AF893021

Methods in Molecular Biology

Series Editor
John M. Walker
School of Life and Medical Sciences
University of Hertfordshire
Hatfield, Hertfordshire, AL10 9AB, UK

For further volumes:
http://www.springer.com/series/7651

Plant Hormones

Methods and Protocols

Third Edition

Edited by

Jürgen Kleine-Vehn

Department for Applied Genetics and Cell Biology, University of Natural Resources and Life Sciences Vienna (BOKU), Austria

Michael Sauer

Department of Plant Physiology, Institute of Biochemistry and Biology, University of Potsdam, Germany

Humana Press

Editors
Jürgen Kleine-Vehn
Department for Applied Genetics and Cell Biology (DAGZ)
University of Natural Resources and Life Sciences Vienna (BOKU)
Austria

Michael Sauer
Department of Plant Physiology
Institute of Biochemistry and Biology
Universität Potsdam
Potsdam, Germany

ISSN 1064-3745 ISSN 1940-6029 (electronic)
Methods in Molecular Biology
ISBN 978-1-4939-6467-3 ISBN 978-1-4939-6469-7 (eBook)
DOI 10.1007/978-1-4939-6469-7

Library of Congress Control Number: 2016955955

© Springer Science+Business Media New York 2017
This work is subject to copyright. All rights are reserved by the Publisher, whether the whole or part of the material is concerned, specifically the rights of translation, reprinting, reuse of illustrations, recitation, broadcasting, reproduction on microfilms or in any other physical way, and transmission or information storage and retrieval, electronic adaptation, computer software, or by similar or dissimilar methodology now known or hereafter developed.
The use of general descriptive names, registered names, trademarks, service marks, etc. in this publication does not imply, even in the absence of a specific statement, that such names are exempt from the relevant protective laws and regulations and therefore free for general use.
The publisher, the authors and the editors are safe to assume that the advice and information in this book are believed to be true and accurate at the date of publication. Neither the publisher nor the authors or the editors give a warranty, express or implied, with respect to the material contained herein or for any errors or omissions that may have been made.

Printed on acid-free paper

This Humana Press imprint is published by Springer Nature
The registered company is Springer Science+Business Media LLC
The registered company address is: 233 Spring Street, New York, NY 10013, U.S.A.

Preface

The field of Plant Hormone Biology is currently advancing at a rapid pace. In little more than a decade we have seen an impressive development that both broadened and deepened our knowledge of how small molecules influence plant physiology and development. Besides the classical hormones, we are now aware of novel compounds that exert hormonal functions, such as strigolactones, karrikins, or signaling peptides, where molecular signaling pathways still unfold. At the same time the understanding of other hormonal pathways has become so detailed that more and more sophisticated questions can be asked. Interactions between members of signaling networks even at the quantitative level, cross-talk between different hormonal pathways, or detailed molecular analyses of activity are now all within the reach of the experimenter.

In this book, we aim to present a representative cross section of modern experimental approaches relevant to Plant Hormone Biology. They range from relatively simple physiological assays, which can be performed in any laboratory with standard equipment, to highly sophisticated methods, which require specialized instrumentation. Some of the chapters describe novel, previously undescribed methods, while others are refined variations of existing protocols. The first four chapters are dedicated to physiological and developmental assays. In line with the increasing demand for high-throughput methods, three chapters on automated phenotyping follow. We tried to cover the wide spectrum of microscopy-based techniques with six chapters ranging from response quantification to four-dimensional tissue reconstruction. Mechanistic insight into hormonal pathways can be gained by interaction studies, and four chapters outline different experimental approaches. Traditionally, the plant hormone field has developed numerous analytical methods to measure hormone contents, and we feature four examples of recent developments. Finally, we conclude with two chapters which outline how the use of heterologous systems can significantly advance the field.

We trust the reader finds this book useful in a twofold way: On the one hand, it can be used as a cookbook, which quickly aids in the setup of a particular experiment directly relevant to the researcher's interest. On the other hand, we encourage the reader to browse through the chapters and explore whether some of the methods may be adapted also for their own research.

We wish our readers the best of luck for their experiments!

Vienna, Wien, Austria — *Jürgen Kleine-Vehn*
Potsdam, Germany — *Michael Sauer*

Contents

Contributors

BIRGIT ABSMANNER • *Cell Biology and Plant Biochemistry, University of Regensburg, Regensburg, Germany*

IOANNA ANTONIADI • *Department of Forest Genetics and Plant Physiology, Umeå Plant Science Centre, Swedish University of Agricultural Sciences, Umeå, Sweden*

ELKE BARBEZ • *Department of Applied Genetics and Cell Biology (DAGZ), University of Natural Resources and Life Sciences (BOKU), Vienna, Austria; Gregor Mendel Institute (GMI) of Molecular Plant Biology, Vienna, Austria*

FRANCIS BARBEZ • *Department of Applied Genetics and Cell Biology (DAGZ), University of Natural Resources and Life Sciences (BOKU), Vienna, Austria*

GEORGE W. BASSEL • *School of Biosciences, University of Birmingham, Birmingham, UK*

EVA BENKOVÁ • *Department of Life Sciences, Institute of Science and Technology Austria, Klosterneuburg, Austria*

CHLOÉ BÉZIAT • *Department of Applied Genetics and Cell Biology (DAGZ), University of Natural Resources and Life Sciences (BOKU), Vienna, Austria*

SJEF BOEREN • *Laboratory of Biochemistry, Wageningen University, Wageningen, The Netherlands*

JAN WILLEM BORST • *Laboratory of Biochemistry, Wageningen University, Wageningen, The Netherlands; Microspectroscopy Center, Wageningen University, Wageningen, The Netherlands*

SIOBHAN A. BRAYBROOK • *The Sainsbury Laboratory, University of Cambridge, Cambridge, UK*

WOLFGANG BUSCH • *Gregor Mendel Institute (GMI), Austrian Academy of Sciences Vienna Biocenter (VBC), Vienna, Austria*

NATHAN CZYZEWICZ • *Division of Plant and Crop Sciences, School of Biosciences, University of Nottingham, Loughborough, UK*

BERT DE RYBEL • *Laboratory of Biochemistry, Wageningen University, Wageningen, The Netherlands; Department of Plant Systems Biology, Flemish Institute of Biotechnology, VIB, Ghent, Belgium; Department of Plant Biotechnology and Bioinformatics, Ghent University, Ghent, Belgium*

IVE DE SMET • *Division of Plant and Crop Sciences, School of Biosciences, University of Nottingham, Loughborough, UK; Department of Plant Biotechnology and Bioinformatics, Ghent University, Ghent, Belgium; Department of Plant Systems Biology, Flemish Institute of Biotechnology (VIB), Ghent, Belgium; Centre for Plant Integrative Biology, University of Nottingham, Loughborough, UK*

ASTRID FASTNER • *Cell Biology and Plant Biochemistry, University of Regensburg, Regensburg, Germany*

ELENA FERARU • *Department of Applied Genetics and Cell Biology (DAGZ), University of Natural Resources and Life Sciences (BOKU), Vienna, Austria*

GAVIN R. FLEMATTI • *School of Chemistry and Biochemistry, University of Western Australia, Crawley, WA, Australia*

ULRICH Z. HAMMES • *Cell Biology and Plant Biochemistry, University of Regensburg, Regensburg, Germany*

Henk W.M. Hilhorst • *Wageningen Seed Lab, Laboratory of Plant Physiology, Wageningen University, Wageningen, The Netherlands*
Jürgen Kleine-Vehn • *Department of Applied Genetics and Cell Biology (DAGZ), University of Natural Resources and Life Sciences Vienna (BOKU), Vienna, Austria*
S. Vinod Kumar • *Cell and Developmental Biology, John Innes Centre, Norwich, UK*
Wilco Ligterink • *Wageningen Seed Lab, Laboratory of Plant Physiology, Wageningen University, Wageningen, The Netherlands*
Jingchun Liu • *Zurich-Basel Plant Science Center, Department of Plant and Microbial Biology, University of Zurich, Zollikerstrasse, Zurich, Switzerland*
Karin Ljung • *Department of Forest Genetics and Plant Physiology, Umeå Plant Science Centre, Swedish University of Agricultural Sciences, Umeå, Sweden*
Víctor Carrasco Loba • *Centro de Biotecnología y Genómica de Plantas (CBGP), Campus de Montegancedo, Pozuelo de Alarcón, Madrid, Spain*
Doris Lucyshyn • *Department of Applied Genetics and Cell Biology, University of Natural Resources and Life Sciences (BOKU), Vienna, Austria*
Charles W. Melnyk • *The Sainsbury Laboratory, University of Cambridge, Cambridge, UK*
Barbara K. Möller • *Laboratory of Biochemistry, Wageningen University, Wageningen, The Netherlands; Department of Plant Systems Biology, VIB Ghent University, Ghent, Belgium*
Thomas D. Montenegro-Johnson • *School of Mathematics, University of Birmingham, Edgbaston, Birmingham, UK*
Bruno Müller • *Zurich-Basel Plant Science Center, Department of Plant and Microbial Biology, University of Zurich, Zollikerstrasse, Zurich, Switzerland*
Maren Müller • *Departament de Biologia Vegetal, Facultat de Biologia, University of Barcelona, Barcelona, Spain*
Sergi Munné-Bosch • *Departament de Biologia Vegetal, Facultat de Biologia, University of Barcelona, Barcelona, Spain*
Richard M. Napier • *School of Life Sciences, University of Warwick, Coventry, UK*
Jennifer L. Nemhauser • *Department of Biology, University of Washington, Seattle, WA, USA*
Ondřej Novák • *Laboratory of Growth Regulators, Centre of the Region Haná for Biotechnological and Agricultural Research, Institute of Experimental Botany AS CR & Faculty of Science, Palacký University, Olomouc, Czech Republic*
Edith Pierre-Jerome • *Department of Biology, University of Washington, Seattle, WA, USA*
Stephan Pollmann • *Centro de Biotecnología y Genómica de Plantas (CBGP), Campus de Montegancedo, Pozuelo de Alarcón, Madrid, Spain*
Justyna M. Prusinska • *School of Life Sciences, University of Warwick, Coventry, UK*
Mussa Quareshy • *School of Life Sciences, University of Warwick, Coventry, UK*
Tatyana Radoeva • *Laboratory of Biochemistry, Wageningen University, Wageningen, The Netherlands*
Alejandra Freire Rios • *Laboratory of Biochemistry, Wageningen University, Wageningen, The Netherlands*
Daniela Ristova • *Gregor Mendel Institute (GMI), Austrian Academy of Sciences, Vienna, Austria*
Hilla Schayek • *Department of Molecular Biology and Ecology of Plants, Life Sciences Faculty, University of Tel-Aviv, Tel-Aviv, Israel*
Eilon Shani • *Department of Molecular Biology and Ecology of Plants, Life Sciences Faculty, Tel-Aviv University, Tel-Aviv, Israel*

DAJO SMET • *Laboratory of Functional Plant Biology, Department of Physiology, Ghent University, Ghent, Belgium*
STEVEN M. SMITH • *School of Biological Sciences, University of Tasmania, Hobart, TAS, Australia; State Key Laboratory of Plant Genomics & National Center for Plant Gene Research (Beijing), Institute of Genetics and Developmental Biology, Beijing, China*
RICHARD SMITH • *Department of Comparative Development and Genetics, Max Planck Institute for Plant Breeding Research, Cologne, Germany*
PETRA STAMM • *School of Biosciences, University of Birmingham, Birmingham, UK*
ELISABETH STES • *Department of Plant Systems Biology, Flemish Institute of Biotechnology VIB, Ghent, Belgium; Department of Plant Biotechnology and Bioinformatics, Ghent University, Ghent, Belgium; Medical Biotechnology Center, Flemish Institute of Biotechnology VIB, Ghent, Belgium; Department of Biochemistry, Ghent University, Ghent, Belgium*
RICHARD STRASSER • *Department of Applied Genetics and Cell Biology, University of Natural Resources and Life Sciences, BOKU, Vienna, Austria*
SOEREN STRAUSS • *Department of Comparative Development and Genetics, Max Planck Institute for Plant Breeding Research, Cologne, Germany*
VESELINA UZUNOVA • *School of Life Sciences, University of Warwick, Coventry, UK*
DOMINIQUE VAN DER STRAETEN • *Laboratory of Functional Plant Biology, Department of Physiology, Ghent University, Ghent, Belgium*
ULRIKE VAVRA • *Department of Applied Genetics and Cell Biology, University of Natural Resources and Life Sciences, BOKU, Vienna, Austria*
CHRISTIANE VEIT • *Department of Applied Genetics and Cell Biology, University of Natural Resources and Life Sciences, BOKU, Vienna, Austria*
MARK T. WATERS • *Australian Research Council Centre of Excellence in Plant Energy Biology, University of Western Australia, Crawley, WA, Australia; School of Chemistry and Biochemistry, University of Western Australia, Crawley, WA, Australia*
DOLF WEIJERS • *Laboratory of Biochemistry, Wageningen University, Wageningen, The Netherlands*
ROY WEINSTAIN • *Department of Molecular Biology and Ecology of Plants, Life Sciences Faculty, Tel-Aviv University, Tel-Aviv, Israel*
JOS R. WENDRICH • *Laboratory of Biochemistry, Wageningen University, Wageningen, The Netherlands*
R. CLAY WRIGHT • *Department of Biology, University of Washington, Seattle, WA, USA*
PETRA ŽÁDNÍKOVÁ • *Institut für Genetik, Heinrich-Heine-University Düsseldorf, Düsseldorf, Germany*
QIANG ZHU • *Department of Life Sciences, Institute of Science and Technology Austria, Klosterneuburg, Austria*

Chapter 1

Real-Time Analysis of the Apical Hook Development

Qiang Zhu, Petra Žádníková, Dajo Smet, Dominique Van Der Straeten, and Eva Benková

Abstract

Mechanisms for cell protection are essential for survival of multicellular organisms. In plants, the apical hook, which is transiently formed in darkness when the germinating seedling penetrates towards the soil surface, plays such protective role and shields the vitally important shoot apical meristem and cotyledons from damage. The apical hook is formed by bending of the upper hypocotyl soon after germination, and it is maintained in a closed stage while the hypocotyl continues to penetrate through the soil and rapidly opens when exposed to light in proximity of the soil surface. To uncover the complex molecular network orchestrating this spatiotemporally tightly coordinated process, monitoring of the apical hook development in real time is indispensable. Here we describe an imaging platform that enables high-resolution kinetic analysis of this dynamic developmental process.

Key words Differential growth, Apical hook development, Hormonal cross talk, Real-time imaging, Phenotype analysis

1 Introduction

To compensate for their sessile lifestyle, plants developed unique mechanisms that provide them with an unusual level of developmental plasticity. Bending of plant organs in response to gravi- and photostimulation is a typical example of such plant-specific adaptation strategies. A particularly intriguing process which comprises the bending and consecutive unbending/straightening of the upper hypocotyl is the development of the apical hook. The apical hook is formed by the folding of the hypocotyl during early seedling development to shield the tender shoot apical meristem and cotyledons from damage while penetrating the soil. It is maintained in a closed stage while seedlings grow through the soil, and rapidly open when exposed to light. Hence, apical hook development gradually progresses through three distinct phases (formation, maintenance, and opening phase), each of them depending on a specific coordination of tissue and cell growth dynamics [1–3]. The common mechanistic basis underlying bending of various

Jürgen Kleine-Vehn and Michael Sauer (eds.), *Plant Hormones: Methods and Protocols*, Methods in Molecular Biology, vol. 1497, DOI 10.1007/978-1-4939-6469-7_1, © Springer Science+Business Media New York 2017

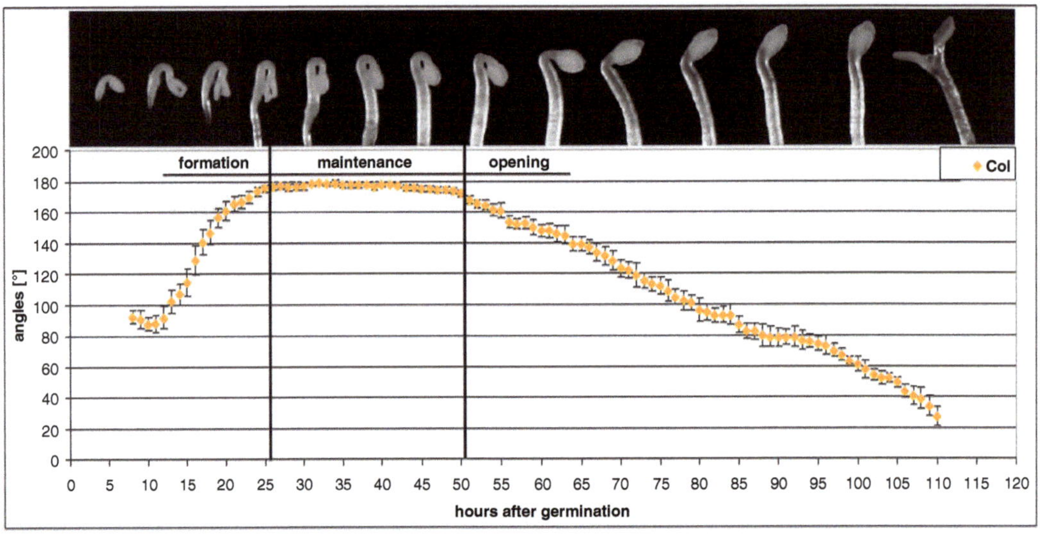

Fig. 1 Real-time monitoring of the apical hook development. Hook growth was continuously recorded from germination on and the angle of curvature was measured. Typically, the hook undergoes three developmental phases: (1) formation, which is the period from seed germination until the hook angle reaches 180°; (2) maintenance, during which the hook remains fully closed; and (3) opening, when it gradually opens to reach an angle of 0° (adapted from [12])

organs, including the apical hook, is differential growth, which, through unequal rates of cell elongation at two opposite organ sides, ultimately results in organ curving [4–6]. To form the hook curvature the elongation rate of cells on the outer side of the upper hypocotyl needs to exceed that of those on the inner side. In contrast the growth rate of cells on the concave (inner side) must exceed that of cells on the convex (outer side) of the hook to straighten the hypocotyl during the opening phase [1–3, 7] (Fig. 1).

Plant hormones are indispensable endogenous regulators of apical hook development. Among them auxin plays a fundamental role. Defects in auxin metabolism, transport, and signaling dramatically affect all phases of apical hook development [8–15]. Particularly the tightly controlled asymmetric auxin distribution is linked with differential cell growth—the driving force of apical hook development [12, 13, 16, 17]. Accumulation of auxin defines the concave side of the apical curvature during formation phase, whereas balancing of the auxin levels between the concave and convex side results in the opening of the apical hook. Besides auxin, a multitude of other hormone signaling pathways including that of ethylene, brassinosteroids, and gibberellins coordinate this developmental process. Increased ethylene levels in ethylene overproducer mutants (*eto1*, *eto2*, *eto3*) result in an enhancement of the apical hook curvature [6, 18, 19] whereas ethylene-insensitive mutants, such as *ethylene resistant1* (*etr1*) and *ethylene-insensitive2* (*ein2*), exhibit a hookless phenotype [6]. Gibberellins and brassinosteroids contribute to hook establishment and their interaction with auxin and ethylene has been described [6, 14, 20–25].

Despite recent progress in dissecting the regulatory pathways and complex hormonal network that guides the development of the apical hook, we are still far from a full understanding of this process. To hasten the elucidation of the molecular components and mechanisms that control the progress of the apical hook through its three phases of development, a reliable high-throughput monitoring of the whole process is one of the major technical prerequisites. Molecular pathways that define (1) the kinetics of the bending and angle of the hypocotyl curvature during the formation phase, (2) the kinetics with which the hypocotyl straightens in response to light stimuli, as well as (3) the transition from formation to maintenance and from maintenance to opening cannot be assessed unless employing continuous monitoring of the whole process over time. Here we describe an imaging platform that enables accurate kinematic analyses of apical hook development in darkness and its opening in response to a light stimulus.

2 Materials and Chemicals

1. Seeds of *Arabidopsis thaliana* and mutants of interest.
2. Square Petri dishes ($120 \times 120 \times 17$ mm).
3. Calcium hypochlorite [$Ca(ClO)_2$].
4. Triton-X-100.
5. Murashige and Skoog (MS) medium.
6. Sucrose (for plant tissue culture).
7. Agar for plant tissue culture.
8. Potassium hydroxide (KOH).
9. Optional: Phytohormones or other biologically active molecules of interest.

2.1 Equipment

1. Dark growth chamber to accommodate plate, camera, and IR light, a cube with 400 mm side length should be adequate (Fig. 2).
2. Infrared light source with 880 nm or alternatively 850 nm wavelength, such as used in conjunction with surveillance cameras (examples are 880 nm IR LED; Velleman, Belgium; alternatively IR LED illuminator 850 nm; ABUS Security Center, Affing, Germany).
3. Spectrum-enhanced digital camera which allows imaging of infrared light and that can be remote controlled by a computer and appropriate software. Frequently, this type of camera is sold as optimized for astro-photography. For guidance, we give the following examples: EOS 600D Canon Rebel T3i, 400DH with built-in clear wideband-multicoated filter, equipped with a standard 18–55 mm f3.5–5.6 lens and standard accessories (Canon), operated by the EOS utility software (for one camera)

Fig. 2 Setup of the infrared imaging platform. Petri dishes are placed into a dark box. Typically two Petri dishes are aligned for monitoring by one camera. As during cultivation water might condense at the lid, Petri dishes are positioned with the lid side facing away from the camera. The manual focus and automatic stabilizer are used for image acquisition. The infrared light (IR-LED) is fixed in the dark box to obtain homogenous illumination. The CCD camera is placed within the dark box and connected to a computer. The camera is operated through DSLR Remote Pro Multi-Camera software with adjustable frequency of picture acquisition and synchronized switching on of the infrared light

or DSLR Remote Pro Multi-Camera software (for one or Multiple Cameras). Another example would be Hercules optical glass USB-type CCD camera without an infrared filter (Guillemot, La Gacilly, France), steered by Active WebCam v.4 software (PY Software, Etobicoke, Canada).

The infrared light is fixed in the dark box as indicated at Fig. 2 to obtain homogenous illumination. The camera is placed within the dark box and connected to a computer. The camera is operated through DSLR Remote Pro Multi-Camera software alternatively EOS utility software with adjustable frequency of picture acquisition and synchronized switching on of the infrared light. Synchronization can be set up using appropriate software (e.g., Q Light Controller).

4. To monitor light-triggered opening of the apical hook a computer-controlled light switch can be installed. This can be quite simply achieved utilizing components frequently used for stage and theatrical lighting adhering to the common DMX standard. With these components, also light pulses of desired spectral quality (by using RGB light strips) can be achieved. A free open-source software solution is the program Qlight controller (http://sourceforge.net/projects/qlc/). We suggest using an LED dimmer controller in combination with an LED Red/Green/Blue strip which is frequently used also for home/ambient lighting applications. As technical advance and product cycles in the lighting field are fast paced, we do not recommend a specific product but rather encourage the reader to research the market for a practical solution.

3 Methods

1. Sterilized seeds are plated on a square Petri dish containing 45 ml of half-strength Murashige and Skoog (MS) medium with 1% sucrose and 0.8% agar (pH 5.7 adjusted with KOH) and sealed with one layer of micropore tape (3M MICROPORE). Optional: According to the experimental design MS medium might be supplemented with plant hormones or other biologically active molecules. For optimal resolution two rows of seeds (15 seeds each) are sown per plate. After stratification for 2 days at 4 °C in darkness, seeds are exposed to light for 6 h at 21 °C (*see* **Note 1**).
2. Petri dishes are placed in a dark box at 21 °C. Typically two Petri dishes are aligned for monitoring by one camera. As during cultivation water might condense on the lid, Petri dishes are positioned with lid facing away from the camera. Manual focus and automatic stabilizer are used for image acquisition. Typical frequency of the image acquisition is every hour for a period of 8–10 days, during which seedlings progress from formation to full opening of the apical hook.
3. To monitor light-triggered opening of the apical hook, the light pulse of desired quality is applied for a defined time period to apical hooks in the maintenance phase and their opening is subsequently recorded.

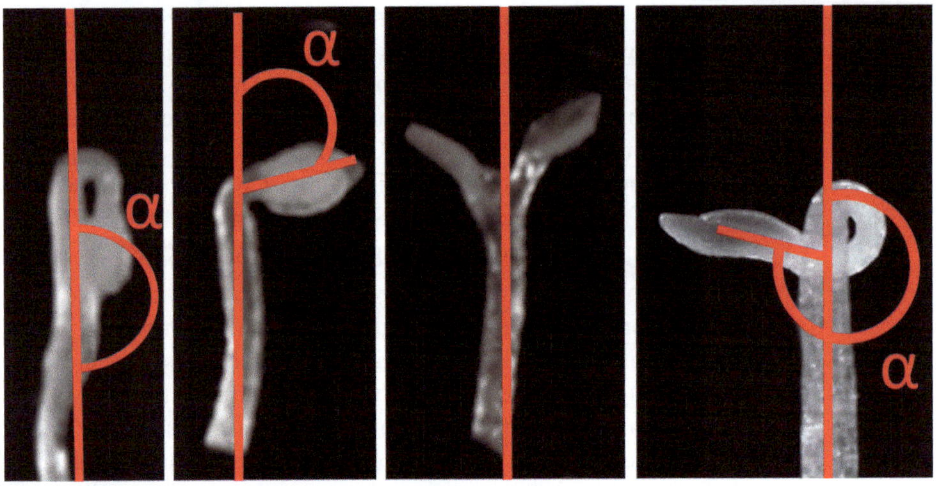

Fig. 3 Kinetic analysis of the apical hook development. The angle of curvature α is defined as 180° minus the angle formed by the tangential of the apical part and the axis of the lower part of the hypocotyl. In case the bending exceeds 180° leading to formation of an exaggerated hook, the angle of curvature is defined as 180° plus α (adapted from [12, 13])

4. The kinetic analysis of the apical hook development is performed by the measurement of the angle between the hypocotyl axis and cotyledons using ImageJ (NIH; http://rsb.info.nih.gov/ij). Typically, 15–20 seedlings are processed. By convention, the angle of curvature α is defined as 180° minus the angle formed by the tangential of the apical part and the axis of the lower part of the hypocotyl (Fig. 3) [12, 13]. In the case the bending exceeds 180°, which leads to the formation of an exaggerated hook, the angle of curvature is defined as 180° plus α (Fig. 3) [13]. In our previous studies [12, 13, 25], the consecutive phases of apical hook development were defined as follows: the hook formation phase is the period from germination until the time point at which the angle of hook curvature reaches 95 % of its maximum value, and the maintenance phase comprises the plateau in which hook angles differ at most 5 % from the maximum angle of curvature. This is then succeeded by the hook opening phase resulting in full straightening of the hypocotyl (*see* **Note 2**).

4 Notes

1. During early germination, the seed coat attached to the germinating seedling might prevent reliable observation of the hook bending. To improve visualization of this early hook formation phase the seed coat can be peeled off. For these purposes seeds are imbibed in water for 6 h at 4 °C in darkness and mature

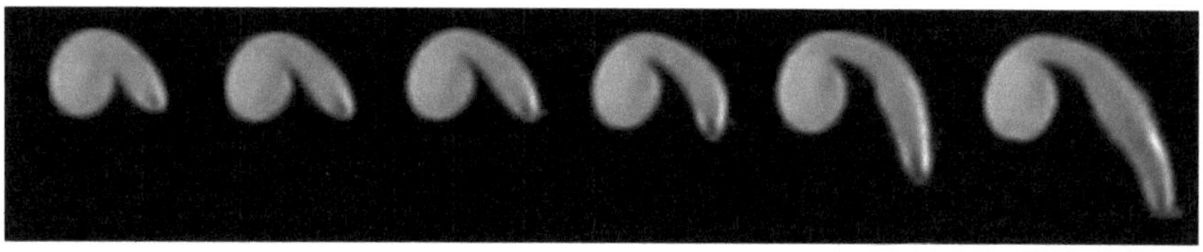

Fig. 4 Monitoring of early phases of apical hook development. To improve visualization of early hook formation phase the seed coat is peeled off and the mature embryo is dissected. The apical hook formation is monitored as described

embryos are cautiously dissected from seed coats using tweezers. For convenient manipulation a stereomicroscope placed in the sterile bench might be used (Fig. 4).

2. Using the real-time platform for examination of apical hook development the heterogeneity in seed germination might be tolerated. This is one of the serious drawbacks of classical "steady-state" assays in which hook curvatures of seedlings grown in darkness are measured for a defined time period. Here, variability in germination might lead to considerable inaccuracies in an assessment of the apical hook developmental phase. In contrast, continuous monitoring of growing seedlings allows to compensate for germination irregularities by setting the time point zero for each individual seedling. Time point zero is considered as the moment when the radicle protrudes through the seed coat.

Acknowledgements

We thank Herman Höfte, Todor Asenov, Robert Hauschield, and Marcal Gallemi for help with the establishment of the real-time imaging platform and technical support. This work was supported by the Czech Science Foundation (GA13-39982S) to Eva Benková. Dominique Van Der Straeten acknowledges the Research Foundation Flanders for financial support (G.0656.13N). Dajo Smet holds a PhD fellowship of the Research Foundation Flanders.

References

1. Silk WK, Erickson RO (1978) Kinematics of hypocotyl curvature. Am J Bot 65:310–319
2. Gendreau E, Traas J, Desnos T, Grandjean O, Caboche M, Höfte H (1997) Cellular basis of hypocotyl growth in Arabidopsis thaliana. Plant Physiol 114:295–305
3. Raz V, Ecker JR (1999) Regulation of differential growth in the apical hook of Arabidopsis. Development 126:3661–3668
4. Darwin C, Darwin F (1881) The power of movement in plants. D. Appleton and Co, New York
5. Boysen-Jensen P (1911) La transmission de l'irritation phototropique dans l'avena. Bulletin Academie Des Sciences et Lettres de Montpellier 3:1–24
6. Guzmán P, Ecker JR (1990) Exploiting the triple response of Arabidopsis to identify

ethylene-related mutants. Plant Cell 2: 513–523
7. Raz V, Koornneef M (2001) Cell division activity during apical hook development. Plant Physiol 125:219–226
8. Boerjan W, Cervera MT, Delarue M, Beeckman T, Dewitte W, Bellini C, Caboche M, Van Onckelen H, Van Montagu M, Inzé D (1995) Superroot, a recessive mutation in Arabidopsis, confers auxin overproduction. Plant Cell 7:1405–1419
9. Lehman A, Black R, Ecker JR (1996) HOOKLESS1, an ethylene response gene, is required for differential cell elongation in the Arabidopsis hypocotyl. Cell 85:183–194
10. Zhao Y, Christensen SK, Fankhauser C, Cashman JR, Cohen JD, Weigel D, Chory J (2001) A role for flavin monooxygenase-like enzymes in auxin biosynthesis. Science 291: 306–309
11. Dharmasiri N, Dharmasiri S, Weijers D, Lechner E, Yamada M, Hobbie L, Ehrismann JS, Jürgens G, Estelle M (2005) Plant development is regulated by a family of auxin receptor F box proteins. Dev Cell 9:109–119
12. Žádníková P, Petrášek J, Marhavý P, Raz V, Vandenbussche F, Ding Z, Schwarzerová K, Morita MT, Tasaka M, Hejátko J et al (2010) Role of PIN-mediated auxin efflux in apical hook development of Arabidopsis thaliana. Development 137:607–617
13. Vandenbussche F, Petrásek J, Zádníková P, Hoyerová K, Pesek B, Raz V, Swarup R, Bennett M, Zazímalová E, Benková E et al (2010) The auxin influx carriers AUX1 and LAX3 are involved in auxin-ethylene interactions during apical hook development in Arabidopsis thaliana seedlings. Development 137:597–606
14. Gallego-Bartolomé J, Arana MV, Vandenbussche F, Zádníková P, Minguet EG, Guardiola V, Van Der Straeten D, Benkova E, Alabadí D, Blázquez MA (2011) Hierarchy of hormone action controlling apical hook development in Arabidopsis. Plant J 67:622–634
15. Mazzella MA, Casal JJ, Muschietti JP, Fox AR (2014) Hormonal networks involved in apical hook development in darkness and their response to light. Front Plant Sci 5:52
16. Kuhn H, Galston AW (1992) Physiological asymmetry in etiolated pea epicotyls: relation to patterns of auxin distribution and phototropic behavior. Photochem Photobiol 55: 313–318
17. Žádníková P, Smet D, Zhu Q, Van Der Straeten D, Benková E (2015) Strategy of seedlings to overcome their sessile nature: auxin in the mobility control. Front Plant Sci 6:218
18. Vogel JP, Schuerman P, Woeste K, Brandstatter I, Kieber JJ (1998) Isolation and characterization of Arabidopsis mutants defective in the induction of ethylene biosynthesis by cytokinin. Genetics 149:417–427
19. Woeste KE, Ye C, Kieber JJ (1999) Two Arabidopsis mutants that overproduce ethylene are affected in the posttranscriptional regulation of 1-aminocyclopropane-1-carboxylic acid synthase. Plant Physiol 119:521–530
20. Achard P, Vriezen WH, Van Der Straeten D, Harberd NP (2003) Ethylene regulates Arabidopsis development via the modulation of DELLA protein growth repressor function. Plant Cell 15:2816–2825
21. Vriezen WH, Achard P, Harberd NP, Van Der Straeten D (2004) Ethylene-mediated enhancement of apical hook formation in etiolated Arabidopsis thaliana seedlings is gibberellin dependent. Plant J 37:505–516
22. De Grauwe L, Vandenbussche F, Tietz O, Palme K, Van Der Straeten D (2005) Auxin, ethylene and brassinosteroids: tripartite control of growth in the Arabidopsis hypocotyl. Plant Cell Physiol 46:827–836
23. An F, Zhang X, Zhu Z, Ji Y, He W, Jiang Z, Li M, Guo H (2012) Coordinated regulation of apical hook development by gibberellins and ethylene in etiolated Arabidopsis seedlings. Cell Res 22:915–927
24. Gendron JM, Haque A, Gendron N, Chang T, Asami T, Wang ZY (2008) Chemical genetic dissection of brassinosteroid-ethylene interaction. Mol Plant 1:368–379
25. Smet D, Žádníková P, Vandenbussche F, Benková E, Van Der Straeten D (2014) Dynamic infrared imaging analysis of apical hook development in Arabidopsis: the case of brassinosteroids. New Phytol 202:1398–1411

Chapter 2

Grafting with *Arabidopsis thaliana*

Charles W. Melnyk

Abstract

Generating chimeric organisms is an invaluable way to study cell-to-cell movement and non-cell-autonomous actions of molecules. Plant grafting is an ancient method of generating chimeric organisms and recently has been used to study the movement of hormones, proteins, and RNAs. Here, I describe a simple and efficient way to graft *Arabidopsis thaliana* at the seedling stage to generate plants with roots and shoots of different genotypes. Using this protocol, success rates of over 80% with up to 80 grafts assembled per hour can be achieved.

Key words *Arabidopsis thaliana*, Micro-grafting, Chimeric plants, Mobile molecules, Graft-transmissible signal

1 Introduction

People have cut and joined together different plant varieties for thousands of years to generate chimeric organisms that have increased stress resistance, increased yields, or improved plant size [1]. This technique, termed grafting, has been used more recently to study the non-cell-autonomous actions of molecules including RNAs, proteins, and hormones [2–4]. Although requiring technical know-how and skill, grafting is far easier and less time consuming than other methods used to generate chimeric plants such as tissue-specific expression of a transgene or through the generation of sectors following transposition, mutagenesis, or recombination. Nonetheless, grafting has been limited to whole organ or tissue chimeras, such as the grafting of a leaf, inflorescence, or root system.

Arabidopsis grafting was first described over 20 years ago [5], and since then, grafting with *Arabidopsis* has proven a very useful and informative technique. Many tissues of *Arabidopsis* are suitable for grafting including cotyledons [6], inflorescence stems [7], developing leaves with the shoot apical meristem [8], the young shoot/young root [2], and the mature shoot/mature root [5]. Most commonly, *Arabidopsis* grafting is performed on young

Jürgen Kleine-Vehn and Michael Sauer (eds.), *Plant Hormones: Methods and Protocols*, Methods in Molecular Biology, vol. 1497, DOI 10.1007/978-1-4939-6469-7_2, © Springer Science+Business Media New York 2017

seedlings between the shoot and root since many grafts can be assembled rapidly and it allows a greater time for molecules to move as the plant matures. Due to the small size of *Arabidopsis* seedlings, this is a technically challenging process requiring the use of a stereomicroscope. The below protocol is adapted from previously published butt-grafting protocols [2, 9, 10] and does not require the use of tubing. With steady hands and sufficient practise, high success rates (over 80% with wild-type Columbia accession) can be obtained with rapid rates (over 80 grafts per hour) for two segment grafts. More advanced techniques such as three segment grafts (where up to three genotypes can be grafted together as a single plant) or Y-grafts (where one shoot is grafted to a second intact plant) are also possible, but with lower success rates and more time required.

Such shoot-root *Arabidopsis* grafting has become routine practice. To date, it has been used to study the movement of small RNAs [4, 10, 11], nutrients [12, 13], secondary metabolites [14], and hormones including jasmonic acid [15], strigolactones [2], gibberellic acid [16], and cytokinin precursors [17]. It has also been informative to study signals associated with flowering time [3], leaf development [18], and disease resistance [19] and to study vascular regeneration [9, 20].

2 Materials

1. Sterile *Arabidopsis* seed.
2. Ultra Fine Micro Knives (manufactured by Fine Science Tools; catalogue number 10315-12; *see* **Note 1**).
3. Fine forceps.
4. Whatman 3MM Chr cellulose chromatography paper, 46 × 57 cm.
5. Hybond N membrane, 20 cm × 3 m.
6. Aluminium foil.
7. Plates with ½ Murashige and Skoog (MS) medium and 0.8–1 % agar.
8. 9 cm Round Petri dishes.
9. Dissecting microscope.
10. Laminar flow hood.
11. 20 °C Growth cabinet.
12. 70% Ethanol for sterilization.
13. Sterile, autoclaved water.
14. Parafilm.

3 Methods

3.1 Two Segment Shoot-Root Graft

1. Sprinkle or pipette out sterilized *Arabidopsis* seed on suitable media such as ½MS plates with 0.8–1 % agar, leaving several millimeters between seeds. Plates without sucrose and without antibiotics work best, but 1 % sucrose can be included if necessary. Leave at 4 °C in the dark for 2–7 days.
2. Move plates from the cold to a growth space set at 20–22 °C with 80–100 μmol/m^2/s of light. Mount plates vertically to ensure correct hypocotyl growth. The growth space can be set for either short-day (8-h light) or long-day (16-h light) conditions. Short day-grown plants are grafted 7 days after moving out. Long day-grown plants are grafted 5 days after moving out. I prefer short day-grown plants, as the hypocotyls are longer and easier to graft.
3. Cut the Whatman paper to ~8.5 cm circles, and Hybond membrane to 2.5 × 4 cm strips. For each grafting Petri dish, two Whatman circles and one Hybond membrane are required. Cut strips of Whatman paper (approximately 3 × 8 cm long) to use for adding or removing water during grafting. Wrap the cut Whatman paper and Hybond membrane in aluminum foil and autoclave to sterilize.
4. Perform grafting in a laminar flow hood under a dissecting microscope that has been wiped clean with 70% ethanol. Forceps and micro knife (*see* **Note 1**) are kept sterile with 70% ethanol, but allowed to dry prior to grafting (*see* **Note 2**). Residual ethanol in the grafting plate will inhibit graft formation.
5. Add sterile water to an empty petri dish, and add two sterile Whatman circles and then one sterile Hybond strip. Sterile water works best for grafting. Liquid ½ MS can also be used with lower efficiency. Sucrose reduces grafting efficiency and should be avoided. Remove the water-soaked Hybond and Whatman with forceps, allow excess water to drip, and then place these in a new petri dish with one Hybond strip on top of two circles of Whatman paper (Fig. 1). During grafting, the Whatman paper maintains the correct moisture, whereas the Hybond ensures that the roots don't anchor and become attached to the Whatman paper.
6. At 5 or 7 days of growth (*see* Subheading 3.1, **step 2**), move 6–12 *Arabidopsis* seedlings from the ½MS plate and place these in a row on the Hybond membrane (Fig. 2a). Select healthy seedlings that have straight hypocotyls and cotyledons that are at right angles to the hypocotyl. Cotyledons that are bent over the hypocotyl make for difficult grafting. The first true set of leaves should be barley visible (Fig. 2b). Be careful

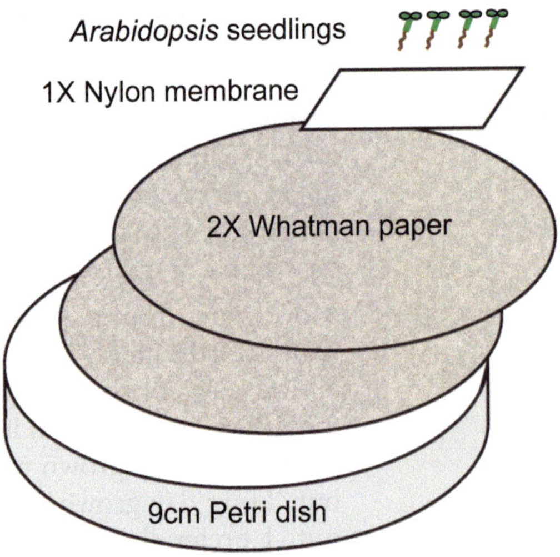

Fig. 1 *Arabidopsis* grafting setup. Two Whatman circles and one Hybond membrane strip are hydrated, excess water left to drip off, and placed in a Petri dish. *Arabidopsis* seedlings are then moved on top of the Hybond membrane

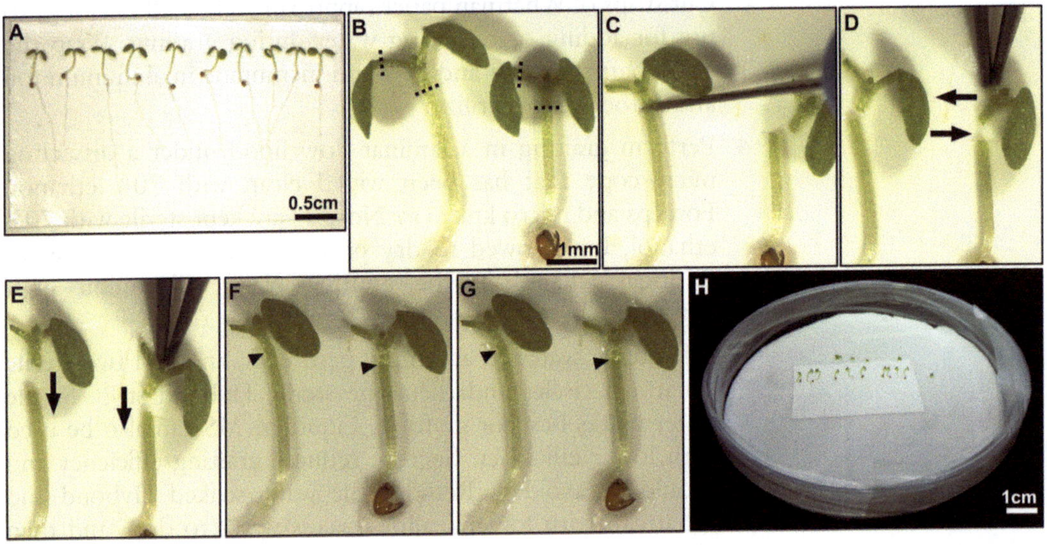

Fig. 2 Two-segment *Arabidopsis* hypocotyl grafting. *Arabidopsis* seedlings are placed on Hybond membrane, cut (*dotted lines*), switched with the desired genotype (*arrows*), and reassembled (*arrows*) (**a**–**f**). *Triangles* denote the graft junction. Before sealing, make sure that water is barely visible around the hypocotyl (**f**) and not excessive (**g**). Plates are then sealed with Parafilm (**h**) and mounted vertically for 7–10 days to recover before transfer to media or soil

when moving seedlings not to damage the hypocotyl, and root or shoot apical meristem. If two genotypes are to be grafted to each other, two rows of seedlings can be made (top row contains the shoot genotype; bottom row contains the root genotype). Alternatively, genotypes can be alternated in the same row such as odd numbers one genotype, even numbers the other.

7. Cut off and discard one cotyledon (Fig. 2b, c), usually the one that is smaller, damaged, or bent suboptimally. This allows the shoot to lie flat on the Hybond membrane. Leave the petiole attached, as later this is useful for picking up the shoot. Make a transverse cut through the hypocotyl close to the shoot (90° to the axis of elongation; a butt-end cut). The cut should be as clean and straight as possible, avoiding crushing the tissue (Fig. 2b, c). Cuts made in the middle or lower portions of the hypocotyl lead to adventitious root formation and graft failure. Ensure that some water is visible around the plants, as this facilitates cutting.
8. Take the cut shoot from one plant and place it close to the cut root from a different plant. This can be accomplished by careful pushing or by picking up the shoot via the exposed petiole (Fig. 2d). The root hypocotyl should lie flat against the Hybond membrane and not be moved. If it is not flat, roll the hypocotyl by carefully pushing the hypocotyl at the root/hypocotyl junction with closed forceps. Be careful not to grab or crush the hypocotyl, as damage to this or the roots will strongly inhibit graft formation.
9. To assemble the graft, keep the forceps closed and gently push on the shoot apical meristem region, cotyledon, or petiole until the cut shoot contacts the cut root (Fig. 2e). Push careful to align and reposition while avoiding grabbing or damaging the tissue. No tubing is used and adhesion of the two cut surfaces is sufficient. The level of moisture here is critical. Some excess water is helpful for cutting, but if the plate is too dry, the plants will stick to the forceps and, in extreme cases, wilt. Too much water will make aligning and adhesion extremely difficult. For aligning and joining the pieces, water should be visible but not excessively pooling on the plate (Fig. 2f, g). Sterile strips of Whatman paper (3 × 8 cm) are useful for removing or adding sterile water (*see* **Note 3**). Alternatively, plates can be left open in the flow hood to dry for a couple minutes.
10. Repeat this procedure until all the plants on the plate have been grafted (*see* **Note 4**). To increase efficiency, cut all plants on the plate at once with the micro knife at low magnification. Discard the cut cotyledons and move the cut shoots. Carefully roll any non-flat root hypocotyls. Use a higher magnification to gently push the cut shoots onto the cut roots. If cutting and alignment become progressively difficult as the plate dries out, add extra water to the plate to facilitate grafting.
11. After grafting, place the lid on the petri dish (Fig. 2h). Water should just be barely visible around some hypocotyls at this stage (Fig. 2f) but not excessive (Fig. 2g). Excessive water will lead to adventitious root formation. Seal the plates with one to

two layers of Parafilm and move these into the growth space at 20–22 °C. Elevated temperatures of 27 °C for 5–7 days are helpful for graft recovery [2], but not necessary. Mount the plates vertically in 80–100 μmol/m^2/s of light in either short- or long-day conditions. Under these conditions, root growth of the grafted plants begins 4–6 days after grafting.

12. 7–10 days after grafting, inspect the plates. Successful grafts have no adventitious roots (roots formed above the graft junction), are well attached, and show signs of new root growth usually in the form of lateral roots. Primary root growth stops after grafting, and does not usually resume. At this point, transfer the grafts to soil or to media. With experience, this method allows success rates of over 80 % and grafting rates of approximately 80 grafts per hour depending on the genotype.
13. Inspect the plants 1–2 weeks after transferring to soil or media to insure that no adventitious roots have formed and the plants are growing normally. Those that have adventitious roots or are not growing should be discarded as the graft has failed. For the first week, plants on soil should have a propagator lid to increase humidity to ensure efficient recovery. Do not bury the graft junction, as this will promote adventitious rooting and make the junction harder to inspect.
14. After the experiment is finished, typically when plants are mature, plants can be inspected for adventitious root formation, though it can be difficult to detect the graft junction in very mature plants. The most reliable indicator is to graft with a shoot or root expressing a visual reporter (i.e., GUS or GFP) [2], or genotype the grafted material.

3.2 Three-Segment Graft (Interstock Graft)

1. Three-segment (or interstock) grafts are set up in a similar manner as two-segment grafts (Subheading 3.1). These types of grafts are informative if the middle segment can block graft formation or a graft-transmissible signal [9] and are used in non-*Arabidopsis* species to improve graft compatibility [1]. The main difference compared to two segment grafts is that two cuts are made in the hypocotyl instead of one (Fig. 3a). The first cut is made in a similar location as for two-segment grafts but the second cut should be approximately 1 mm below the first cut. Longer segments can be used and are easier to move but reduce the grafting success rates.
2. After cutting, you should end up with segments from the desired shoot genotype, middle genotype, and root genotype (Fig. 3b). Three-segment grafting is facilitated by having multiple rows of plants (up to three), or by alternating genotypes in a row and discarding the tissues not required. Care should be taken not to mix up tissues of the various genotypes, so it is recommended not to graft too many plants on one plate.

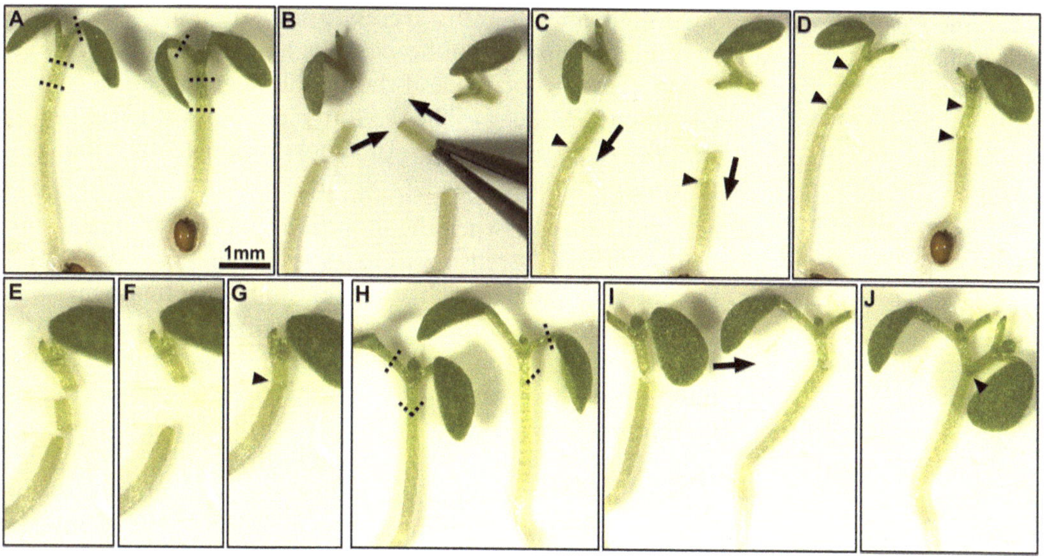

Fig. 3 Three-segment, self-, and Y-grafts. Three-segment (interstock) grafts are made by making two cuts in the hypocotyl (*dotted lines*) and moving the segments as desired (*arrows*) (**a** and **b**). Note that middle segment is moved by pushing either end with forceps (**b**). The lower junction is attached first (**c**), followed by the upper junction (**d**). Self-grafts (two segment) are also made by cutting the hypocotyl twice (**e**), but the middle segment is discarded (**f**) before joining (**g**). Y-grafts involve cutting one hypocotyl halfway through, and the other in a V pattern (**h**). The V segment is then attached to the partially cut hypocotyl (**i** and **j**). *Triangles* denote the graft junction

3. After cutting, do not move root segments except for a gentle roll at the root/hypocotyl junction to get the hypocotyl lying flat if necessary. Move the middle segment by closing the forceps and pushing at one cut end (Fig. 3b). Pushing at the side (the epidermis) should be minimized as this damages the tissue. Push the middle segment (using either cut end) to the cut root until the two segments join (Fig. 3c). Maintain the correct orientation of the middle segment, as upside-down segments will not graft. This forms the lower junction.
4. Move the cut shoot into place using the petiole or by gentle pushing of the cotyledon or meristem. Then close the forceps and gently push the cut shoot onto the middle segment (Fig. 3d). Care should be taken not to dislodge the lower junction.
5. The remaining grafts are assembled, paying careful attention to water levels, before the plate is sealed as in Subheading 3.1. Transfer grafts to media or soil after 10 days since grafts take longer to heal. Expect a lower grafting efficiency with three-segment grafts, as adventitious roots can form above the upper or lower junction. Those plants with adventitious roots should be discarded. It is best to practise with two-segment grafts before attempting three segment.

3.3 Two-Segment Self-Graft

1. Two-segment self-grafts are essentially the same protocol as two-segment grafts (Subheading 3.1), except that two cuts are made in the hypocotyl and the middle segment (~0.5 mm) is discarded (Fig. 3e–g). The 0.5 mm piece is discarded to ensure that the same surfaces are not simply realigned, and that new tissue is used for the graft.
2. Self-grafts are used when the shoot and root need to be the same genotype but the seeds are from a segregating population or the phenotype is variable such as expression from a transgene. Self-grafting has been useful for studying gene expression changes at the graft junction and for dissecting the genetic requirements of graft formation [9]. Self-grafts are also useful controls (*see* **Note 4**).

3.4 Two-Shoot Y-Graft

1. Y-grafts involve adding an additional shoot to an intact plant. This type of graft is useful to test if the molecule of interest moves within the shoot. For instance, Y-grafting was used to demonstrate that the protein FT was mobile from shoot to shoot [3].
2. Prepare plants and the grafting setup identical to that in Subheading 3.1. Different genotypes can be set up in different rows, or beside one another. On one side of the recipient (intact) plant, cut off one cotyledon and make a diagonal cut in the hypocotyl near the shoot that does not cut completely through the hypocotyl (Fig. 3h). The cut should be approximately halfway through. Push gently the shoot with forceps after cutting to slightly widen the hypocotyl cut (Fig. 3i).
3. For the donor shoot, cut off one cotyledon and then cut the hypocotyl in a V-fashion near the upper part of the hypocotyl (Fig. 3h, i). Using the petiole, move the donor shoot close to the recipient hypocotyl. Using closed forceps, push the donor shoot into place. Ideally, the donor shoot's cotyledon should face the opposite direction of the recipient shoot's cotyledon to ensure that there is sufficient room (Fig. 3j).
4. These grafts are time consuming and to avoid the plate from drying out, fewer grafts should be made per plate. Once finished, the plates are sealed with Parafilm and mounted vertically as in Subheading 3.1.

4 Notes

1. Using an Ultra Fine Micro Knife is important for success, but if unavailable, sharp double-sided razor blades will work, though with lower success rates. The Ultra Fine Micro Knives are fragile and take care not to bend or damage the tips. Usually after several hundred grafts the knives need to be replaced to maintain clean cuts.

2. Periodically dip the forceps and micro knife in 70% ethanol and leave to dry. Use aseptic technique, avoid or minimize sources of contamination, and sterilize the hood and all materials with ethanol prior to grafting. Contamination of the grafted plants reduces the grafting efficiency or leads to a complete loss of grafted material.
3. The amount of water is critical to success or failure. You should see some water on the Hybond membrane when cutting and grafting as it facilitates both processes. If the plate starts to dry out while grafting, add a bit more sterile water as this makes grafting easier. After making all grafts, the water needs to be reduced usually by waiting a minute or two for the plate to partially dry out. You should see a shimmer of water that is hardly noticeable under a couple of the hypocotyls (Fig. 2f). Do not have not much water before sealing (Fig. 2g). One sign of too much water is that you will have a lot of adventitious roots formed above the graft junction several days after grafting. For cutting there should be visible water, for grafting an intermediate level of water, and prior to sealing the plates, a low level of water that is barely visible around some of the hypocotyls.
4. Controls should be included in all experiments. In addition to the grafted samples between different genotypes, genotypes grafted to themselves are normally included. This control is necessary to ensure that the effect observed is due to movement of molecules, rather than an effect from grafting itself. Occasionally ungrafted controls are also useful. These have one cotyledon removed but are not cut in the hypocotyl region. Controls can be included in a separate plate, or placed on a second strip of Hybond membrane above or below the Hybond membrane with grafted samples.

Acknowledgement

I thank Elliot Meyerowitz and Raymond Wightman for critical reading. This work was funded by a Clare College Junior Research Fellowship and through Gatsby Charitable Trust grants GAT3272/C and GAT3273-PR1.

References

1. Melnyk CW, Meyerowitz EM (2015) Plant grafting. Curr Biol 25:R183–R188
2. Turnbull CG, Booker JP, Leyser HM (2002) Micrografting techniques for testing long-distance signalling in Arabidopsis. Plant J 32:255–262
3. Corbesier L, Vincent C, Jang S, Fornara F, Fan Q, Searle I, Giakountis A, Farrona S, Gissot L, Turnbull C et al (2007) FT protein movement contributes to long-distance signaling in floral induction of Arabidopsis. Science 316:1030–1033

4. Molnar A, Melnyk CW, Bassett A, Hardcastle TJ, Dunn R, Baulcombe DC (2010) Small silencing RNAs in plants are mobile and direct epigenetic modification in recipient cells. Science 328:872–875
5. Rhee SY, Somerville CR (1995) Flat-surface grafting in Arabidopsis thaliana. Plant Mol Biol Rep 13:118–123
6. Yoo SJ, Hong SM, Jung HS, Ahn JH (2013) The cotyledons produce sufficient FT protein to induce flowering: evidence from cotyledon micrografting in Arabidopsis. Plant Cell Physiol 54:119–128
7. Nisar N, Verma S, Pogson BJ, Cazzonelli CI (2012) Inflorescence stem grafting made easy in Arabidopsis. Plant Methods 8:50
8. Huang NC, Yu TS (2015) A pin-fasten grafting method provides a non-sterile and highly efficient method for grafting Arabidopsis at diverse developmental stages. Plant Methods 11:38
9. Melnyk CW, Schuster C, Leyser O, Meyerowitz EM (2015) A developmental framework for graft formation and vascular reconnection in Arabidopsis thaliana. Curr Biol 25:1306–1318
10. Brosnan CA, Mitter N, Christie M, Smith NA, Waterhouse PM, Carroll BJ (2007) Nuclear gene silencing directs reception of long-distance mRNA silencing in Arabidopsis. Proc Natl Acad Sci U S A 104:14741–14746
11. Pant BD, Buhtz A, Kehr J, Scheible WR (2008) MicroRNA399 is a long-distance signal for the regulation of plant phosphate homeostasis. Plant J 53:731–738
12. Green LS, Rogers EE (2004) FRD3 controls iron localization in Arabidopsis. Plant Physiol 136:2523–2531
13. Widiez T, El Kafafi S, Girin T, Berr A, Ruffel S, Krouk G, Vayssieres A, Shen WH, Coruzzi GM, Gojon A et al (2011) High nitrogen insensitive 9 (HNI9)-mediated systemic repression of root NO^{3-} uptake is associated with changes in histone methylation. Proc Natl Acad Sci U S A 108:13329–13334
14. Andersen TG, Nour-Eldin HH, Fuller VL, Olsen CE, Burow M, Halkier BA (2013) Integration of biosynthesis and long-distance transport establish organ-specific glucosinolate profiles in vegetative Arabidopsis. Plant Cell 25:3133–3145
15. Gasperini D, Chauvin A, Acosta IF, Kurenda A, Stolz S, Chetelat A, Wolfender JL, Farmer EE (2015) Axial and radial oxylipin transport. Plant Physiol 169:2244–2254
16. Ragni L, Nieminen K, Pacheco-Villalobos D, Sibout R, Schwechheimer C, Hardtke CS (2011) Mobile gibberellin directly stimulates Arabidopsis hypocotyl xylem expansion. Plant Cell 23:1322–1336
17. Matsumoto-Kitano M, Kusumoto T, Tarkowski P, Kinoshita-Tsujimura K, Vaclavikova K, Miyawaki K, Kakimoto T (2008) Cytokinins are central regulators of cambial activity. Proc Natl Acad Sci U S A 105:20027–20031
18. Van Norman JM, Frederick RL, Sieburth LE (2004) BYPASS1 negatively regulates a root-derived signal that controls plant architecture. Curr Biol 14:1739–1746
19. Xia Y, Suzuki H, Borevitz J, Blount J, Guo Z, Patel K, Dixon RA, Lamb C (2004) An extracellular aspartic protease functions in Arabidopsis disease resistance signaling. EMBO J 23:980–988
20. Yin H, Yan B, Sun J, Jia P, Zhang Z, Yan X, Chai J, Ren Z, Zheng G, Liu H (2012) Graft-union development: a delicate process that involves cell-cell communication between scion and stock for local auxin accumulation. J Exp Bot 63:4219–4232

Chapter 3

Tips and Tricks for Exogenous Application of Synthetic Post-translationally Modified Peptides to Plants

Nathan Czyzewicz, Elisabeth Stes, and Ive De Smet

Abstract

The first signaling peptide discovered and purified was insulin in 1921. However, it was not until 1991 that the first peptide signal, systemin, was discovered in plants. Since the discovery of systemin, peptides have emerged as a potent and diverse class of signaling molecules in plant systems. Peptides consist of small amino acid sequences, which often act as ligands of receptor kinases. However, not all peptides are created equal, and signaling peptides are grouped into several subgroups dependent on the type of post-translational processing they undergo. Here, we focus on the application of synthetic, post-translationally modified peptides (PTMPs) to plant systems, describing several methods appropriate for the use of peptides in *Arabidopsis thaliana* and crop models.

Key words Post-translationally modified peptide, Synthetic peptide, *Arabidopsis*, Cereal crops, In vitro growth

1 Introduction

In addition to phytohormones, such as auxin, cytokinin, gibberellin, and abscisic acid (see other chapters in this book), peptides also impact on various aspects of plant growth and development [1–3]. The terms "peptide," "peptide signal," "signaling peptide," and "peptide hormone" (not to be confused with "signal sequences" which are small domains of larger proteins that dictate cellular localization) are often used synonymously, although all of these are relatively vague terms. In mammalian models, there are three main classes of hormones, comprising peptides, steroids, and amino acid derivatives. Each of these classes are further grouped by mechanism of signaling, namely autocrine (self), paracrine (adjacent/nearby tissue), or endocrine (systemic) signaling [4]. In this regard, "classical" plant hormones bear similarity to steroids (as isoprenoid derivatives) or amino acid derivatives, and act systemically. Using this system of classification, most plant peptides can be classified as "paracrine peptide hormones", although there are exceptions

Jürgen Kleine-Vehn and Michael Sauer (eds.), *Plant Hormones: Methods and Protocols*, Methods in Molecular Biology, vol. 1497, DOI 10.1007/978-1-4939-6469-7_3, © Springer Science+Business Media New York 2017

(e.g., *Glycine max* RHIZOBIA-INDUCED CLE1 and 2) [5], which act in an endocrine manner. Peptides consist mainly of relatively short chains of amino acid monomers linked by amide bonds (<100 amino acids), and are—so far—generally acting as ligands for receptor kinases, initiating signaling cascades that steer the development of the plant or a specific organ, in response to environmental, developmental or physiological stimuli. Most (but not all) peptides are proteolytic cleavage products of larger precursor sequences [1–3], and are found in two main families, post-translationally modified peptides (PTMPs) and cysteine-rich peptides. Cysteine-rich peptides, as the name suggests, have multiple cysteine residues, which form disulfide bridges as the peptide folds, and are critical for maintaining the three-dimensional shape of the peptide [1, 2]. Here, we focus on one class, namely PTMPs. PTMPs are usually shorter than cysteine-rich peptides, and are subjected to an array of modifications prior to proteolytic cleavage, which are able to alter the bioactivity of the peptide, presumably by altering the shape or affinity of the peptide for its receptor [6]. Since PTMPs tend to exist as short 12–15 amino acid sequences, it is relatively easy to procure synthetic forms of these peptide signals (even with some modifications) and examine their activity on plant models.

Functional analysis of PTMPs requires multiple approaches, including (but not restricted to) loss-of-function (if available) and gain-of-function (overexpression lines or application of synthetic peptides) studies, analysis of expression patterns, and biochemical ligand-receptor studies. Whilst not an ideal approach, in part due to the off-target effects, exogenous application of synthetic variants of PTMPs can still give an idea of the peptide function; and can similarly facilitate the study of downstream effects and the mode-of-action.

In this chapter, we describe several PTMP-associated experimental setups and highlight important aspects and considerations in the use of synthetic peptide application. Key points to the use of synthetic peptides that are described are:

- Pure peptide—check for contamination
- Define a relevant biological assay
- Use correct solvent for dissolution/stock creation
- Determine optimum working concentration experimentally (and peptide should be active at low concentrations to increase biological relevance)
- Determine best route of application for required experiment
- Complement with genetic loss- and gain-of-function approaches and expression analysis

2 Materials

2.1 Reconstitution and Dilution of Peptide

1. Synthetic peptide (several companies provide these).
2. Water (ideally MilliQ 18 MΩ cm at 25 °C) (*see* **Note 1**).
3. −20 °C Freezer.

2.2 Analysis of Peptide Effect

1. Seed stocks (wild-type, reporters, and mutants of interest) (acquired from for example NASC (http://arabidopsis.info/) or ABRC (https://abrc.osu.edu/)).
2. Microcentrifuge tubes.
3. 70% ethanol.
4. Sterilization solution: 4% hypochlorite (NaOCl), 0.05% Tween 20.
5. Growth medium: 2.15 g/l Murashige and Skoog salts, 0.1 g/l myo-inositol, 0.5 g/l, 2-(*N*-Morpholino)ethanesulfonic acid hydrate. Adjust pH to 5.7 using 1 M NaOH (*see* **Note 2**).
6. 200 ml conical flasks.
7. Square petri dishes, 120 × 120 × 17 mm.
8. Glass test tubes 20 mm diameter × 150 mm height and appropriate sterilisable caps.
9. 50 ml sterile centrifuge tubes.
10. Pipette filler and sterile 25 ml pipettes.
11. Micropore tape.
12. Laminar flow cabinet.
13. 10, 200, and 1000 μl micropipettes and sterile tips.
14. Forceps.
15. Growth chamber (21 °C, continuous light).
16. −20 °C freezer.
17. Rotary shaker (21 °C, continuous light).

3 Methods

3.1 Initial Reconstitution and Dilution

1. Determine the p*I* of the peptide using the "Compute pI/Mw" tool via ExPASy Bioinformatics Resource Portal.
2. To dilute the peptide to a consistent concentration, accounting for purity of the peptide (*see* **Note 3**), calculate the total volume of solvent required using the following equation (Eq. 1) prior to dissolution.

$$\text{Volume required (L)} = \left(\frac{\text{Mol}}{\text{concentration required (mol / L)}}\right) \times \left(\frac{\text{Purity (\%)}}{100}\right) \quad (1)$$

3. Reconstitution should always first be attempted in a small volume of sterile, distilled water, (ideally MilliQ 18 MΩ cm at 25 °C), but other chemicals may be required to adjust pH in order to fully dissolve the peptide (*see* **Note 4**).

3.2 Sterilization of Arabidopsis Seeds

1. Distribute seeds into microcentrifuge tubes.
2. Add 1 ml of 70% ethanol, invert to ensure full coverage of seeds and incubate at ambient temperature for 30 s. A rotary shaker can be used to keep large amounts of seeds suspended for even coverage.
3. Replace the ethanol with 1 ml of sterilization solution, invert to ensure full coverage of seeds and incubate at ambient temperature for 15 min. A rotary shaker can be used to keep large amounts of seeds suspended for even coverage.
4. In a laminar flow cabinet, rinse the seeds three times in 1 ml sterile, distilled water to remove sterilization solution.
5. Vernalize the seeds by storage at 4 °C in the dark for two nights.

3.3 Sterilization of Wheat/Tomato/Oilseed Rape/Brachypodium Seeds

1. Distribute seeds into 50 ml centrifuge tubes.
2. Add 30 ml of 70% ethanol, invert to ensure full coverage of seeds and incubate at ambient temperature for 30 s. A rotary shaker can be used to keep large amounts of seeds suspended for even coverage.
3. Replace the ethanol with 30 ml of sterilization solution, invert to ensure full coverage of seeds and incubate at ambient temperature for 30 min. A rotary shaker can be used to keep large amounts of seeds suspended for even coverage.
4. In a laminar flow cabinet, rinse the seeds three times in 30 ml sterile, distilled water to remove sterilization solution.
5. Vernalize the seeds by storage at 4 °C in the dark for 1 week.

3.4 Inclusion of Synthetic Peptide in Solid Medium for Arabidopsis

1. Melt solid growth medium in a covered 100 °C water bath or microwave (or use directly from the autoclave).
2. Cool media to <50 °C.
3. Working in a laminar flow cabinet, use a 50 ml centrifuge tube to dilute peptide to required working concentration in 50 ml cooled media. Invert five times to mix.
4. Pour media into square petri dishes and allow to set.

5. Add approximately 10–20 surface-sterilized seeds (*see* Subheading 3.2) using a 10 μl micropipette and sterile tips.
6. Seal plates with Micropore tape.
7. Place in growth chamber until required for analysis.

3.5 Inclusion of Synthetic Peptide in Solid Medium for Crop Plants

1. Melt solid growth medium in a covered 100 °C water bath or microwave (or use directly from the autoclave).
2. Cool media to <50 °C.
3. Working in a laminar flow cabinet, aliquot required amount of peptide stock for dilution in 20 ml media into sterile glass test tubes.
4. Use a pipette filler and sterile 25 ml pipettes to dilute peptide to required concentration in 20 ml cooled media. Pipette up and down to mix, and allow to set.
5. Add a single surface-sterilized seed (*see* Subheading 3.3) to the surface of media in each test tube using sterilized forceps.
6. Replace the sterile cap on each test tube and seal with Micropore tape.
7. Place in growth chamber until required for analysis.

3.6 Inclusion of Synthetic Peptide in Liquid Medium for Arabidopsis

1. After preparing 100 ml aliquots of ½ MS liquid medium (*see* **Note 5**) and sterilizing in conical flasks to be used for the subsequent experiment, cool to ambient temperature.
2. Working in a laminar flow hood, remove and retain aluminum foil seal.
3. Add surface-sterilized seeds (approximately 50 mg dry weight seeds per 100 ml media) using a 1000 μl micropipette and sterile tips.
4. Peptide can be added at this stage by diluting it to the required concentration in the liquid medium (*see* **Note 6**).
5. Reseal culture vessel with aluminum foil (sterilized with flask).
6. Place on rotary shaker at 90–100 rpm in optimal growth conditions until required for analysis.

3.7 Direct Application of Peptide to Tissue

1. Dilute peptide to appropriate working concentration—determined by dose–response analysis (*see* Subheading 3.8).
2. Autoclave or melt growth medium, cool to <50 °C.
3. Working in a laminar flow cabinet, measure out 50 ml of medium using a sterile 50 ml centrifuge tube, aliquot into square petri dishes and allow to set.
4. Add approximately 10–20 surface-sterilized seeds (*see* Subheading 3.2) using a 10 μl micropipette and sterile tips.
5. Seal plates with Micropore tape.

6. Place in growth chamber until required for treatment.
7. In a laminar flow cabinet, remove seal, and apply peptide directly to the tissue of interest.
8. Reseal plate and replace in growth chamber until required for analysis.

3.8 Dose–Response Analysis

1. Create stocks of peptide at a range of concentrations between 10 mM and 1 nM by serial dilution.
2. These stocks can be included in medium (usually 1:1000) as detailed in Subheadings 3.3 and 3.4, giving a final concentration range of 10 μM to 1 pM.
3. *Arabidopsis* or crop plants can be grown on this medium as detailed in Subheadings 3.3 and 3.4, respectively.

4 Applications

4.1 Determination of Relevant Biological Assay for Measuring Peptide Bioactivity

To determine the effect of a synthetic peptide on plants, it must be applied to a tissue, organ or even the whole plant to induce a biological effect. Since peptides are able to affect a range of developmental, physiological and biochemical processes [2], it is important to establish a relevant biological assay for determining bioactivity either prior to, or upon first application of the peptide. This can be at the level of a growth, physiological or biochemical read-out or assessing the response of relevant markers [7, 8]. Transcriptional fusions can illustrate the type of tissue in which the peptide acts or is secreted from, allowing for some idea of whether any observed phenotype is relevant. Reporter lines can give an indication of how peptides can affect the downstream response [7]. With respect to biological assays, examples include (but are not limited to) root growth, meristem size, medium acidification, membrane depolarization, and measuring oxidative burst [8–10]. Additionally, when studying a large peptide family (e.g., CLAVATA3/EMBRYO-SURROUNDING REGION (CLE) peptides), it is important to factor for functional redundancy within the family. This can be achieved by screening multiple peptides using the same biological assay, and determining any difference in peptide bioactivity. However, the best indication of relevance is given by comparison with expression pattern analysis and loss-of-function mutants.

4.2 Screening for Homologous Signaling Modules Across Multiple Species

A continuing question surrounding studies conducted in *Arabidopsis*, is whether any discovery made in this model organism is transferrable to other species. Utilizing synthetic peptides, it is possible to treat other model organisms with synthetic peptides in the same way as *Arabidopsis*, with some minor adaptations for the size and amount of growth of the species.

4.3 Amino Acid Substitutions

It is possible to obtain a range of synthetic peptides with modified amino acid sequences from most peptide suppliers. Substitution of amino acids with inert residues can be useful in providing insight into peptide conformation, and also can reveal information on exposed side chains important for interaction of peptide and receptor. Replacement of amino acids in this manner can also occasionally identify peptides that inhibit the function of the native peptide via competitive inhibition—known as antagonistic peptides [11–13].

Alanine scanning is a useful technique for determining critical amino acids for peptide function, and can additionally provide insight into peptide secondary structure. Since peptides can be synthesized directly, it is relatively simple and far less time consuming to apply synthetic peptides where every amino acid is sequentially replaced by an alanine (or another inert amino acid) to probe for bioactivity than it is to create and transform constructs into *Arabidopsis*. Occasionally, alanine scanning (and subsequent physiological or biochemical analysis) can also lead to identification of mutant peptides that are able to produce the opposite phenotype compared to the unmodified peptide (putative antagonistic peptides). Antagonistic peptide technology should be applied with caution however, since there is no strict single modification or substitution of amino acid residues that works for every peptide, as different peptides will interact with their individual receptors via a different combination of exposed side chains and overall 3D structure [11, 13].

4.4 Modification of Peptides

Proteins are known to undergo multiple forms of modification, and peptides are no exception, often undergoing multiple forms of modification in vivo prior to proteolytic cleavage. There are several common post-translational modifications that are able to affect peptide bioactivity, for example tyrosine sulfation and/or hydroxyprolination (and subsequent l-arabinosylation), are known to alter the bioactivity of the peptide by adjusting the shape and/or binding affinity of the peptide to its receptor [6]. Peptides can be synthesized with a number of modifications, which can be useful if the user desires to determine if amino acid modification affects structure and/or bioactivity.

5 Caveats

5.1 Nonspecific Effects

Perhaps the most obvious caveat to peptide application is that the applied peptide is made available to cell types and tissues that would not normally be exposed, and may produce nonspecific effects in the studied plant. Since the mature sequence of many peptides is reasonably short, there is often a relatively large amount of similarity in the mature peptide sequences within a family, which may activate receptors in the incorrect tissue and mislead the analyst with an unrelated phenotype. From this point of view, it is critical that any data generated by peptide application is supported by a corresponding expression analysis, and—if possible—comparison with a phenotypic or biochemical analysis of knockout and (mild) overexpression lines to determine if observed effects to the plant correspond accordingly. Comparison with activity of similar peptides can also help to negate any concern over nonspecificity.

5.2 Contamination

There are multiple routes by which contamination can affect experiments utilizing peptide application, including (but not restricted to) contamination of the peptide during synthesis, stock solution generation, and preparation of medium. It is important to confirm that any observed effects on a biological model are indeed the result of applying the peptide of interest and not some form of contamination [e.g., *see* ref. 14 on flg22]. HPLC chromatogram data (provided with synthesized peptide) should indicate any contamination, however this is not infallible if the contaminant is also a similarly sized/charged peptide or present in extremely low (but sufficient for a response) quantities. If there is concern as to the specificity of a peptide, obtaining a peptide of high purity decreases the risk of contamination-related effects, and checking the supplied LC chromatograph for additional peaks can be used to determine whether any potential contaminant is present in the peptide stock. Working with ultrapure water, using analytical grade reagents, and observing sterile technique while preparing solutions and media will further minimize any user-introduced contamination.

6 Notes

1. Other chemicals may be required for optimizing pH of reconstitution solution, according to peptide provider's instructions.
2. For solid medium add 10 g/l plant tissue culture agar prior to autoclaving.

   For liquid medium, autoclave individual 100 ml aliquots in 200 ml conical flasks topped with an aluminum foil cap.
3. Peptide purity is important for minimizing contamination. When purchasing synthetic peptides, first consider for what purpose they are needed.
   - For initial screening of peptide function, aliquots of approximately 70% purity are relatively inexpensive and can be used to determine whether any of the peptides are able to induce a change to morphology/biochemical processes.
   - For further experiments, it is important to use peptides of higher purity (at least >85%, ideally higher), to safeguard against contamination either from a biological or chemical

source within the synthesized peptide aliquot. If there is particular concern over the biological effect of the peptide, aliquots of >99 % purity can be obtained to ensure that any observed effect is specific to the peptide.

4. If the peptide does not fully dissolve in water, further chemicals may be added to adjust pH, allowing dissolution of the peptide.
   - Basic peptides are best dissolved in acidic solutions (acetic acid/TFA).
   - Acidic peptides are best dissolved in basic solutions (NH_4OH).
   - Uncharged peptides are best dissolved in organic solvents (acetonitrile/methanol/isopropanol) according to the peptide provider's instructions.
   - If any chemical other than water is used in the preparation of peptide, it is prudent to additionally create a blank solution, to determine any effect of these chemicals on the organism, tissue, or reporter system downstream of treatment.
5. It is important not to overfill the culture flasks, or the medium will not remain in motion on a rotary shaker and will become stagnant. Use 100 ml media in a 200 ml conical flask for up to 50 mg (dry weight) seeds. This can be scaled up for larger weights of seeds.
6. Synthetic peptide can alternatively be added at a later stage if analysis focuses on more short-term response to peptide treatment.

References

1. Murphy E, Smith S, De Smet I (2012) Small signaling peptides in Arabidopsis development: how cells communicate over a short distance. Plant Cell 24(8):3198–3217
2. Czyzewicz N, Yue K, Beeckman T et al (2013) Message in a bottle: small signalling peptide outputs during growth and development. J Exp Bot 64(17):5281–5296
3. Tavormina P, De Coninck B, Nikonorova N et al (2015) The plant peptidome: an expanding repertoire of structural features and biological functions. Plant Cell 27(8):2095–2118
4. Petraglia F, Florio P, Nappi C et al (2013) Peptide signaling in human placenta and membranes: autocrine, paracrine, and endocrine mechanisms. Endocr Rev 17(2):156–186
5. Reid D, Li D, Ferguson B et al (2013) Structure–function analysis of the GmRIC1 signal peptide and CLE domain required for nodulation control in soybean. J Exp Bot 64(6):1575–1585
6. Shinohara H, Matsubayashi Y (2013) Chemical synthesis of Arabidopsis CLV3 glycopeptide reveals the impact of hydroxyproline arabinosylation on peptide conformation and activity. Plant Cell Physiol 54(3):369–374
7. Czyzewicz N, Shi CL, Vu LD et al (2015) Modulation of Arabidopsis and monocot root architecture by CLAVATA3/EMBRYO SURROUNDING REGION 26 peptide. J Exp Bot 66(17):5229–5243
8. Qi Z, Verma R, Gehring C et al (2010) Ca^{2+} signaling by plant Arabidopsis thaliana Pep peptides depends on AtPepR1, a receptor with guanylyl cyclase activity, and cGMP-activated Ca^{2+} channels. Proc Natl Acad Sci U S A 107(49):21193–21198
9. Pearce G, Strydom D, Johnson S et al (1991) A polypeptide from tomato leaves induces wound-inducible proteinase inhibitor proteins. Science 253(5022):895–897

10. Butenko M, Wildhagen M, Albert M et al (2014) Tools and strategies to match peptide-ligand receptor pairs. Plant Cell 26(5):1838–1847
11. Czyzewicz N, Wildhagen M, Cattaneo P et al (2015) Antagonistic peptide technology for functional dissection of CLE peptides revisited. J Exp Bot 66(17):5367–5374
12. Song XF, Guo P, Ren SC et al (2013) Antagonistic peptide technology for functional dissection of CLV3/ESR genes in Arabidopsis. Plant Physiol 161(3):1076–1085
13. Xu T, Ren S, Song X et al (2015) CLE19 expressed in the embryo regulates both cotyledon establishment and endosperm development in Arabidopsis. J Exp Bot 66(17):5217–5227
14. Mueller K, Chinchilla D, Albert M et al (2012) Contamination risks in work with synthetic peptides: flg22 as an example of a pirate in commercial peptide preparations. Plant Cell 24(8):3193–3197

Chapter 4

Assaying Germination and Seedling Responses of *Arabidopsis* to Karrikins

Mark T. Waters, Gavin R. Flematti, and Steven M. Smith

Abstract

Karrikins are a small family of naturally occurring plant growth regulators present in the smoke and char produced from burning plant material in wildfires. They can stimulate germination of dormant seed and can influence seedling morphogenesis. Although *Arabidopsis thaliana* is not considered to be a smoke-responsive species, karrikins will stimulate seed germination under the appropriate circumstances and will cause repression of hypocotyl elongation in low light. This chapter describes how to conduct assays of the activity of karrikins on *Arabidopsis* seeds and seedlings. The methods presented can potentially be modified for use in a range of *Arabidopsis* genotypes or in other plant species.

Key words Karrikins, *Arabidopsis thaliana*, Seed germination, Dormancy, Hypocotyl elongation, Photomorphogenesis

1 Introduction

Originally isolated from smoke from burning plant material, karrikins (KAR) stimulate the germination of seed in a wide variety of plant species. In *Arabidopsis thaliana*, freshly harvested seed exhibits primary dormancy, such that the seed will not germinate readily under conditions that are otherwise permissive for germination. The depth of primary dormancy depends on the ecotype in question and the growth conditions experienced by the parental plant. However, in many cases KAR at micromolar concentrations can overcome this primary dormancy [1]. Besides germination, KAR also influence seedling development and light sensitivity: KAR-treated seedlings show increased inhibition of hypocotyl elongation, and greater cotyledon expansion [2]. Consistent with these effects, genetic studies indicate that KAR act through a signaling pathway that promotes germination and seedling responses to light [3, 4]. Thus, KAR are potent exogenous plant growth regulators that have revealed a distinct signaling mechanism of plant development.

Jürgen Kleine-Vehn and Michael Sauer (eds.), *Plant Hormones: Methods and Protocols*, Methods in Molecular Biology, vol. 1497, DOI 10.1007/978-1-4939-6469-7_4, © Springer Science+Business Media New York 2017

Here we describe two generic methods for measuring the effects of KAR on *Arabidopsis*—though they could equally be applied to other plant growth regulators, and to other species. A key aspect of detecting an effect of KAR on seed germination is obtaining suitably dormant seed: if the seeds are too dormant, KAR will not be effective, and if the seeds have lost their primary dormancy through prolonged storage (after-ripening) then no further stimulation of seed germination will be detectable. As dormancy is highly variable across genotypes, ecotypes, and growth environments, there is no "one-size-fits-all" approach. However, we believe that the method we describe here will assist researchers by serving as a guide for optimizing experimental conditions. Observing effects of KAR on hypocotyl elongation requires attention to the light conditions but is potentially more transferable between different genotypes or plant species.

2 Materials

2.1 Seed Germination Equipment

1. Round 6 cm diameter Petri dishes.
2. Growth cabinet, temperature controlled.
3. Low-magnification stereomicroscope.
4. Ultralow-temperature freezer for seed storage.
5. Microfuge tubes or seed envelopes for storage.

2.2 Seed Germination Reagents and Materials

1. *Arabidopsis* seed of desired genotype.
2. Bacto-agar or phytoagar.
3. Plant growth regulator stock solutions (1000× or greater).
4. Silica gel desiccant.

2.3 Hypocotyl Elongation Assay Equipment

1. Round 6 cm diameter Petri dishes.
2. Growth cabinet, temperature controlled with LED red light source (*see* **Note 1**).
3. High-resolution digital camera on a camera stand (SLR with macro lens is recommended).
4. PC running ImageJ software (http://imagej.nih.gov/ij/) or similar (*see* **Note 2**).

2.4 Hypocotyl Elongation Assay Reagents

1. 0.05 % (v/v) Triton-X100 in 70 % (v/v) ethanol.
2. 100 % (v/v) ethanol.
3. *Arabidopsis* seed of desired genotype.
4. Half-strength (0.5×) Murashige and Skoog (MS) basal medium, pH 5.9 adjusted with KOH, solidified with 0.7 % (w/v) agar.
5. Plant growth regulator stock solutions (1000× or greater, in appropriate solvent).

3 Methods

3.1 Generating Seed for Germination

1. Grow *Arabidopsis* under consistent, controlled conditions; artificial growth facilities that maintain light levels, temperature, and humidity are highly recommended. Recommended conditions for enhancing primary seed dormancy and rapid growth are long (16 h) days, 22 °C day/18 °C night, 60% relative humidity, and white light intensity 100–150 μmol/m^2/s. Recent data indicate that shifting the plants to a lower temperature (16 °C) after transition to flowering dramatically increases seed dormancy [5]; this technique may be beneficial for generating seed of *Arabidopsis* accessions with otherwise little to no primary dormancy (e.g., Columbia-0).
2. Grow sufficient maternal plants to yield at least three independent seed batches for each genotype, with 3–4 plants contributing to a single batch. The plants should be randomized with respect to position within a tray; ideally, rotate the positions weekly to ensure that all plants grow at approximately similar rates. If your genotypes vary substantially in flowering time, then you may need to determine this in advance, and stagger the growth such that flowering occurs synchronously among all genotypes.
3. Allow the maternal plants to set seed and the siliques to mature. Inevitably, early siliques will mature first, and thus the seed collected from any given plant will be heterogeneous in age. In our view, it is impractical to collect seed from individual siliques as they mature, but for very subtle germination phenotypes, a staged collection strategy might be necessary. In general, the longer the seed is allowed to dry on the plant, the sooner the primary dormancy is lost. Therefore, the time of harvesting is a compromise between obtaining a suitable amount of mature seed and maintaining a useable level of primary dormancy. Finding the right time to harvest is critical and depends on the particular growth conditions. It must be determined empirically, perhaps by harvesting seed from different plants at varying levels of maturity over a period of 4 weeks. Typically we stop watering once flowering has ceased (the primary shoot apex has terminated), but before all siliques have turned brown. The youngest siliques will still be maturing at this stage, but these can be removed before seed harvesting. We then harvest seed about 1 week later.
4. Harvest the seed and pool from 3 to 4 plants to form one seed batch. Collect the seed in small envelopes or in 1.5 ml microfuge tubes. Envelopes are easier to label and will assist subsequent drying, but are perhaps less convenient to handle and store in the freezer.
5. Place the seed under silica gel desiccant at room temperature for 48–72 h, to improve viability of seed under storage. A plas-

tic food container containing anhydrous silica gel is suitable for this purpose (*see* **Note 3**).

6. Store the seed in an ultralow-temperature freezer (−70 to −86 °C). This reduces the rate of loss of primary dormancy. If the seeds are stored at higher temperatures, after-ripening will gradually occur and the effects of KAR will be hard or impossible to observe (*see* **Note 4**).

3.2 Conducting Germination Tests

Before setting up a very large experiment, we suggest testing one genotype—typically wild type—under different incubation conditions. Consider the gelling agent (**step 1**), the incubation temperature, and the diurnal period (**step 6**). When you have something that appears to work well, set up the full experiment.

1. Prepare 1% (w/v) agar in milliQ water by autoclaving. We have used bacto-agar with success in the past, but we have found that certain batches of agar inhibit germination. Phytoagar is more consistent; agarose could also be considered. If required, add chemical compounds using 1000× stocks in acetone (e.g., 1 mM KAR_1), and mix well. When pouring the plates, 6 cm petri dishes hold 7–8 ml of agar. Allow plates to dry out for about 30 min in a laminar flow hood, to avoid condensation later. Use fresh plates each time you set up an experiment. When labeling plates, we recommend a "blind" approach.
2. Surface sterilize the seed. A convenient method when dealing with many samples is by exposure to gaseous chlorine: Dispense sufficient seed into 1.5 ml microfuge tubes, and place in a desiccator jar that contains 100 ml 10% (w/v) sodium hypochlorite solution. In a fume cupboard, add 3 ml concentrated HCl to the solution, and quickly close the jar. Leave for about 3–4 h. For an alternative sterilization method, *see* Subheading 3.3 below.
3. Sow seed, about 100 per plate. Punch a small hole in the top of the tube with a pin or needle, invert the tube, and tap to sprinkle the seed on the plates. Try to avoid clumps of seed.
4. Seal the plates with gas-permeable tape, often referred to as "surgical tape" or "Leukopor"®.
5. On the underside of the plate mark out with a small black dot, four groups of 25 seeds (100 in total) to make germination counting easier. Because the seeds can move or clump together while imbibing, it is best to mark the seeds a couple of hours after sowing, but before germination starts.
6. Incubate the plates under white light, 100–150 μmol/m^2/s, at 20–22 °C. If the seeds are not very dormant, you can slow down germination by increasing the temperature to 28 °C (thermoinhibition). This often makes the effect of germination stimulants easier to detect. We typically use constant light conditions, but others report 12-h light:12-h dark with success [5].

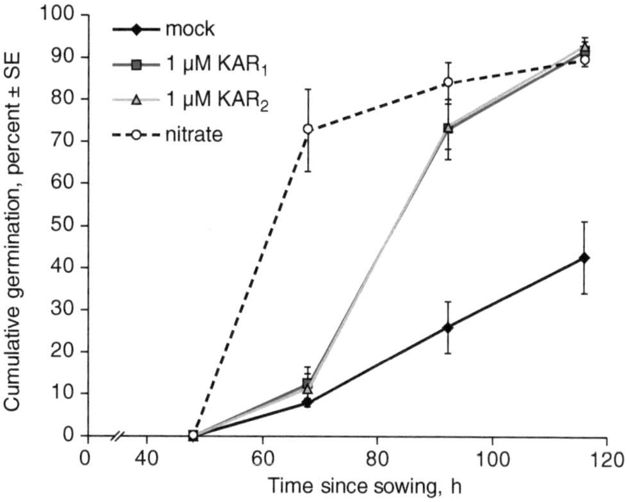

Fig. 1 Germination response of primary dormant *Arabidopsis* seed (ecotype Landsberg *erecta*) to karrikins. Seed were sown on water-agar supplemented with 0.1 % (v/v) acetone (mock) or with 1 μM of two different karrikins, KAR_1 and KAR_2. For "nitrate," seeds were sown on 0.5× MS medium + 1 % (w/v) agar, stratified in the dark for 72 h, and transferred to the same growth conditions as the other treatments. Note that the germination response to karrikins is not as strong as to stratification in the presence of nitrates. Germination was inspected every 24 h. Error bars are mean ± SE of $n = 3$ seed batches

Count germinated seeds at least every 24 h using a stereomicroscope. Classify a seed as germinated when you can see unambiguous radicle emergence (e.g., 1 mm), and mark such seeds with a different color. Be aware that some inks fade rapidly under fluorescent light in incubators! Failure to inspect plates regularly will make it hard to detect germination, as the first seeds to germinate can quickly obscure the others. Length of time to germination depends on many factors—ecotype, incubation temperature, and seed age. With Landsberg *erecta*, germination normally starts after 48 h, and Col-0 within 24 h. Most experiments have run their course within 6 days; cumulative germination expressed as a percentage typically follows a sigmoidal curve (Fig. 1).

For statistical analysis, we treat each seed batch within a genotype/treatment as an independent replicate. Because the germination data are expressed as percentages, standard procedure is to use the arcsine transformation before statistical analysis. The entire experiment should be repeated—ideally with different batches of seed—to ensure that a conclusion is robust.

3.3 Hypocotyl Elongation Assays

For this assay, seedlings are grown in the presence of KAR under red light (*see* **Note 1**). Seedlings grown under red light have much longer hypocotyls than under white (or blue) light, making the effect of KAR on seedling development easier to observe.

1. Surface sterilize seed using chlorine as above (Subheading 3.2, **step 2**), ethanol (below), or your preferred method. Determine how much seed you will require for all treatments, assuming that you will sow approximately 40 seedlings per plate. We dispense seed into 1.5 ml microfuge tubes, add 70% (v/v) ethanol containing 0.05% (v/v) Triton-X100, and mix for 5 min by occasional inversion. Allow the seed to settle, remove the solution by pipetting, and replace with approximately 1 ml of 100% (v/v) ethanol. Mix briefly, allow the seed to settle, and aspirate as much of the ethanol as possible. Close the lid, flick the tube to disperse the seed, and leave the tube open and on its side in the laminar flow hood to allow residual ethanol to evaporate. After about 10 min, the seed will be dry and ready to sow.
2. In the meantime, prepare the MS-agar plates by adding KAR at the desired concentration, as described for seed germination assays above.
3. Sow seed, seal the plates with permeable tape, and stratify at 4 °C for 72 h in the dark.
4. Transfer the plates under white light (~100–150 μmol/m^2/s) for 3 h at 20–22 °C. This exposure to broad-spectrum light will ensure homogeneous germination.
5. Transfer the plates into darkness for 21 h at 20–22 °C, placing them in the same location as the red light source. This period of darkness is not essential but it means that red light treatment will begin at a consistent 24 h after removal of the seed from cold.
6. Incubate the plates at 20–22 °C under continuous red light for 4 days.
7. Remove the plates from the incubator and photograph them on a blue or black background to enhance contrast. If there are too many seedlings to clearly see each one individually, you can remove them manually and arrange them on another agar plate prior to photography. (IMPORTANT: Also photograph an object of known size (e.g., a coin, or a grid pattern). If you do not have a camera stand that allows each photograph to be uniformly composed, this object should be in every photograph and the image dimensions calibrated for each photograph.)
8. Import each image into ImageJ software, one at a time.
9. Calibrate the image dimensions (to convert pixels into mm) as follows. Draw a line of known physical length across your calibration image taken in **step 7**. Use the "Set Scale" command under the "Analyze" menu to define the length in millimeter of the calibration object.

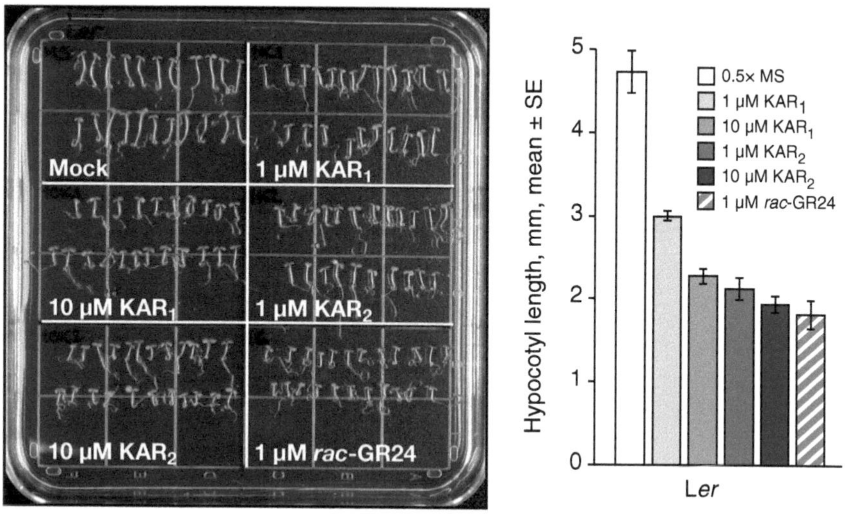

Fig. 2 Example of KAR-treated seedlings from the hypocotyl elongation assay. Approximately 20 wild-type Landsberg *erecta* seedlings from each treatment group were arranged flat on an agar plate after 4 days of growth under continuous red light. KAR_1 and KAR_2 are two different karrikins; *rac*-GR24 is a synthetic strigolactone analogue that has similar effects on seedling growth as KAR (but note that *rac*-GR24 does not promote cotyledon expansion). "Mock" corresponds to seedlings grown on 0.5× MS media supplemented with 0.1 % (v/v) acetone as a treatment control. The grid pattern, with 13 mm spacing, serves as a size calibration for the image. This image has been processed in Adobe Photoshop to enhance contrast for reproduction in grayscale. The chart on the *right* depicts data from three such images derived using ImageJ, each representing an independent experimental replicate

10. Measure hypocotyl lengths manually using the "segmented lines" tool. Measure from the root-hypocotyl junction to the apical meristem, following the line of the hypocotyl. Do this for 15–20 seedlings, and copy the data into a spreadsheet.

The procedure described above, performed once, comprises one experimental replicate. Because of sampling error, we perform this procedure three times or more (on different dates) for a given experiment, and use each occasion as a separate experimental replicate for statistical analysis. Thus, individual seedlings can be nested within a replicate, but each replicate should be considered the independent sample. In practise, we simply take the mean hypocotyl length for each of the three replicates, and compare the mean of these replicates across genotypes and treatments (Fig. 2).

4 Notes

1. Suitable LEDs can be found at electronics supply stores and online, often sold as self-adhesive strips with red, green, and blue LEDs, complete with power supply and remote control (e.g., www.eaglelight.com). Ideally these should be characterized

with a spectrometer to determine the wavelength and intensity of each light source. The peak output of the red LED should lie between 600 and 660 nm, with integrated intensity of between 2 and 20 μmol/m^2/s, but these parameters are not critical for this assay.

2. ImageJ is a public domain, Java-based image processing program developed at the National Institutes of Health (http://imagej.nih.gov/ij/).
3. Seeds harvested for germination assays may be equilibrated in an atmosphere of 15 % relative humidity if you have facilities to do this, instead of incubation in a box containing anhydrous silica gel, prior to cold storage.
4. While −70 to −86 °C is preferred for storage of seeds to maintain dormancy from many months to years, −20 °C may be adequate for several months.

References

1. Nelson DC, Riseborough J-A, Flematti GR, Stevens J, Ghisalberti EL, Dixon KW et al (2009) Karrikins discovered in smoke trigger Arabidopsis seed germination by a mechanism requiring gibberellic acid synthesis and light. Plant Physiol 149:863–873
2. Nelson DC, Flematti GR, Riseborough J-A, Ghisalberti EL, Dixon KW, Smith SM (2010) Karrikins enhance light responses during germination and seedling development in Arabidopsis thaliana. Proc Natl Acad Sci U S A 107: 7095–7100
3. Nelson DC, Scaffidi A, Dun EA, Waters MT, Flematti GR, Dixon KW et al (2011) F-box protein MAX2 has dual roles in karrikin and strigolactone signaling in Arabidopsis thaliana. Proc Natl Acad Sci U S A 108: 8897–8902
4. Waters MT, Nelson DC, Scaffidi A, Flematti GR, Sun YK, Dixon KW et al (2012) Specialisation within the DWARF14 protein family confers distinct responses to karrikins and strigolactones in Arabidopsis. Development 139:1285–1295
5. Chen M, MacGregor DR, Dave A, Florance H, Moore K, Paszkiewicz K et al (2014) Maternal temperature history activates Flowering Locus T in fruits to control progeny dormancy according to time of year. Proc Natl Acad Sci U S A 111:18787–18792

Chapter 5

Low-Cost Microprocessor-Controlled Rotating Stage for Medium-Throughput Time-Lapse Plant Phenotyping

Francis Barbez, Jürgen Kleine-Vehn, and Elke Barbez

Abstract

Here we provide the instructions to build a cost-friendly rotating stage, which enables time-lapse phenotyping of seedlings, grown vertically on in vitro plates, in a medium-throughput manner.

Key words Rotating stage, Plant phenotyping, Arduino

1 Introduction

Due to their sessile lifestyle, plants feature an enormous capacity for dynamic growth regulation in response to developmental as well as environmental signals. To improve our understanding of these dynamic developmental processes, the use of time-lapse plant phenotyping approaches is becoming increasingly important. Nowadays, a high range of excellent time-lapse phenotyping platforms have been developed and are custom-made as well as commercially available [1–3]. Unfortunately, the development and construction or the commercial purchase of these platforms are expensive and may not be cost effective for certain projects. Here, we present an in-house-built rotating stage which enables customized time-lapse plant phenotyping in a medium-throughput manner. Using the open-source Arduino prototyping platform, we developed and programmed a microcontroller that steers a rotating stage carrying up to eight in vitro-grown plates (Fig. 1). In combination with an ordinary consumer-grade single-lens reflex camera, the micro-controlled rotating stage significantly increases the number of seedlings that can be monitored over time. Below are comprehensive instructions, a complete electric schematic diagram, and the adjustable Arduino script for building a micro-controlled rotating platform for less than €200.

Jürgen Kleine-Vehn and Michael Sauer (eds.), *Plant Hormones: Methods and Protocols*, Methods in Molecular Biology, vol. 1497, DOI 10.1007/978-1-4939-6469-7_5, © Springer Science+Business Media New York 2017

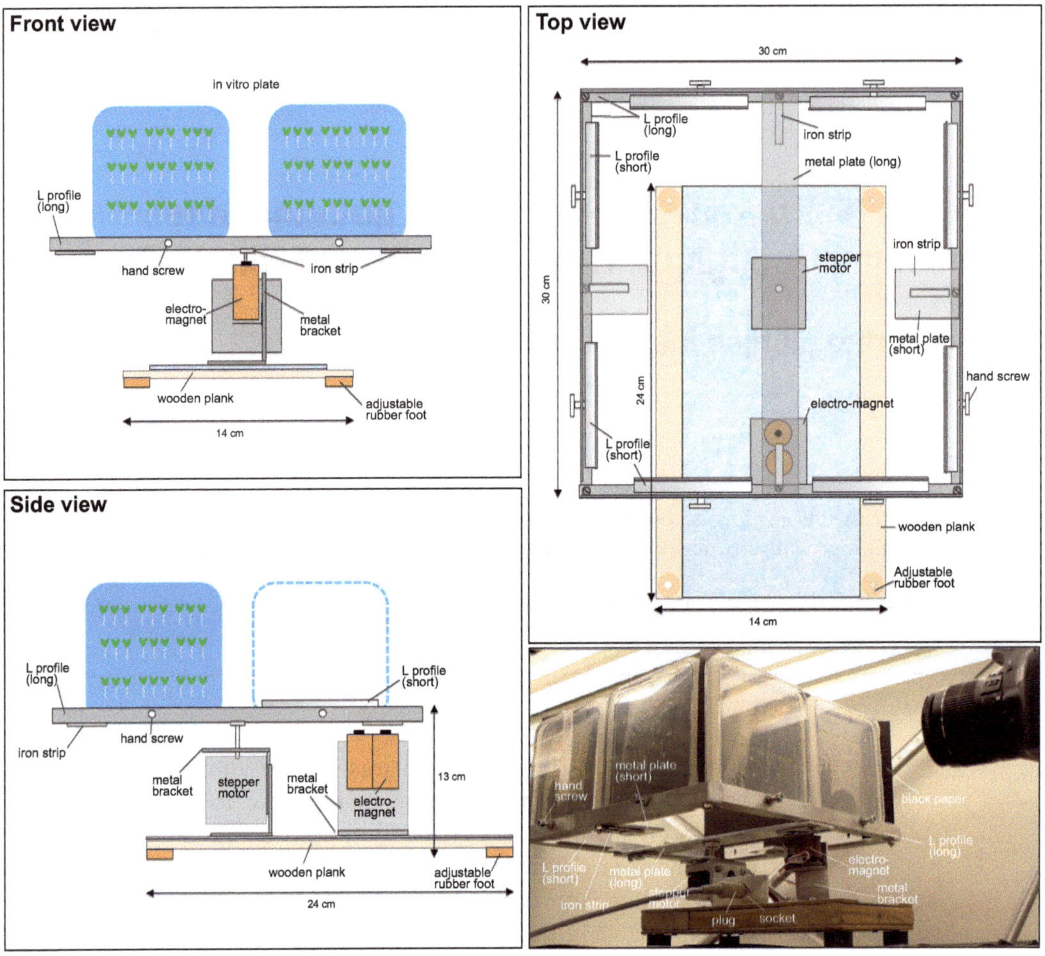

Fig. 1 Schematic drawing of construction

2 Materials

2.1 The Rotating Stage

Table 1 depicts the materials needed to construct the rotating stage.

2.2 The Microcontroller

Table 2 depicts the materials needed (*see* **Note 1**) to construct the microcontroller.

2.3 Software

The software used is free of charge.

1. Arduino to upload the provided script on the Arduino-Uno microcontroller: download from https://www.arduino.cc/
2. Fiji for image analysis: download from http://fiji.sc/Fiji

2.4 Imaging

1. Digital, single-lens reflex (or mirrorless-type system) camera including imaging software to control the camera from a com-

Table 1
Material needed for constructing the rotor

#	Symbol	Name	Specifications
1×		Stepper motor	6 wires (SanyoDENKI 103-547-52500)
1×	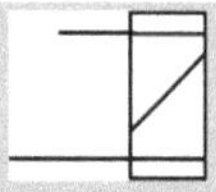	Electromagnet	100 N 12 V/DC 3.8 W
#	Name	Specifications	Comments
1×	Wooden plank	14 cm×24 cm	The basis of the stage
4×	Adjustable rubber feet		To keep the stage in a stable position
4×	L-shaped metal bracket		To attach the motor and electromagnet on the plank
4×	Aluminum L profile (long)	2.5 cm×2.5 cm×30 cm (2 mm thick)	For the frame
8×	Aluminum L profile (short)	2.5 cm×2.5 cm×30 cm (2 mm thick)	For the plate holders
1×	Metal plate (long)	29 cm×4 cm (2 mm thick)	To attach the frame to the motor
2×	Metal plate (short)	6 cm×4 cm (2 mm thick)	To attach the magnetic strips
8×	Hand screws and nuts		To tighten the in vitro plates on the frame
4×	Magnetic strip		To position the stage exactly above the electromagnet after turning
	⊘ Additional screws and nuts		To construct the frame and attach the different parts
4×	Black paper	30 cm×12 cm	To provide a black background behind the in vitro plates

puter. There are multiple options for equipping this setup with a camera and the pace of technical advance in this field is fast; therefore, we do not recommend any specific brand or make. Just make sure that the camera/lens combination can bring two in vitro culture plates side by side into focus (*see* Fig. 1) and can be remote controlled by a computer.

2. Laptop to control the camera.

2.5 Tools

No specialized tools are required to build the rotating stage.

Standard tools include Soldering iron, screwdrivers, metal glue, drill, and saw.

Table 2
Electric components needed for constructing the control unit

#	Symbol	Name	Specifications
1×		Arduino microcontroller	Arduino-UNO (US only)/ Genuino-UNO (outside of US)
1×		Big Easy Driver Stepper motor driver board	Sparkfun, Big Easy Driver ROB-12859
2×		Diode	1000 V/1 A
2×	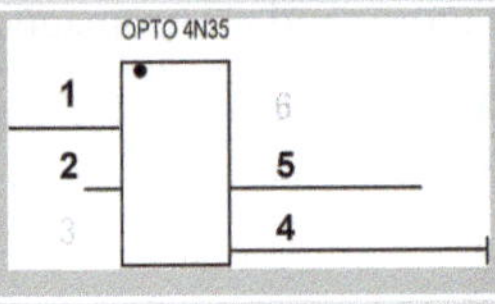	Optocoupler	4N35
2×	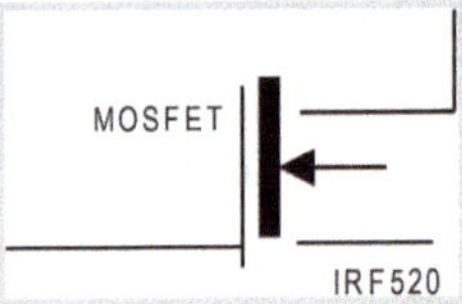	Mosfet/hexfet	IRF520
5×		Led	5 mm, round, 20 mA, 2.5 V
1×		Electrolytic capacitor	2.5 mm 100 µF 16 V/DC 20% (Ø×l) 6.3 mm×11 mm SY 100 µF/16 V 6.3×11 mm
	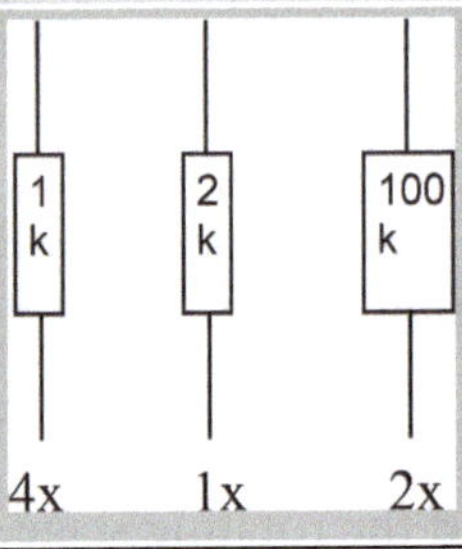	Resistor	Carbon film-fixed resistors

(continued)

Table 2 (continued)

#	Symbol	Name	Specifications
1×	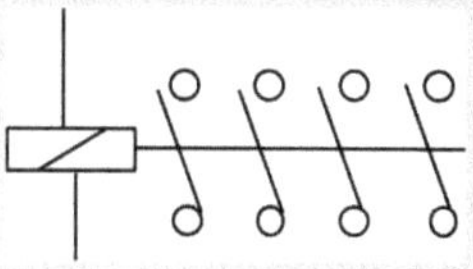	Relais	4 Pole relais 12 V/DC 7 A
1×	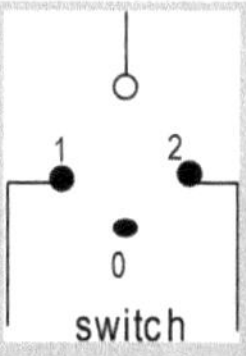	Round rocker switch	250 V/AC 6 A
1×	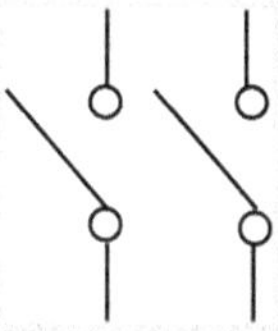	Rocker switch	250 V/AC 10 A

#	Name	Specifications
1×	Panel-mounted socket with flange	6 contacts, 4 A, 34 V
1×	Plug with insulated handle	6 contacts, 4 A, 34 V
1×	6-Wire control cable	6×0.25 mm^2
	Electric wire	0.14–0.25 mm^2
	Perfboard, dot-PCB	For example 75 mm × 100 mm, epoxy, Cu 35 μm
1×	USB cable	
1×	Adaptor/power supply	12 V/3 A
1×	Adaptor/power supply	5 V/1 A
	Soldering tin	

3 Methods

3.1 The Stage

Figure 1 gives a detailed representation of how the stage can be built.

1. Drill two holes (with a diameter slightly bigger than the hand screws) in one flank of every long L-profile. The holes are located 8 cm from each end.

2. To construct the frame, screw the long L-profiles tightly together at the corners so that the perforated flanks stand upright and the horizontal flanks face the inside of the frame.
3. Screw the long metal plate in between two sides of the frame. Drill a hole, exactly in the middle of the metal plate, to later on attach the frame on the axle of the stepper motor.
4. Screw on the two short metal plates in the middle of the two sides that were not connected with the long metal plate.
5. Attach the iron strips to the lower sides of the metal plates using metal glue.
6. Glue the eight short L-profiles onto the horizontal flanks of the frame to create a groove, which is slightly wider than the thickness of the in vitro plates used. Attach a profile facing every perforation.
7. Glue the eight nuts belonging to the hand screws over the perforations on the outside of the frame. When the glue is thoroughly dry, screw the hand screws loosely in the nuts. The hand screws will enable the in vitro plates to be fixed tightly on the frame (*see* **Note 2**).
8. Screw the adjustable rubber feet onto the bottom of the wooden plank.
9. Use 2×2 metal brackets to attach the stepper motor and the electromagnet to the plank as shown in Fig. 1 (*see* **Note 3**).
10. The presence of the electromagnet will keep the stage in the right position after every 90° turn (*see* **Note 4**). The optimal distance between the iron strip and the electromagnet (3–6 mm) may have to be optimized by trial and error.
11. Install the socket in a leftover piece of L-profile and attach this to the wooden plank. The six contacts of the socket will be later connected to the four cables of the stepper-motor and the two cables of the electromagnet according to the electric diagram given in Fig. 2.

3.2 Microcontroller

Figure 2 gives a detailed representation of how the microcontroller can be built (*see* **Note 5**).

1. Build the microcontroller on a perfboard, also called dot-PCB (or, depending on your electronics skills, a breadboard, stripboard, or even a printed circuit board of your own design) using a soldering iron according to the electric diagram in Fig. 2 (*see* **Notes 6** and 7).
2. We chose to build the microcontroller in the shell of an old external hard drive (Fig. 2). It can, however, be built into any handmade or commercially available heat-resistant electronic housing.

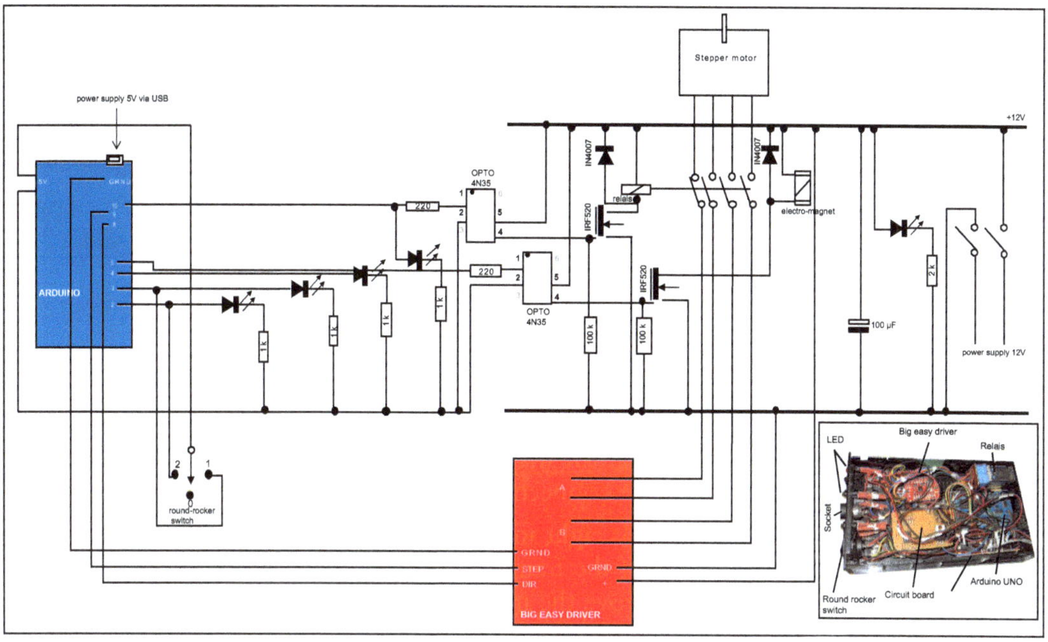

Fig. 2 Electric circuit for the control unit

3.3 Uploading the Arduino Script

Here we provide an Arduino script, which allows steering of the stage's rotation every desired time interval (in our case 15 min) over an angle of 90°. The script also contains a manual paragraph enabling a single 90° turn of the stage on command, depending on the configuration of the switches (explained in Subheading 3.4).

1. Download the open-source Arduino software IDE from https://www.arduino.cc/.
2. Download the Arduino script to steer the rotating stage from http://www.dagz.boku.ac.at/pgz/kleine-vehn/tools/supplements-rotating-stage/.
3. Connect the Arduino-UNO to the computer using the USB connection.
4. Open the Arduino script.
5. Make sure that the software is compatible with the Arduino-Uno hardware. Tools > board > Arduino/Genuino UNO.
6. Adjust the time interval if necessary.

   `X = desired time between the start of two 90°-turns in milli-seconds - 3000`

   `Y = desired time between the start of two 90°-turns in milli-seconds - 18000`

   For example, in case of the given program, we want the stage to turn 90°, every 15 min.

```
X= 900 000 msec(15 min)-3000 = 897 000
Y= 900 000 msec (15 min) -18000 = 882 000
```

7. Click upload to upload the script on the Arduino-board (*see* **Note 8**).
8. Disconnect the Arduino-UNO from the computer.
9. The Arduino-UNO is now able to steer the stage according to the status of the switches (manual vs. automatic, Subheading 3.4).

3.4 How to Use the Stage

The rotating stage can be operated in the absence of a computer by using the rocker switch (on/off) and the round rocker switch to change between the manual (a single turn of 90°) and the automatic mode (the stage turns 90° every time interval as required).

1. Note that the main power (rocker switch) should be switched off while connecting or disconnecting any cables.
2. Place the eight in vitro plates on the stage, the lid facing the inside of the frame, and tighten them using the hand screws. To improve the image quality, placing black paper behind the plates is recommended (Fig. 1).
3. To use the stage: Switch on the main power (rocker switch). The main power control LED will glow.
4. In the manual mode (I), the stage will turn 90° every time the switch is activated. This allows for correct positioning of the turning stage carrying the in vitro plates in front of the camera.
5. In the automatic mode (II), the stage will turn 90° at the end of every time interval as programmed (in our case, every 15 min) until the switch is turned off (0).
6. Set the camera to take a picture every time interval (in our case, 15 min) using the remote control software of your choice. Start the time lapse around 30 s after the start of the rotating stage to avoid pictures being taken during a stage turn.
7. In our case, every 15 min, one side of the stage will be imaged so that every plate will be imaged once an hour.

4 Notes

1. In Table 2, we provide the electric components which we used to build the microcontroller. However, most of the electric components can be purchased from any other company as long as they meet the technical specifications provided in the online data sheets which can be downloaded from http://www.dagz.boku.ac.at/pgz/kleine-vehn/tools/supplements-rotating-stage/.

2. The in vitro plate holders can be replaced by a horizontal stage enabling the imaging of plants in pots.
3. To ensure that the components are tightly attached to the wooden plank, covering the plank with a perforated metal plate before attaching the motor and the electromagnet is recommended. The perforation allows for the easy attachment of the metal brackets.
4. The frame can be built out of any light and strong material, but note that the material used must not be magnetizable by the electromagnet.
5. Data sheets for the components used can be downloaded from http://www.dagz.boku.ac.at/pgz/kleine-vehn/tools/supplements-rotating-stage/.
6. The stepper motor used is wired with six cables. However, only four are operational. The unused cables can be identified by measuring the resistance between the different cable pairs. The resistance between two used cables will be double then between a used/unused cable pair. (In the case of the stepper motor used, the unused cables are black and white.)
7. The active state of several modes (main switch, manual mode, automatic mode, and relays) is indicated by glowing LED lamps. To ease the interpretation of the different glowing LEDs, we decide to use LED lamps in different colors.
8. If it is not possible to upload the Arduino script on the Arduino-UNO board, the default export-PORT in the Arduino software may be incorrect. Select another PORT under TOOLS > PORT and try again.

Acknowledgements

We would like to thank the Arduino community for setting up this wonderful and easily accessible prototyping platform. We are grateful to David Whittaker for help with the manuscript. This work was supported by the Vienna Science and Technology Fund (WWTF) (Vienna Research Group), Austrian Science Fund (FWF) (P26568-B16; P26591-B16), and the European Research Council (ERC) (Starting Grant 639478-AuxinER).

References

1. Araus JL, Cairns JE (2014) Field high-throughput phenotyping: the new crop breeding frontier. Trends Plant Sci 19:52–61
2. Fahlgren N, Gehan MA, Baxter I (2015) Lights, camera, action: high-throughput plant phenotyping is ready for a close-up. Curr Opin Plant Biol 24:93–99
3. Fiorani F, Schurr U (2013) Future scenarios for plant phenotyping. Annu Rev Plant Biol 64:267–291

Chapter 6

Genome-Wide Association Mapping of Root Traits in the Context of Plant Hormone Research

Daniela Ristova and Wolfgang Busch

Abstract

Genome-wide association (GWA) mapping is a powerful method for the identification of alleles that underlie quantitative traits. It enables one to understand how genetic variation translates into phenotypic variation. In particular, plant hormone signaling pathways play a key role in shaping phenotypes. This chapter presents a protocol for genome-wide association mapping of root traits of *Arabidopsis thaliana* in the context of hormone research. We describe a specific protocol for acquiring primary and lateral root trait data that is appropriate for GWA studies using FIJI (ImageJ), and subsequent GWA mapping using a user-friendly Internet application.

Key words Natural variation, Phenotyping, GWAS, Hormones, Root development

1 Introduction

Understanding how genotypic variation translates into phenotypic variation remains one of the fundamental challenges in biology. In recent years, tremendous progress has been made in this regard mainly through the application of genome-wide association studies (GWAS) [3]. GWAS is a method in which variation of a phenotype of interest is associated with sequence polymorphisms throughout the genome in a large number of genetically distinct individuals. In *Arabidopsis thaliana*, natural variation in conjunction with GWAS can be used to map alleles responsible for different quantitative traits (Fig. 1). This is enabled by the large collection of *Arabidopsis* accessions (naturally occurring strains), which were collected from a broad range of habitats around the world and genotyped at high resolution [1]. Different accessions of *Arabidopsis thaliana* exhibit high levels of phenotypic variation [2, 6], which can be used to relate diverse phenotypes to the genetic variation. Thus, GWAS have become a very powerful tool for the identification of genes and their alleles that underlie different quantitative traits [3]. Moreover, recent progress in addressing the potential confounding

Jürgen Kleine-Vehn and Michael Sauer (eds.), *Plant Hormones: Methods and Protocols*, Methods in Molecular Biology, vol. 1497,
DOI 10.1007/978-1-4939-6469-7_6, © Springer Science+Business Media New York 2017

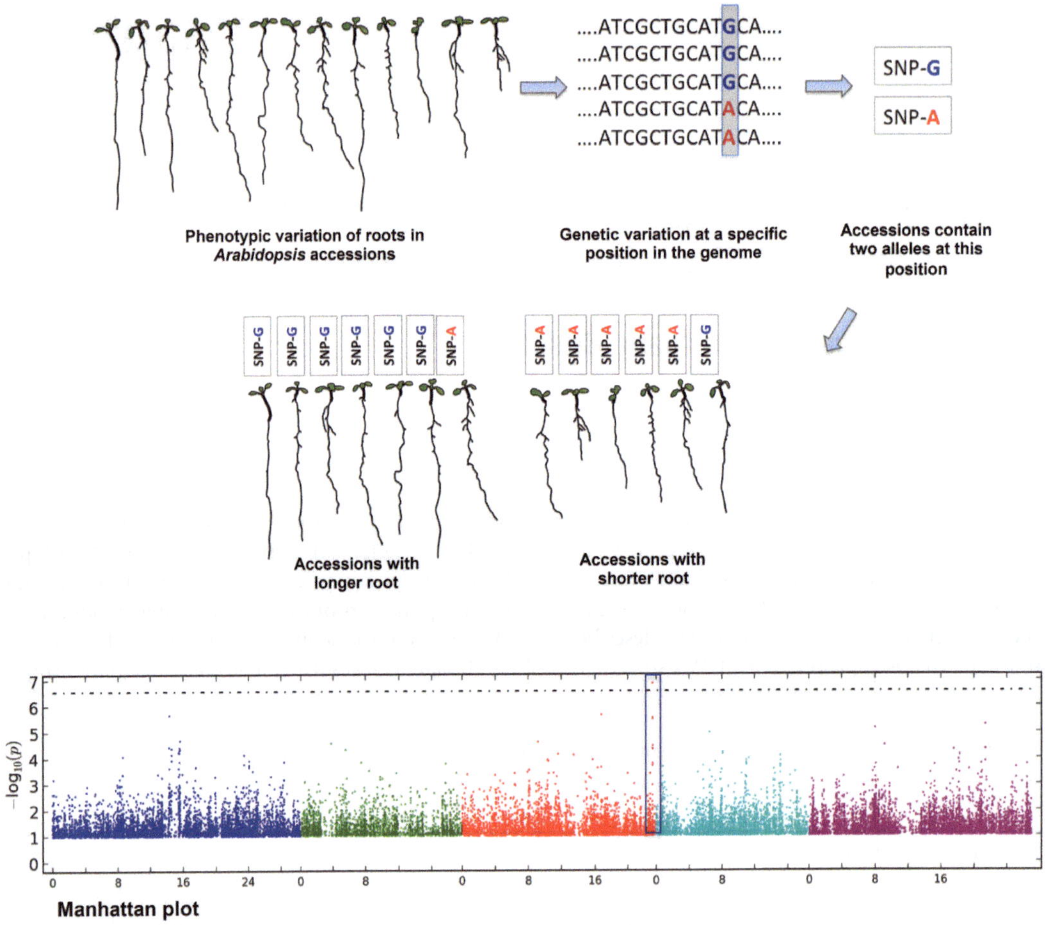

Fig. 1 Schematic representation of a GWAS in *Arabidopsis* roots. Variation of root phenotypes in diverse accessions is quantified. This phenotypic variation is associated with genetic variation at a single-nucleotide level. In this example, accessions with a short root show a highly significant difference to accessions from long roots at one position of the genome which results in a significant association above the threshold in the third chromosome (Manhattan plot)

effect of population structure [3] has made it even more powerful and practical [4, 5].

In this chapter, we present a protocol for performing GWAS on root traits acquired from *Arabidopsis thaliana* accessions grown on media supplemented with phytohormones. In principle the same protocol is generally applicable for root phenotypes, as GWAS can be performed on any trait of interest that can be described as a numeric value. We describe how to culture *Arabidopsis* accessions in sterile conditions, and acquire root phenotypes. Further we explain how to quantify root traits using the image analysis software FIJI or ImageJ. Finally, we describe how to perform GWAS by submitting the quantified phenotype to a free online Web application, the GWA-Portal [7].

2 Materials

2.1 Materials Needed for Culturing Plants

1. *Arabidopsis thaliana* seeds from multiple accessions, approximately 15 seeds per accession. For one experimental set ten seeds are used for plating. However, start with 15 seeds to have few extra during the sterilization and handling.
2. Other materials needed: Tube rack for 1.5 ml microcentrifuge tubes and 1.5 ml microcentrifuge tubes; 250 ml glass beaker; household Bleach containing 5–8% sodium hypochlorite; 37% hydrochloric acid; 15 ml polypropylene conical centrifuge tube; polycarbonate lockable airtight box; sterile square Petri dishes, 12 × 12 × 1.7 cm; transfer pipette, tips (1000 μl), parafilm; sterile pipettes (50 ml) and electronic pipette; non-woven ventilating tape; custom-made support rack to hold plates vertically (Fig. 2a); leveling table, for filling media into plates (Fig. 2c); 0.22 μm filter,

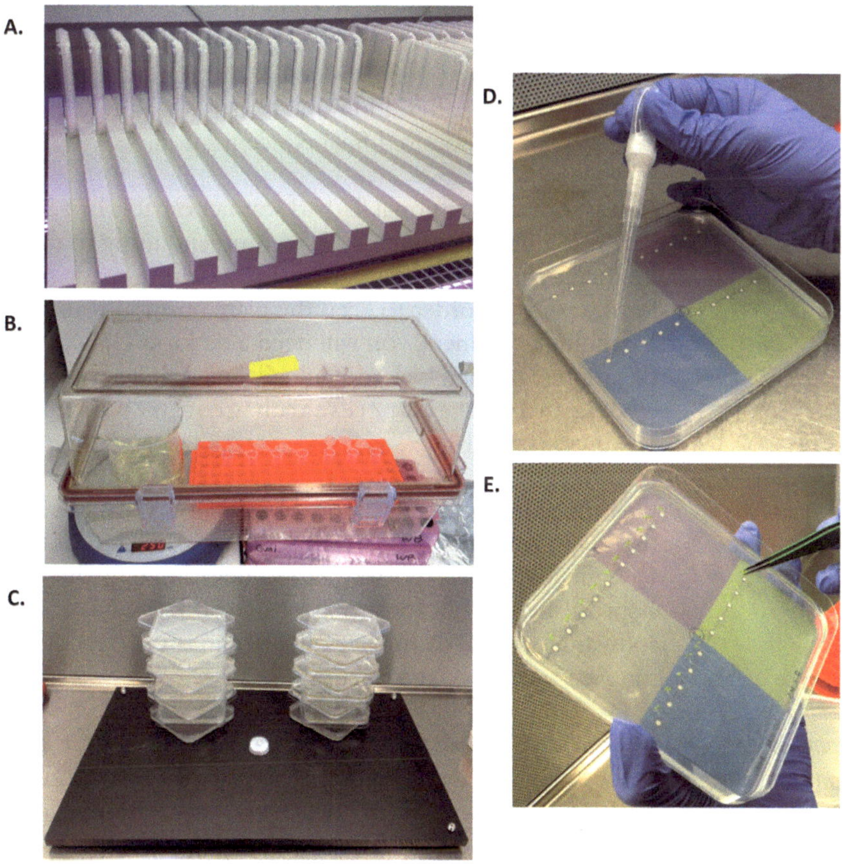

Fig. 2 Culturing plants and acquisition of root trait data. (**a**) Custom-made support racks for vertically positioned plates. (**b**) Lockable airtight box for gas sterilization of seeds. (**c**) Culture media plate production using a leveling table to ensure identical growth conditions. (**d**) Seed distribution on culture plates using plating model. (**e**) Seedling transfer to MS media supplemented with hormone

for sucrose sterilization; water bath, for media cooling; Murashige and Skoog including MES plant growth medium, sucrose, plant agar, pH meter, KOH solution (1 M).

3. Hormones: 3-Indoleacetic acid (IAA), kinetin (CK), and (+)-abscisic acid (ABA) are all solved in dimethyl sulfoxide (DMSO).

2.2 Media Preparation

Preparing media (1 l): Prepare sterile sucrose (1 M) solution, and filter-sterilize it in the laminar hood, using a 0.22 μm filter. Weigh an appropriate amount of Murashige and Skoog including MES, into 966.67 ml of sterile water into a beaker containing a magnetic stirrer bar, and mix well. Adjust pH to 5.7 with 1 M KOH solution. Add the media into a glass bottle, and add 8 g (0.8 %) of plant agar. Sterilize the media at 125 °C for 15 min. After sterilization, cool the bottle/media to 60 °C in water bath, and move the bottle to the laminar hood. Add 33.33 ml of 1 M sucrose solution mix well and pour 57 ml of media in each plate. Leave the plates open for 45 min, then close them, wrap the plates in sterile plastic bags, and store upside-down at room temperature until plating. Media should be prepared 1–3 days before the plating. For preparation of hormone media, after addition of sucrose, add hormones diluted in DMSO to the desired final concentration.

2.3 Image Data Acquisition

At least one conventional flatbed scanner capable of 1200 dpi image data resolution is required. We recommend using multiple scanners for increasing throughput, and using a custom-made support frame for the plates in order to keep the plate position constant. To operate multiple scanners, you will need a desktop UNIX computer and the multiscan interface (download link and instructions: http://www.gmi.oeaw.ac.at/research-groups/wolfgang-busch/resources/brat) [8].

2.4 Quantification of Root Traits

Quantification and extraction of root traits are conducted using the Fiji software (a distribution of ImageJ; http://fiji.sc/Downloads). The exported trait values can be further processed in a spreadsheet application (e.g., Microsoft Excel) or any statistical software (e.g., R) to further calculate mean and median trait value for each accession/treatment.

2.5 Genome-Wide Association Mapping

The GWA-Portal is an online Internet application (http://gwas.gmi.oeaw.ac.at/) that requires a browser supporting HTML5. The input file for GWA-Portal is a text with comma-separated columns of accession IDs (unique number assigned for each *Arabidopsis* accession, for example for Col-0 is 6909) and trait values. Accessions IDs can be retrieved on the GWA-Portal Web (http://gwas.gmi.oeaw.ac.at/#/taxonomy/1/passports?alleleAssayId=0).

3 Methods

3.1 Culturing Plants

1. Surface sterilization of seeds: For each *Arabidopsis* accession place 15 dry seeds in an open 1.5 ml microcentrifuge tube in a tube rack, and place the rack into an airtight lockable box along with a beaker containing a magnetic stirrer bar and 100 ml of bleach. Place the lid partially on the box. Put 3.5 ml of 37 % hydrochloric acid in a 15 ml conical tube, lift the lid on the side, quickly decant the HCl into the beaker to generate chlorine gas (*see* **Note 1**), and then quickly close and secure the lid. Place the box on a magnetic stirrer (speed = 2.5 rpm) for 1 h of sterilization time (Fig. 2b).
2. After 1 h, unclamp the box, remove the beaker from the box, and slightly tilt the lid in order to vent the rest of the chlorine gas for approximately 15–20 min.
3. Close the lid. Transfer the box with the sterilized seeds inside to a laminar hood and leave the tube rack with open tubes to vent for 30 min.
4. To each tube add 2–3 drops of sterile water, close the tubes, vortex, and centrifuge briefly. Place the rack in the dark at 4 °C for stratification for 3 days.
5. Plating seeds: Plating is performed under sterile conditions (laminar hood) to avoid contamination. Use a regular pipette or a sterile custom-made pipette for speeding the plating process (Fig. 2d). To make the custom-made pipette cut one transfer pipette just below the base, and secure a sterile tip (1000 μl) to the transfer pipette with parafilm. Plate four accessions per plate (use background design for easy plating (Fig. 2d)). First plate ten seeds/accession on MS media without hormone (control media), and on day 7 transfer the five most common-appearing plants from each accession to MS media supplemented with hormone or to the control media (Fig. 2e). Final hormone concentrations depend on the research project and are best established during pilot experiments prior to the GWAS. In our studies we used IAA (0.5 μM), CK, and ABA (1 μM).
6. After plating, carefully seal the edges of the plates with non-woven ventilating tape impermeable to bacteria (Fig. 2a).
7. Place plates vertically into the custom-made rack, and place the rack into growth chamber (Fig. 2a).

3.2 Image Data Acquisition

1. Transfer the rack with petri plates to the image acquisition room (*see* **Note 2**).
2. Place each plate (without removing the tape) onto the horizontally oriented scanners. They should fit adequately into the support frames (Fig. 3a).

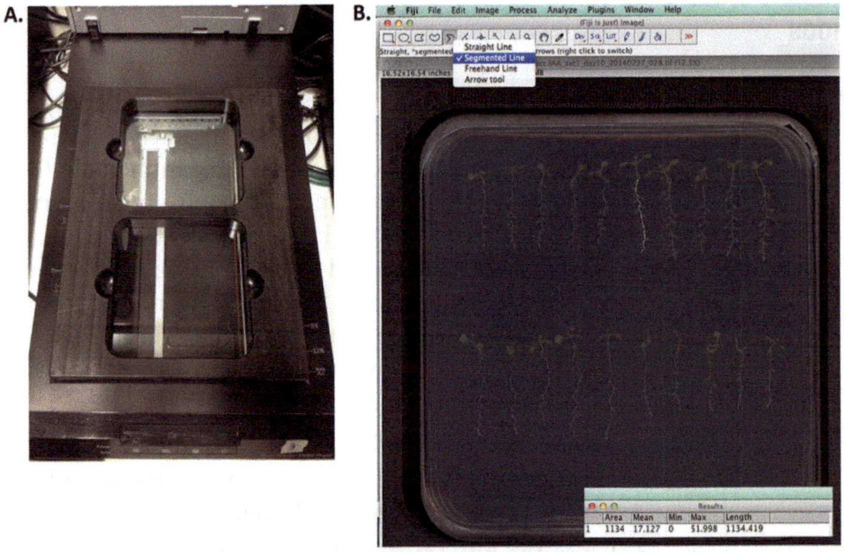

Fig. 3 Image acquisition and root trait quantification. (**a**) CCD scanner with a custom frame to allow for identical plate positions during scanning. (**b**) Example of root length extraction in FIJI (ImageJ). The example plate has twenty 10-day-old Arabidopsis seedlings (from four accessions) in two rows, grown on 500 nM IAA for 3 days

3. Scan the plates with the following parameters: 1200 dpi 8-bit RGB TIFF files, and name the plates. For optimal image quality, scan in a darkroom with scanner lid open. After image acquisition, return the rack with plates to the growth chamber (*see* **Note 3**).

3.3 Trait Quantification

1. Open Fiji, and then open the image of one plate. Choose "Segmented Line" to draw the length of the primary root (Fig. 3b), and run the "Measure" function by selecting from the "Analyze" pull-down menu bar on top of the window or alternatively press cmd + M (Mac) (length shown in pixels). Repeat for each length, then copy the values, and export them into Excel-file format.
2. For performing GWAS use mean and/or medium of all individuals of one accession for each root trait.
3. Quantified traits (example): length of primary root on transfer day (P1), length of primary root after transfer to hormone (P2), total length of lateral roots (TLRL), and number of lateral root (LR.No). After obtaining these values, calculate the following traits: total length of primary root (P = P1 + P2), average lateral root length (LRL = TLRL/LR.No), total root length (TRL = TLRL + P), lateral root density (LRd = LR.No/P), and length ratio (LRR = TLRL/P).

3.4 Genome-Wide Association Mapping

To identify alleles associated with the phenotype (root trait) of interest, perform genome-wide association mapping by submitting the quantified phenotype to GWA-Portal. GWA-Portal allows one to quickly run GWAS on the four available datasets (250k SNP, 1001 Fullsequence, Swedish genomes, or Imputed Fullsequence) (*see* **Note 4**).

1. Open the GWA-Portal web page (http://gwas.gmi.oeaw.ac.at) and create an account. Press "Take a tour" to familiarize yourself with the web page (*see* **Note 5**) [7].
2. To perform GWAS on your traits, click on "Create" in the "New GWAS analysis." A new page will open where the user needs to fill in information for six steps (left column). Step 1 is "Study"; here the user can choose already uploaded phenotypes or can create a new one by clicking "Create new study." When "Create new study" is chosen, a new, smaller window will open to add information for the new experiment, including name, originator, design, and comments. After each step Click "Next" (in the lower right side) to move to the next one.
3. Step 2, "Phenotype." Click "Upload phenotype" and upload your trait of interest. Use the .csv file format, where the first column is "Accession_ID" and the second column is the trait values (*see* **Note 6**). After uploading your file click "SAVE." Select your newly created phenotype and then click "Next."
4. Step 3, "Genotype." Here you can select which sequencing data to use. Use the 250k SNP chip data set (first option) [5].
5. To transform the phenotype data distribution, go to step 4, "Transformation." Choose the desired type of transformation or no transformation.
6. Step 5 is where the user chooses the type of analysis. We recommend using the "AMM" method (accelerated mixed model) that efficiently addresses the confounding effects of population structure. The other two methods (KW and LM) might produce population structure confounded association mappings [4–6]. Finally, go to step 6, "Summary," where all of the previous user choices are shown. Press "Finish" to proceed to the next page where the analysis is performed.
7. A new page will open where the summary is shown on the left side. In the middle under "GWAS Status" click on the drop-down arrow "N/A," and choose "Run analysis," after which the user can monitor the progress of the analysis indicated in percentage (*see* **Note 5**). When finished, click on "Plot (AMM)" to visualize the interactive Manhattan plots for each chromosome (Fig. 1). A Manhattan plot is a type of scatter plot where genomic coordinates of the five chromosomes are displayed on the *X*-axis, while the negative logarithm of the association *P*-value for each SNP is displayed on the *Y*-axis. On the interactive Manhattan plot the user can zoom into the region of interest by selecting the region above the association.

To save the plot click "Download plot"; a new window will open where one can choose the "mac" (minor allele count) threshold, chromosomes, and the format.

4 Notes

1. Chlorine is a strong oxidizing agent with a yellow-green color, and it is a very toxic gas that irritates the respiratory system. Therefore wear gloves and safety goggles, and work in a well-ventilated chemical hood.
2. To prevent condensation of droplets on the plate lids, the temperature in the image acquisition room should be the same as in the growth room or slightly higher.
3. Images acquired on scanners appear in a mirrored orientation. This must be accounted for that when quantifying the root traits and assigning accession number (i.e., flip the image to obtain the positions as visible from top view during plating).
4. "250k SNP" refers to sequencing data genotyped using a custom Affymetrix single-nucleotide polymorphism (SNP) chip containing 250,000 SNPs [5], "1001 Fullsequence" dataset contains full sequence data (~10 million SNPs) for 1135 accessions, while the "Imputed Fullsequence" dataset has the same amount of SNPs for 2029 accessions (unpublished). "Swedish genomes" are 259 Swedish accessions (6 million SNPs) [10].
5. At times the GWA-Portal server is quite busy; thus the page can take some time to load.
6. The phenotype file has to be in the .csv file format (which can be created in excel) and contain at least two columns: the first column should contain the GWAS IDs of accessions (a number which has already been assigned by the GWA-Portal, see web page), while the second column contains the mean and/or median values for the trait of interest for the respective accession. Additional traits can be added in additional columns. The phenotype file must contain headers with unique trait names. Do not use space characters in trait names. Missing data must be denoted as "NA.".

Acknowledgement

We are grateful to Elke Barbez, Takehiko Ogura, and Santosh Satbhai for comments and suggestion for this protocol, Ümit Seren for information about the GWA-Portal, and Matthew Watson for editing.

References

1. Horton MW, Hancock AM, Huang YS et al (2012) Genome-wide patterns of genetic variation in worldwide Arabidopsis thaliana accessions from the RegMap panel. Nat Genet 44:212–216
2. Koornneef M, Alonso-Blanco C, Vreugdenhil D (2004) Naturally occurring genetic variation in Arabidopsis thaliana. Annu Rev Plant Biol 55:141–172
3. Weigel D (2012) Natural variation in Arabidopsis: from molecular genetics to ecological genomics. Plant Physiol 158:2–22
4. Platt A, Horton M, Huang YS et al (2010) The scale of population structure in Arabidopsis thaliana. PLoS Genet. doi:10.1371/journal.pgen.1000843
5. Segura V, Vilhjálmsson BJ, Platt A et al (2012) An efficient multi-locus mixed-model approach for genome-wide association studies in structured populations. Nat Genet 44:825–830
6. Atwell S, Huang YS, Vilhjálmsson BJ et al (2010) Genome-wide association study of 107 phenotypes in Arabidopsis thaliana inbred lines. Nature 465:627–631
7. Seren Ü et al (2012) GWAPP: a web application for genome-wide association mapping in Arabidopsis. Plant Cell 24(12):4793–4805
8. Slovak R et al (2014) A scalable open-source pipeline for large-scale root phenotyping of Arabidopsis. Plant Cell 26(6):2390–2403
9. Long Q et al (2013) Massive genomic variation and strong selection in Arabidopsis thaliana lines from Sweden. Nat Genet 45(8):| 884–890

Chapter 7

High-Throughput Scoring of Seed Germination

Wilco Ligterink and Henk W.M. Hilhorst

Abstract

High-throughput analysis of seed germination for phenotyping large genetic populations or mutant collections is very labor intensive and would highly benefit from an automated setup. Although very often used, the total germination percentage after a nominated period of time is not very informative as it lacks information about start, rate, and uniformity of germination, which are highly indicative of such traits as dormancy, stress tolerance, and seed longevity. The calculation of cumulative germination curves requires information about germination percentage at various time points. We developed the GERMINATOR package: a simple, highly cost-efficient, and flexible procedure for high-throughput automatic scoring and evaluation of germination that can be implemented without the use of complex robotics. The GERMINATOR package contains three modules: (I) design of experimental setup with various options to replicate and randomize samples; (II) automatic scoring of germination based on the color contrast between the protruding radicle and seed coat on a single image; and (III) curve fitting of cumulative germination data and the extraction, recap, and visualization of the various germination parameters. GERMINATOR is a freely available package that allows the monitoring and analysis of several thousands of germination tests, several times a day by a single person.

Key words *Arabidopsis thaliana*, Automatic scoring, Curve-fitting, Germination, High-throughput analysis, Image analysis

1 Introduction

Fundamental and applied seed biology research relies heavily on accurate quantification of seed germination. Nowadays, large-scale experiments using large genetic populations or mutant collections are popular tools to unravel molecular aspects of seed development, germination, dormancy, and seed performance [1, 2]. Detailed analysis of these traits requires the complete cumulative germination curve and extraction of parameters such as start, rate, and uniformity of germination. *Arabidopsis thaliana* is one of the most used model species for plant science and it has also become a very useful model plant to study seed biology. However germination of the very small *A. thaliana* seeds (200–400 μm) is often evaluated by using a binocular microscope, which makes this a very

Jürgen Kleine-Vehn and Michael Sauer (eds.), *Plant Hormones: Methods and Protocols*, Methods in Molecular Biology, vol. 1497, DOI 10.1007/978-1-4939-6469-7_7, © Springer Science+Business Media New York 2017

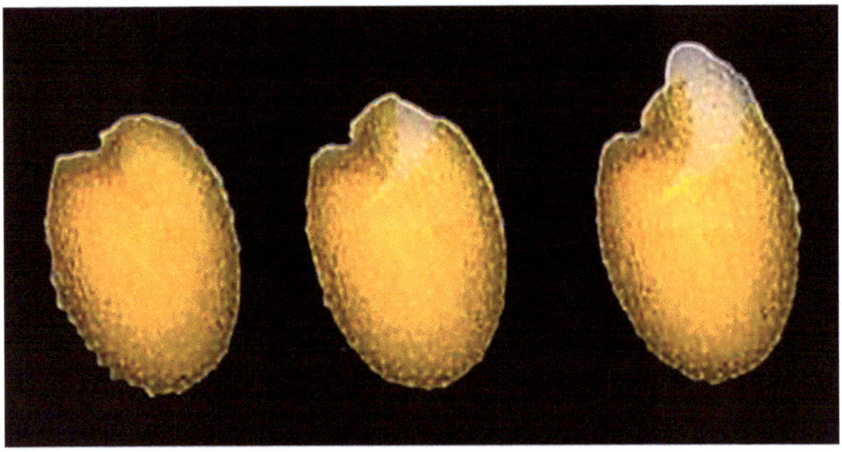

Fig. 1 Photographs of different stages of *Arabidopsis* seed germination. From *left* to *right*: fully imbibed seed (6 h after start of imbibition), seed coat rupture, and radicle protrusion

laborious task. This often hampers the collection of cumulative germination data in large-scale experiments and therefore often only the less informative percentage of germination after a nominated period of time is used to describe the germination performance of a seed lot. With this in mind, the GERMINATOR was developed, a cost-efficient and flexible procedure for high-throughput automatic scoring of germination that can be implemented without the use of complex robotics [3].

The uptake of water during seed imbibition is triphasic, and consists of a rapid initial uptake (phase I), followed by a plateau phase (phase II) and a further increase (phase III). During this last phase, the embryo axis elongates and breaks through the seed coat, which consists of dead tissue. The next step is the morphological completion of germination, which is characterized by the protrusion of the radicle through the endosperm (Fig. 1).

1.1 GERMINATOR

Most approaches to the automatic evaluation of germination based on image analysis extract information from a time series of images. For example, change in seed size over time can be used to evaluate the progress of germination very precisely. However, this analysis requires fully aligned images, which are often obtained by using fixed setups in specialized cabinets, flat-bed scanners, or camera systems over Jacobsen tables [4–7]. This type of setup has important consequences not only for the amount of samples that can be measured and the flexibility to follow germination under various environmental conditions, but also for the costs of the equipment. Especially this flexibility and level of throughput are essential characteristics of the GERMINATOR.

Evaluation of germination by using image analysis requires good contrast between background, seed, and protruding radicle. Therefore, a blue filter paper specifically designed for germination tests is used. By using two different color threshold levels on a single image we can distinguish between background and seed coat, or background, seed coat, and radicle. This allows germinated and non-germinated seeds to be distinguished. To ensure the most accurate scoring, the number of non-germinated seeds to monitor germination is used. With this approach it is not necessary to obtain fully aligned images between the various time points at which germination is evaluated. The GERMINATOR package consists of three separate modules, based on the components of our procedure: (I) experiment design, (II) image analysis, and (III) curve fitting. Module I uses Microsoft Excel combined with Visual Basic scripts to create a database of experiments, and assists in experiment design. Germination is documented by photographs taken manually at various intervals during germination. Automated scripts were developed for Adobe Photoshop and ImageJ to automatically analyze the images and to minimize user input as much as possible. With the help of ImageJ, a two-color threshold analysis is performed on each individual image: first, a color threshold that selects the seed coat only, and second, a color threshold that selects both the seed coat and the protruding radicle. The specific settings for the color thresholds can be adjusted for each seed type during a system calibration. The output of the image analysis consists of four tables with information about the total number of seeds and the position and size of each individual seed that has been detected. In module II, these tables are then coupled with the experiment setup tables from module I, and cumulative germination curves are calculated. The calculation uses position and seed size in the various color threshold analyses to determine whether a seed has germinated. Module III performs a curve-fit analysis of the cumulative germination data. This module uses data generated by the image analysis of module II, or cumulative germination data of other sources (for example derived from manual counting). This module executes a curve fit using the four-parameter Hill function, which can accurately fit the typical sigmoid curve that describes the course of germination [8]. As soon as the best-fitted curve has been determined, typical germination parameters are extracted from the curve. Generally Gmax (maximum percentage of germination), t50 (time to reach 50% germination), tx (time to reach x% germination), uniformity (time interval between, e.g., 16 and 84% of viable seeds to germinate), and AUC (integration of area under the germination curve until a certain point of time) are determined. As described by El-Kassaby and co-workers [8] the AUC can also be used to calculate a dormancy index (DI), by extracting the area under the curve after dormancy release by cold stratification, by the area under the curve without cold stratification. By the same analogy the AUC can be used to measure the effect of any stress treatment and calculate a stress index (SI). The GERMINATOR script will summarize the results by calculating averages and standard errors for repeated samples, performing a student-t test, and provide a clearly formatted output including graphs for the different germination parameters.

With the GERMINATOR package, a single person can easily handle more than 1000 germination tests at a time and monitor germination precisely over time. This greatly facilitates large-scale screens of mutant collections or genetic populations, such as recombinant inbred line populations [9, 10]. The package was tested and optimized for *Arabidopsis* seeds and it has been shown to also work for other species including *Brassica* spp. and rice. The whole GERMINATOR package, including a manual and video tutorials, is available from http://www.wageningenseedlab.nl/germinator.

2 Materials

2.1 Germination Assays

1. Germination trays (*see* **Note 1**).
2. Colored germination paper that fits in the germination trays (*see* **Note 2**).
3. A plastic flexible mold/template with the positions for the different germination tests in the same size as the germination paper (Fig. 2).
4. A germination cabinet (*see* **Note 3**).

2.2 For Image Acquisition

1. A digital single-lens reflex camera with more than ten million pixels and a macro lens.

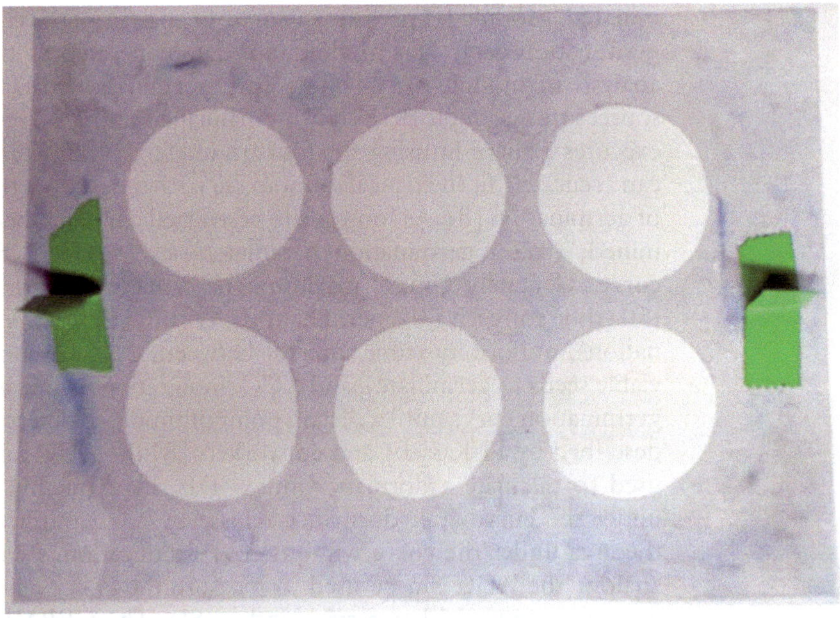

Fig. 2 Plastic mold/template to put in the germination trays to ensure that the different germination tests will be on the correct positions

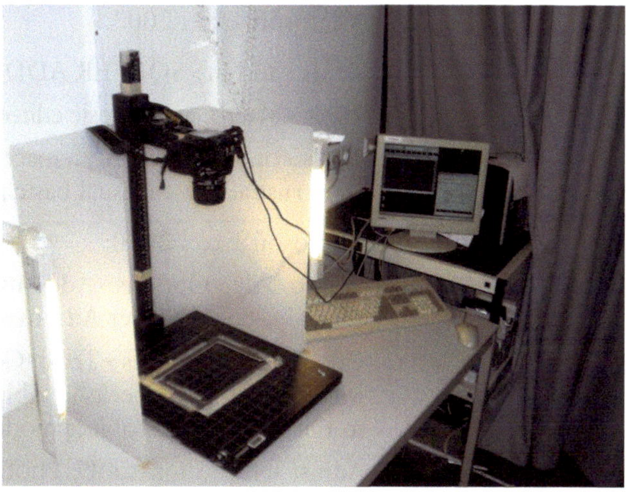

Fig. 3 Camera setup for the GERMINATOR

2. A power adaptor and USB connector cable for the camera.
3. A repro-stand.
4. Proper, preferably indirect, lightning. A Perspex screen is used around the camera setup to soften the light (Fig. 3) in a room without daylight in order to keep the light conditions as constant as possible. Our Perspex screen is 60 cm high, 55 cm wide, and 35 cm deep. As light source two fluorescent tubes (TL) of 30 cm long on both sides of the camera and one of 115 cm shining towards the ceiling of 115 cm are used (*see* **Note 4**).
5. A Windows-based PC (XP or Windows 7).
6. Software to control the camera. Nikon Control Pro version 2.0 is used in our setup.

2.3 For Image Analysis

1. A Windows-based PC.
2. Microsoft Excel (version 2003 or 2010 and English language version).
3. Adobe Photoshop CS3 or newer (English language version).
4. The GERMINATOR package (can be downloaded from www.wageningenseedlab.nl/germinator).
5. ImageJ version 1.40g with Montpellier RIO Imaging plug-ins (included in the GERMINATOR package).

3 Methods

3.1 Installing Scripts and Macros

1. Unzip the whole "Germinator.zip" file to a specific directory (keeping the folder structure of the file).
2. Adjust settings in Microsoft Excel to allow macros and install the Solver add-in.

For Microsoft Excel 2003:

1. Tools > add in > SOLVER ADD IN.
2. Tools > Macro > visual basic editor > tools > references > SOLVER.
3. Tools > options > security > macro security > trusted publishers > trust access to visual basic project.

For Microsoft Excel 2010:

1. Go to File > Options > Add-Ins. Choose Excel Add-ins and click Go. Select Solver Add-in and click OK.
2. Go to File > Options > Trust Center > Trust Center Settings > Macro Settings > Check "Enable all macros" and "Trust access to the VBA project object model" and click OK.
3. Adjust settings for Adobe Photoshop.
   (a) Select the window "Actions" with |Alt F9|. Open a menu on the right top of this window and select "load actions." Browse to the GERMINATOR package, select folder "Photoshop," and select "germinator.atn" (*see* **Note 5**).
   (b) In Photoshop, open one image of a typical seed germination test. In the "Actions" panel go to Germinator > crop and save and double-click the first "crop." Adjust the crop position that appears on the image to encompass the first germination test (*see* **Note 6**), and press enter and subsequently Ctrl + z to go back to the original image. Then do the same steps for the following five "crops" that are shown in the "Actions" panel and close Photoshop.
4. Install and adjust settings for Image J.
   (a) Install Image J 1.40g from the GERMINATOR package (*see* **Note 7**).
   (b) Extract Zip file "mri-plugins-base.zip" to the ImageJ 1.40g directory.
   (c) Extract Zip file "mri-all-plugins.zip" to the ImageJ 1.40g directory.
   (d) Copy the Germinator folder in the ImageJ folder of the Germinator.zip file to the ImageJ_applications directory.
   (e) Open ImageJ program, go to Plugins > Montpellier RIO Imaging > MRI VisualScripting. Go to Applications > Applications > Germinator > germination score + white, and click on the O. Make sure that the macro file points to the correct location of the "particle analysis macro.txt" file. Close the window, right click on the top white frame (just above the blue part), and click save. Close the script and do the same thing for germination score – white.

3.2 Prepare for a Germination Experiment

1. Open Germinator_menu 1.01.xls and set defaults: Provide a name for the overview and click Browse to select a default directory in to store the germination results. Click Save twice.
2. Click Browse and select the germinator table file from the GERMINATOR package (Germinator_table 1.05.xls) by double-clicking and subsequently click save (*see* **Note 8**).
3. Click on the "add new experiment" button, fill in the different parameters for the germination experiment that should be performed, and finish by clicking Add (*see* **Note 9**).
4. Click button "Add tables," select "New tables," and click the button "Make Tables" in the newly opened Excel file.
5. Fill in the parameters for the germination experiment like details, the number of samples that should be analyzed, and the number repetitions and treatments. The number of treatment option is there for when a germination test should be performed at, e.g., 22 and 25 °C or in water and in a certain concentration of ABA.
6. By checking the "Optional sample file name" box, sample names can be extracted from an Excel file (according to the format as shown in the "sample file name example.xls" file in the Excel directory of the GERMINATOR package). If biological replicates should have different names, these can be placed in different columns in this file. Select "separate biological replicates" to have the biological replicates in separate trays in the experiment setup.
7. After clicking start, the tables that show the experiment setup in which for each sample the tray number and position in the tray are indicated can be printed (*see* **Note 10**). The Excel file will be saved in the default directory under the experiment's name.
8. The printed table is brought to the lab to start the germination experiment.

3.3 Start a Germination Experiment

1. Take a tray and insert two sheets of blue blotter paper (*see* **Note 11**).
2. Mark the sheets with a plate number (depending on the amount of germination tests that will be performed) in the upper-left corner with a pencil.
3. Add 50 ml of demi water (*see* **Notes 12** and **13**).
4. Place the plastic mold/template in the tray to ensure that the different germination tests will be on the correct positions (Fig. 2) (*see* **Notes 14** and **15**).
5. Add around 50 *A. thaliana* seeds in each well (Fig. 4, *see* **Note 16**) according to the printed Germinator table and note the time of the start of imbibition in the table.

Fig. 4 A picture of a typical germination experiment with *A. thaliana* seeds

6. Pile up to 17 filled trays, add 1 tray with filter paper and water at the bottom of the pile and 2 at the top + a lid, and pack the whole pile in a big plastic bag (*see* **Note 17**).
7. Place the pile in a germination cabinet and set the desired temperature (*see* **Note 18**).

3.4 Optimize Image Acquisition

1. Attach the camera in a repro-stand and adjust height to ~53 cm (*see* **Note 19**).
2. Switch camera to full manual control and connect with USB to a PC.
3. Place the Perspex screen around the camera setup to soften the light, place the lights, and try to avoid interference of other, unstable, light sources (Fig. 3).
4. Place a mask that fits the germination trays underneath the camera to fix the imaging position (Fig. 3) (*see* **Note 20**).
5. Put a germination tray as described in Subheading 3.1 under the camera and focus.
6. Use Nikon camera Control Pro software to set resolution to maximum and file format to jpg with normal compression.
7. Adjust camera settings (aperture, ISO, and shutter speed) with Nikon camera Control Pro software and if necessary light conditions to optimize the pictures with respect to contrast between germination paper, seed coat, and radicle (*see* **Note 21**).
8. Save the optimal settings via Settings > Save Control Settings (as *.ncc file) (*see* **Note 22**). Also the tone compensation

curves can be adjusted manually to fine-tune the pictures via Camera > Edit Camera Curves. Optimal tone compensation curves can be saved by clicking Save... under the curve (as *.ntc file).

3.5 Take Pictures of the Germination Experiment

1. Open the Camera Control Pro program, select settings > settings > load control settings, and select the previously saved optimal settings (*.ncc file). Go to image processing > tone comp > edit > load and select previously saved optimal settings for tone compensation curves (*.ntc file).
2. Go to tools > download options and select a folder to store the pictures. Subsequently go to edit file name. In prefix, first month, then the day followed by the year and time that the pictures are taken are indicated in a mmddyyyy-hhmm format followed by # (e.g., 10212015-0923#). Set start numbering at 1 and length of number at 3 digits. Finish by clicking OK.
3. Start taking pictures of all germination trays in the experiment starting with tray number 1 and adjust time settings every 5–10 min as indicated under Subheading 3.5, **step 2**. An experienced user can easily take pictures of 30 plates in 10 min (*see* **Note 23**).
4. Repeat taking pictures in certain intervals. For *Arabidopsis* routinely pictures are taken two or three times a day (*see* **Note 24**).

3.6 Determining the Optimal ImageJ Settings to Score Germination

1. Open a photo of germinating seeds (some already germinated, others not) with ImageJ (*see* Fig. 4 for an example).
2. Go to Process > Enhance Contrast and adjust contrast of the picture to 0.2 % saturated pixels and click OK (*see* **Note 25**).
3. Go to Plugins > Segmentation > Threshold Colour, select YUV (at the bottom of the window) (*see* **Note 26**) and adjust color settings of the three channels until settings are found that selects all seeds and all radicles but no background (Fig. 5, Filter +).

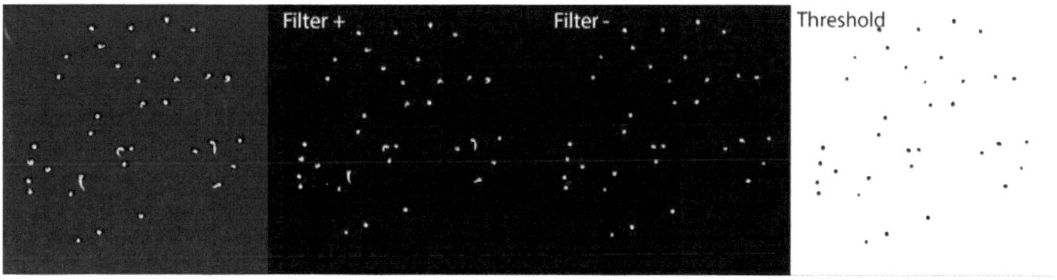

Fig. 5 A photograph of a typical germination experiment and the two color thresholds used to count seed germination. Filter +: an example of filtering everything but the background. Filter –: an example of filtering only for the seed coat color. Threshold: example of the Filter—picture after binarization and inversion and ready for particle analysis

Write these settings down. For the example image, these settings are Y 50-255, U 0-130, and V 0-255.

4. Adjust color settings until the settings are found that select the seed coat, but not the radicles and the background (Fig. 5, Filter –). For the example image, these settings are Y 50-255, U 0-90, and V 0-255. Write these settings down (*see* **Note 27**).
5. Select Threshold and close the Colour Threshold window.
6. Go to Process > Binary > Make Binary followed by Edit > Invert. This results in a black-and-white image with black seeds on a white background (Fig. 5, Threshold).
7. To count the black spots on such a photo go to Analyse > Set Measurements and select Area, Circularity, Centroid, Perimeter and click OK. Go to Analyse > Analyse Particles, select size 0–Infinity, Circularity 0.00–1.00, Show Outlines, Display Results, Summarize, Exclude on Edges, and Include Holes, and click OK.
8. In the Summary table that will appear the total number of seeds that are counted can be found (Count) and in the Results table parameters for the individual seeds are shown. The Area is a measure for size of the individual seeds. This area will be used in the Excel Germinator script.
9. To prevent counting of clustered seeds or other artifacts, such as remnants of siliques, a limit needs to be set for the size of particles that are still measured and considered as single seeds. The lower size limit is typically set at approximately 0.5 × the smallest seed that is counted and the upper limit at approximately 2 × the biggest seed that is counted (*see* **Note 28**).
10. Open the "particle analysis macro.txt" file (from Subheading 3.1, **step 4e**). Adjust the line "size = **40–175**" by replacing the bold numbers with the limits obtained under Subheading 3.6, **step 9**, and save the file.
11. Open the visual scripting tool of ImageJ by going to Plugins > Montpellier RIO Imaging > MRI VisualScripting. Go to Applications > Applications > Germinator > germination score + white and click on the O. Fill in the values obtained under Subheading 3.6, **step 3**, for the three different channels in the fields min/max channel 1–3. Close the window, right click on the top white frame (just above the blue part), and click save.
12. Close the script and do the same thing for germination score – white with the settings obtained under Subheading 3.6, **step 4**.

3.7 Analyze a Germination Experiment

1. When germination is completed pictures are analyzed for automatic scoring of germination in batch. The first step is to divide the pictures into the separate germination tests. To do so, first

Adobe Bridge is opened, all pics to be analyzed are selected, and if the pictures are upside down they are rotated first: go to Edit > Rotate 180° (*see* **Note 29**). Subsequently go to Tools > Photoshop > Batch and select the following settings: Set: Germinator; Action: Crop and Save; Source: Bridge; Destination: Folder. Click Source and select the folder in which the cropped separate pictures should be saved. Click override action "save as" and click OK (*see* **Note 30**).

2. Open ImageJ and go to Plugins > Montpellier RIO Imaging > MRI VisualScripting. Go to Applications > Applications > Germinator > germination score + white and click the big blue button. Select file type as all images and select all images to be analyzed. Click Open. When ImageJ is finished, there are two windows. One named Summary, which should be saved as summary+.txt, and one named Results, which should be saved as results+.txt (*see* **Note 31**). Perform the same steps on the same pictures for Applications > Applications > Germinator > germination score – white (*see* **Note 32**) and save the two windows as summary-.txt and results-.txt, respectively.
3. The first time the GERMINATOR is used, the optimal settings for variances that are allowed for the position of seeds and differences in seed area will have to be determined and thereby the performance of the system will be optimized. To do so first count the number of germinated seeds from some samples and save them in an Excel file according to the format as shown in the "manualcount-example file.xls" file in the Excel directory of the GERMINATOR package.
4. Open the Excel table that was made under Subheading 3.2, **step 7**, click Optimize parameters, and select the txt files that were created under Subheading 3.7, **step 2** (*see* **Note 33**). Under Name manual results file, select the Excel file created under Subheading 3.7, **step 3**. Select Area and Absolute (*see* **Note 34**) and select the range in which to look for the optimal settings for both the variance in Area and XY position. Use first a rough screen (e.g., for our system with *Arabidopsis* for Area from 10 to 100 in steps of 10 and for *XY* variance from 0 to 2 in steps of 0.2) to get an idea about the optimal range and do subsequent analyses to zoom in to the best settings (*see* **Notes 35** and **36**).
5. To measure germination for the whole germination experiment with the optimal settings, click Calculate; select the txt files that were created under Subheading 3.7, **step 2**; fill in the optimized parameters as obtained under Subheading 3.7, **step 4**; and click Calculate. The results file with the cumulative germination table of the experiment will be automatically saved in the default directory.

3.8 Curve Fitting and Interpretation of the Data

1. A cumulative germination table in itself does not provide much information. Therefore the GERMINATOR also has a curve-fitting module to convert the cumulative germination data into useful parameters that can be used to compare germination characteristics of seed batches. After the cumulative germination table has been calculated, there will be an option to fit curves. Click Yes and select the Germinator_curve-fitting 1 29.xls file in the Excel directory of the GERMINATOR package and click Save.
2. In the OUTPUT sheet of the curve-fitting file, several parameters for the curve fitting can be adjusted: the minimal germination at which a germination curve will be fitted through the data, the number of hours until which the area under the curve (AUC) should be calculated (*see* **Note 37**), the critical *p*-value that should be used for the statistical analysis (default on 0.05), which additional parameter should be calculated apart from t50 (e.g., t10 or t90), a lower limit for the *r*-square of the fit, the uniformity parameter that should be calculated (e.g., u8416 or u7525), and if letters should be added above the bars in the graphs to indicate statistical differences between the different measurements.
3. Click GO in the OUTPUT or INPUT sheet to fit all germination curves and calculate the different parameters per curve. These will be shown in the OUTPUT sheet. By placing 1 in column R behind a specific germination sample and clicking Graph, the curves of the selected samples will be plotted.
4. For germination samples with the same names, averages will be calculated with standard errors and statistical differences between samples will be calculated and shown in the statistics sheet.
5. Also after the curves are fitted, some parameters as mentioned under Subheading 3.8, **step 2**, can be adjusted and statistical analysis will be repeated after clicking the "recalculate statistics" button in the OUTPUT sheet.

4 Notes

1. Our system is optimized for stackable 15 × 21 cm trays (ref 109, DBP plastics, Belgium. www.dbp.be), but any other germination tray can also be used. The advantage of the DBP trays is that they are stackable and allow for convenient storage in germination incubators in which each tray functions as the lid of the tray below and as such avoids evaporation as much as possible.
2. The germination paper should have a color (in wet conditions) which is sufficiently contrasting to the color of the seed coat

and radicle. We use 14 × 19.5 cm blue blotter germination paper from Anchor (Anchor paper company, St Paul, MN, USA; www.seedpaper.com). It can be ordered in any desired size if it should be used in different size germination trays.

3. Any germination cabinet that is used for normal germination tests can be used. For working with *Arabidopsis* it is important to have light and preferably from the side, so that all trays get equal exposure to light.
4. As alternative, A3 paper sheets can be placed between the light and the trays to get a similar effect of indirect lightning and also the strength and position of the lights can be changed to optimize the light conditions.
5. There are also specific Photoshop action files for use with 3, 20, and 24 germination tests per tray. Other formats can be requested from the author.
6. It is wise to select the cropping areas as broad as possible with as little as possible chance to select seeds of the neighboring germination tests.
7. It is important to use ImageJ version 1.40. Newer versions unfortunately do not support Visual Scripting.
8. Select Germinator_table 2.01.xls if a different number of germination tests on one tray is used. Version 2.01 allows the formation of germination tables with any number of germination tests per tray.
9. The content of the drop-down selectors can be changed or items can be added to it in the "Experiment descriptors" sheet.
10. Make sure that for the date format in cell B10 "-" is used as separator instead of "/". This can be changed in the Region and Language control panel of the used Windows system under the Short date format.
11. It is important to use two sheets of germination paper to obtain a big enough buffer of solution. This will make the system less sensitive to influences on germination due to evaporation.
12. The amount of solution added to the trays is very important. Too little will have influence on the germination, but too much will hamper the image analysis: too much water will cause a light-reflecting meniscus around the seed that will be mistakenly detected as a radicle by the system.
13. Instead of adding water, also salt, mannitol, PEG, ABA, or any other compound can be added to the trays to analyze the effect of these compounds on seed germination.
14. The mold is made manually by cutting holes in a plastic sheet. In our system, a mold for 6 germination tests per tray is used, but as mentioned before, also molds for, e.g., 2, 3, or 24 germination tests per tray can be made.

15. The mold is reused for a next tray, after sowing one tray.
16. The amount of seeds depends on the size of the used seeds and of the holes in the used mold. Take care to have the seeds as evenly distributed as possible since too many seeds too close together will hamper reliable automatic scoring of germination.
17. The trays on top and bottom of the pile are there to prevent unequal evaporation as much as possible.
18. For dormancy breaking purposes, the piles can also be incubated at 4 °C for 3–4 days before incubation in the germination cabinet.
19. The 53 cm is the optimal distance for the 15 × 21 cm trays, but should obviously be adapted for trays with different sizes.
20. The mask does not need to fit the trays very accurately, since our system is not sensitive to slight changes in position of the trays at the different time points. Changes up till 5 mm will not have an effect on the automatic scoring.
21. Low light with longer shutter speed works in general very well to enhance contrast.
22. In our setup, ISO 400, F/18, and 1/3 s are used, but these settings have to be defined depending on the specific light conditions. It is advised to be very punctilious in optimizing the camera settings. The better the pictures, the better the automatic scoring will perform.
23. The first picture should be taken after the initial imbibition, but before the first seeds begin to germinate, since taking a picture before initial imbibition (imbibition phase I) is completed will have an effect on the size of the seeds and the count of non-germinated seeds on the first picture will determine the total amount of seeds that is counted in the germination test.
24. The frequency of taken pictures depends on the germination speed. In general it is advised to have at least three pictures in the steep part of the germination curve. After all germination tests within an experiment have reached t50, the frequency can also go down. In general, pictures are taken till the moment that no further germination is expected.
25. In our setup 0.2 % saturated pixels is used to enhance contrast, but slightly other settings might be optimal for other setups.
26. In our setup the best results are obtained with YUV channels, but HSB, RGB, or CIE channels might work better in other setups.
27. Make sure that there is no, or as little as possible, size difference between both settings for non-germinated seeds, but that there is an obvious size difference for germinated seeds.

28. The upper size limit will also filter out seeds with long radicles, but this will not hamper the analysis, because seeds not present in the + white filter will be automatically counted as germinated seeds.
29. In our setup pictures are automatically taken upside down due to the orientation of the camera on the repro-stand.
30. The system is designed for the Photoshop version in the English language. The names of the files for the cropped separate germination tests should end with "copy" followed by a number (e.g., 10212015-0923#001 copy 4.jpg).
31. It is important to save the files with the .txt extension and not with the default .xls extension due to compatibility issues with the subsequent analysis by the Excel scripts.
32. Both the germination score + white and germination score – white analysis can be performed simultaneously, which saves PC calculation time. To do so, ImageJ has to be started twice and the subsequent steps have to be followed in each open instance of ImageJ for germination score + white and germination score – white, respectively. In this case it is important to keep the two appearing Summary and Results windows apart. One way to do so is to drag the two windows of the germination score – white analysis to the left of the desktop and the two windows of the germination score + white to the right of the desktop.
33. When the suggested naming of the txt files is used, only the first txt file (summary+.txt) has to be selected and Excel automatically will fill in the other three files.
34. In our system with *Arabidopsis* the Area variance and absolute variances are used. For other setups and seed types, other settings might work better. Also the Perimeter can be used instead of the Area of the seeds, or relative differences (%) as compared to absolute differences.
35. The optimal settings for our system with *Arabidopsis* seeds are 48 for Area and 0.8 for *xy* variance.
36. The maximum number of combinations that can be tested at one time is 255.
37. The AUC is determined until a user-defined time point. This time point is preferably chosen shortly after most germination tests have just reached maximum germination. When using an earlier time point, the AUC will mainly correlate with the t50 and using a much later time point will result in AUCs that are highly correlated with the maximum germination. If chosen at the right position, AUC will be a combinational measurement reflecting aspects of both the t50 and Gmax.

References

1. Penfield S, King J (2009) Towards a systems biology approach to understanding seed dormancy and germination. Proc R Soc B Biol Sci 276:3561–3569
2. Ligterink W, Joosen RVL, Hilhorst HWM (2012) Unravelling the complex trait of seed quality: using natural variation through a combination of physiology, genetics and -omics technologies. Seed Sci Res 22:S45–S52
3. Joosen RVL, Kodde J, Willems LA, Ligterink W, van der Plas LH, Hilhorst HW (2010) Germinator: a software package for high-throughput scoring and curve fitting of Arabidopsis seed germination. Plant J 62:|148–159
4. Dell'Aquila A (2009) Digital imaging information technology applied to seed germination testing. A review. Agron Sustain Dev 29: 213–221
5. Ducournau S, Feutry A, Plainchault P, Revollon P, Vigouroux B, Wagner MH (2005) Using computer vision to monitor germination time course of sunflower (Helianthus annuus L.) seeds. Seed Sci Technol 33:329–340
6. Teixeira PCN, Coelho Neto JA, Rocha H, De Oliveira JM (2007) An instrumental set up for seed germination studies with temperature control and automatic image recording. Braz J Plant Physiol 19:99–108
7. Wagner M-H, Demilly D, Ducournau S, Dürr C, Léchappé J (2011) Computer vision for monitoring seed germination from dry state to young seedlings. Seed Sci 142:49–51
8. El-Kassaby YA, Moss I, Kolotelo D, Stoehr M (2008) Seed germination: mathematical representation and parameters extraction. Forest Sci 54:220–227
9. Nguyen T-P, Keizer P, van Eeuwijk F, Smeekens S, Bentsink L (2012) Natural variation for seed longevity and seed dormancy are negatively correlated in Arabidopsis thaliana. Plant Physiol 160(4):2083–2092
10. Joosen RVL, Arends D, Willems LAJ, Ligterink W, Jansen RC, Hilhorst HWM (2012) Visualizing the genetic landscape of Arabidopsis seed performance. Plant Physiol 158:570–589

Chapter 8

Histochemical Staining of β-Glucuronidase and Its Spatial Quantification

Chloé Béziat, Jürgen Kleine-Vehn, and Elena Feraru

Abstract

Microscope images of plant specimens showing expression of GUS markers, besides being very beautiful, provide useful information regarding various biological processes. However, the information extracted from these images is often purely qualitative, and in many publications is not subjected to quantification. Here, we describe a very simple quantification method for GUS histochemical staining that enables detection of subtle differences in gene expression at cellular, tissue, or organ level. The quantification method described is based on the freely available image analysis software ImageJ that is widely used by the scientific community. We exemplify the method by quantifying small and precise changes (at the cellular level) as well as broad changes (at the organ level) in the expression of two previously published reporter lines, such as the *pPILS2::GUS* and *pPILS5::GUS*. The method presented here represents an easy tool for converting visual information from GUS histochemical staining images into quantifiable data and is of general importance for plant biologists performing GUS activity-based evaluation of reporter genes.

Key words GUS staining, Reporter gene expression, PILS, Auxin, Quantification, ImageJ

1 Introduction

The β-glucuronidase (GUS)-based reporter system, proposed by Jefferson and co-authors several decades ago for the analysis of gene expression in transformed plants, remains a very powerful and highly sensitive technique to study gene regulation and protein stability [1–4]. Around 5000 citations recorded in PubMed show that the GUS reporter system is used in hundreds of labs, making it one of the most widely used tools in molecular plant biology [http://www.ncbi.nlm.nih.gov/pubmed/?term=beta-glucuronidase+plant]. The reporter system contains the *Escherichia coli GUS* gene as a fusion marker and provides quantitative information or visualization of gene expression activity in vivo and in situ [1–4]. The expression of GUS can be accurately quantified using methods such as fluorometric assays on tissue extracts, while its histochemical staining provides visual information on the

Jürgen Kleine-Vehn and Michael Sauer (eds.), *Plant Hormones: Methods and Protocols*, Methods in Molecular Biology, vol. 1497, DOI 10.1007/978-1-4939-6469-7_8, © Springer Science+Business Media New York 2017

localization of gene activity in cells, tissues, or organs [2, 5]. The methods available for quantification of GUS activity are often laborious and time consuming, whereas histochemical staining protocols usually are less labor intensive. However, the latter are normally employed only in a qualitative manner and, in many publications, not subjected to quantification.

Here, we describe a simple method for quantification of GUS activity in histochemically stained specimens, using the open-access image processing and analysis software ImageJ ([6]; http://rsb.info.nih.gov/ij/). We exemplify this method by visualizing and quantifying the expression of GUS activity in roots of transgenic *pPILS2::GUS* and *pPILS5::GUS* [7]. PILS (PIN-like) proteins are a family of putative auxin carriers that localize to the endoplasmic reticulum and regulate intracellular auxin homeostasis [7, 8]. *pPILS2* and *pPILS5* activity depends on their putative substrate [7].

The most common substrate for GUS histochemical staining is 5-bromo-4-chloro-3-indolyl-beta-d-glucuronide (X-Gluc) that, in the final reaction, produces a blue precipitate which shows the localization of transgene expression [2]. The method presented here requires the conversion of this blue color intensity into gray values quantifiable with ImageJ software, and encompasses two main steps: (1) conversion of GUS histochemical staining images from RGB (stands for *R*ed *G*reen *B*lue) color mode to HSB (stands for *H*ue *S*aturation *B*rightness) mode, and (2) quantification of GUS dye present in the sample by measuring the signal intensity in the *S*aturation channel. Saturation is a value depicting the relative bandwidth of the color information. Since the intensity of the blue color (from GUS staining) correlates with the saturation value only in largely transparent tissues, a thorough clearing of the tissue is strictly required. Notably, a similar approach has been previously proposed [9], propounding the suitability of measuring the saturation for GUS quantification. The method allows precise quantification of subtle intensity differences in GUS histochemically stained samples that are otherwise used only for the visualization of gene expression.

2 Materials

2.1 Plant Medium and Seedling Growth

1. *Arabidopsis* seedlings are grown in a plant cabinet under long-day photoperiod for 5 days: *pPILS2::GUS* and *pPILS5::GUS* in *Arabidopsis thaliana Col-0* background have been previously published [7].
2. Plant growth medium contains 0.5 g/l MES, 2.3 g/l Murashige and Skoog salt, 10 g/l sucrose, and 8 g/l plant agar (pH 5.9 KOH).

3. In vitro growth is carried out in standard, sterile square Petri dishes (120 × 120 × 17) with vents.
4. 70 % Ethanol is used for seed sterilization.

2.2 Auxin Treatment

1. 24-Well culture plates.
2. Featherweight entomological tweezers.
3. 1-Naphthaleneacetic acid (NAA) solved in dimethyl sulfoxide (DMSO).

2.3 GUS Staining

1. 24-Well culture plates.
2. Cold acetone 90 % (optional) (*see* **Note 1**).
3. Na-phosphate buffer (0.1 M; pH 7) made of 39 ml (2.76 g to 100 ml H_2O) $NaH_2PO_4 \cdot H_2O$ (0.1 M final concentration), 61 ml (5.35 g to 100 ml H_2O) $Na_2HPO_4 \cdot H_2O$ (0.1 M final concentration), and 100 ml water. Adjust pH to 7 with $NaH_2PO_4 \cdot H_2O$ solution. Add 0.1 % Triton.
4. Ferro-ferricyanide buffer (5 mM) made of $K_3Fe(CN)_6$ (0.08 g) and $K_4Fe(CN)_6$ (0.105 g) dissolved in 50 ml Na-phosphate buffer (*see* **item 3**).
5. Dimethylformamide (DMF) solvent.
6. X-Gluc dissolved freshly as 1 mg X-Gluc in 10 μl DMF (*see* **Note 2**).
7. GUS staining buffer made of seven parts Na-phosphate buffer (*see* **item 3**), three parts ferro-ferricyanide buffer (*see* **item 4**) and 10 mg (1 mg/ml) freshly dissolved X-Gluc (*see* **item 6**).
8. Ethanol:acetic acid (3:1).
9. 70, 50, and 20 % ethanol.
10. Clearing solution made of 90 ml H_2O, 240 g chloral hydrate, and 30 ml glycerol.
11. Thermo incubator set at 37 °C.
12. Stereomicroscope to monitor the staining.

2.4 Sample Preparation, Imaging, and GUS Quantification

1. Tweezers.
2. Microscope slides and cover glasses.
3. Transmitted light microscope with attached digital color camera.
4. ImageJ 1.41 software (http://rsb.info.nih.gov/ij/).

3 Methods

3.1 Seedling Growth

1. Sterilize the seeds. We soaked the seeds for 1–5 min in 70 % ethanol.
2. Remove ethanol and allow the seeds to dry well in a laminar flow hood.

3. Sow the seeds in a Petri dish containing solid plant medium.
4. Stratify the seeds for a minimum of 2 days (up to 4 days) by keeping the plates (wrapped in aluminum foil) at 4 °C in a fridge or cold room.
5. Grow the seedlings in a growth chamber or cabinet for 5 days (*see* **Note 3**).

3.2 Auxin Treatment

1. We incubated *pPILS2::GUS* and *pPILS5::GUS* seedlings in MS liquid medium containing 500 nM NAA or the same amount of DMSO (solvent control) for 12 h.

3.3 Histochemical Staining with X-Gluc

1. Fixation (optional) (*see* **Note 1**):
   (a) Put about 20 seedlings in 1 ml cold acetone for at least 30 min.
   (b) Wash the seedlings at room temperature in 1 ml Na-phosphate buffer for minimum 30 min (2–3 times). Shake occasionally.
2. Histochemical staining with X-Gluc:
   (a) Put or transfer the fixed seedlings into 1 ml prepared staining buffer (*see* Subheading 2.3, **item 7**), wrap the incubation plate in aluminum foil, and place it at 37 °C.
   (b) Observe the samples regularly for optimal staining (*see* **Note 4**). Avoid over-staining the zone of interest. For the here used *pPILS2::GUS* and *pPILS5::GUS* we stained for 6 h and 2 h, respectively.
3. Clearing:
   (a) Transfer the stained seedlings into 1 ml mixture of ethanol:acetic acid (*see* Subheading 2.3, **item 8**) until complete disappearance of chlorophyll.
   (b) Wash the seedlings with 1 ml 70%, 50%, and 20% ethanol, each step for 10 min or longer.
   (c) Transfer the seedlings into an appropriate volume of (all seedlings should be submerged) clearing solution (*see* Subheading 2.3, **item 10**) and keep them overnight or longer at 4 °C. It is crucial to perform a thorough clearing for optimal reduction of optical density and because all residual (non-blue) color needs to be removed to allow its reliable ImageJ-based quantification.

3.4 Sample Preparation and Imaging

1. Place the overnight-cleared seedlings on microscope slides. Add few drops of chloral hydrate as a mounting solution. Carefully place the cover slip on the seedlings.
2. Observe the samples with a light microscope using bright-field illumination (*see* **Note 5**).
3. Take photos of the GUS-stained regions of interest (*see* **Notes 6** and **7**).

3.5 ImageJ-Based Quantification of Histochemical GUS Staining

1. Prior to quantification, convert the color mode of light microscope images from RGB to HSB by using ImageJ software function: Image → Type → HSB stack.
2. Select Saturation channel in the HSB stack (scroll the bar to the second position).
3. Measure the intensity of GUS staining in the Saturation channel. An increase of saturation depicts more "pure" color, while a decrease denotes a more "washed-out" signal. We thoroughly cleared the tissue and therefore the color information correlates with the degree of blue staining. Thus, this channel reflects the intensity of GUS dye present in the sample.
4. Display the results, taking into consideration the biological process that is being investigated (various ImageJ software options are available for this). Here, we (1) compared different cells (Fig. 1c, d), cell files (Fig. 1c, d), or regions (Fig. 1a, b) of the same sample; (2) compared different samples, such as NAA-treated and -untreated (Fig. 2a–c); and (3) used various *r*egion *o*f *i*nterest (ROI) selections, such as rectangular (Figs. 1a and 2a, b) or linear (Fig. 1c) to define where to measure the GUS intensity. Moreover, we described two very basic quantification methods, focusing on intensity profile [rectangular profile (*see* **step 5**; Fig. 1a, b) or line profile (*see* **step 6**; Fig. 1c, d)] and mean intensity depiction (*see* **step 7**; Fig. 2a–c).
5. For quantifying differences in the intensity of GUS staining observed in the same sample, we used roots of *pPILS5::GUS* reporter seedlings that showed weak GUS staining in the meristem and strong GUS staining above the meristem (Fig. 1a). To depict these differences we performed an intensity profile of the GUS signal present in a defined region:
   (a) Define a rectangular ROI by using "Rectangular selection" drawing tool from ImageJ toolbar (Fig. 1a).
   (b) Plot the intensity values measured within the selected region: Analyze → Plot profile (Fig. 1b).
   (c) To obtain the plot values, press "List" in the Plot window (*see* **Note 8**).
6. To quantify subtle differences (or differences at cellular level) in GUS staining we used roots of *pPILS2::GUS* reporter line and performed an intensity profile of the GUS signal measured along defined linear ROIs drawn crossing cells or crossing cell files (Fig. 1c):
   (a) Draw lines crossing the specific cell files (blue and gray lines) by using the ImageJ drawing tool "Straight line selection" (Fig. 1c).
   (b) Plot the values measured along the linear ROIs: Analyze → Plot profile (Fig. 1d).

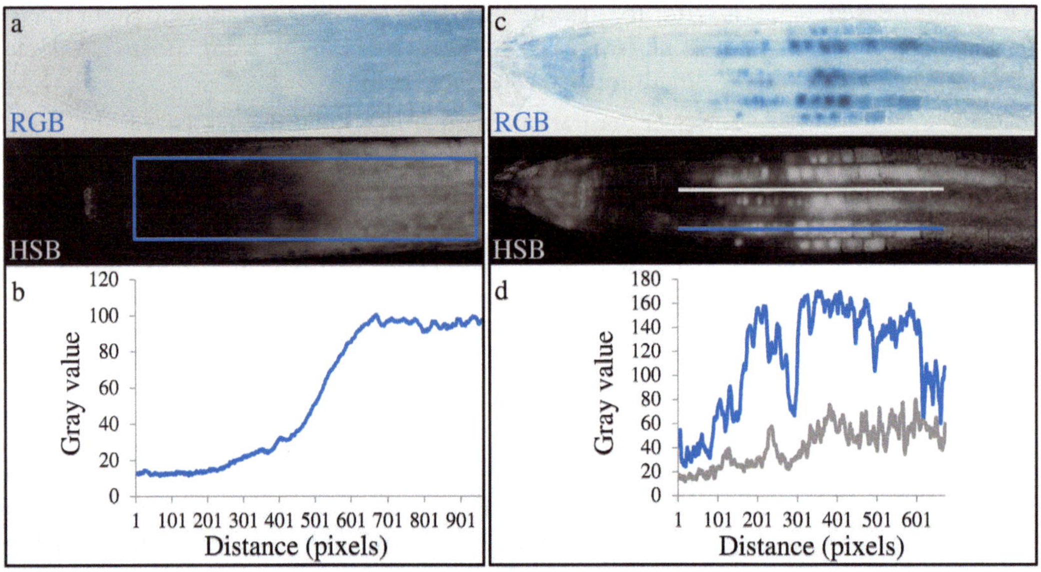

Fig. 1 Quantification of GUS histochemical staining in roots of *pPILS5::GUS* (**a** and **b**) and *pPILS2::GUS* (**c** and **d**) marker lines. (**a**) RGB (*upper*) or HSB (*lower*) color mode images showing histochemical staining of graded *pPILS5::GUS* expression in root tip. The *blue box* in the HSB image depicts the rectangular ROI for measuring the GUS staining intensities. (**b**) Graph showing the profile of GUS intensity measured within the rectangular ROI depicted in the HSB image (**a**). Note the weak expression of *pPILS5::GUS* in the meristem and strong expression above the meristem. (**c**) RGB (*upper*) or HSB (*lower*) color mode images showing cell- and cell file-specific histochemical staining of *pPILS2::GUS* expression in root tip. Drawings in HSB image represent linear ROIs crossing neighboring root epidermal cell files (*blue* and *gray lines*), showing distinct GUS staining. (**d**) Graph depicts the profile of GUS intensity measured along the linear ROIs depicted in HSB image (**c**). Note the gradual increase in *pPILS2::GUS* expression (*blue profile*). Moreover, *pPILS2::GUS* expression is distinct in neighboring epidermal cell files showing strong (*blue profile*) and weak (*gray profile*) staining

(c) To visualize and analyze the signal intensity differences between the two distinct cell files, copy the measured plot values from the Plot window (press "List" in the Plot window) into Excel or any other graphing program of choice (*see* **Note 9**).

7. For quantifying changes in GUS activity of a certain reporter gene, measure the signal intensity across an entire image or a defined region. We used *pPILS5::GUS* seedlings treated with NAA, and quantified the mean gray value in a selected region of the root, showing an increase after NAA treatment (Fig. 2a–c):

(a) Define the shape and size of ROI, where the signal should be measured by using the ImageJ drawing tool "rectangular selection" (Fig. 2a, b).

(b) Measure the mean gray value of the defined ROI: Analyze → Measure.

(c) Display the measured values in graphs (Fig. 2c). Compare signal intensity between auxin-treated and non-treated seedlings (*see* **Note 8**).

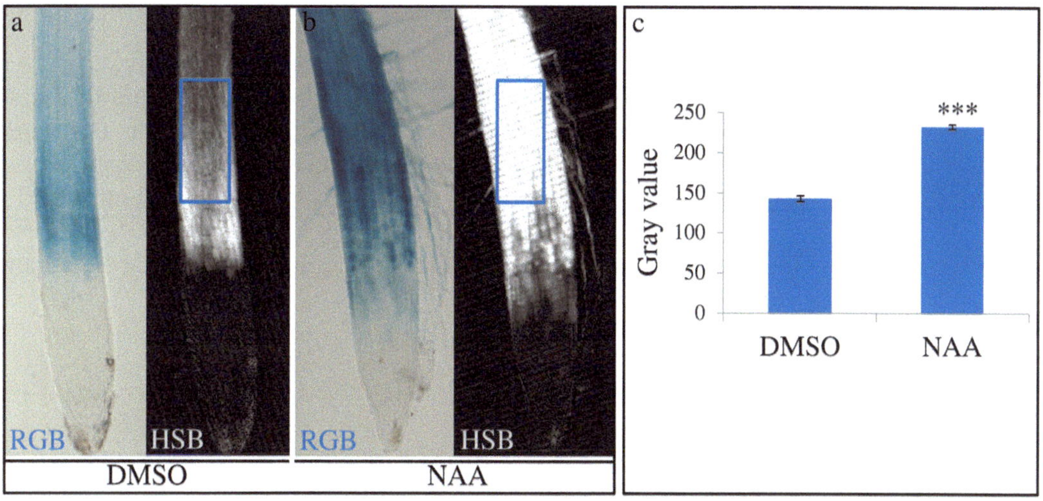

Fig. 2 Quantification of GUS histochemical staining of *pPILS5::GUS* in roots treated with auxin. (**a** and **b**) RGB (*left*) or HSB (*right*) color mode images showing histochemical staining of *pPILS5::GUS* expression in roots treated with DMSO (**a**) or NAA (**b**). Drawings in HSB images represent identical rectangular ROIs defined in DMSO-treated (**a**) and NAA-treated (**b**) *pPILS5::GUS* expressing roots. (**c**) Graph showing the mean GUS intensity measured within the rectangular ROIs depicted in HSB images (**a** and **b**). $n = 10$ seedlings, ***$p < 0.0001$

4 Notes

1. Fix your seedlings because the activity of the investigated promoter could be responsive to heat or dark stresses.
2. Commercially available X-Gluc salts deviate in their chemical properties and hence may show distinct tissue perfusion that could affect staining outcome.
3. Seedlings can be younger or older, depending on the biological processes and reporter genes that are being investigated.
4. Some GUS reporter markers show strong GUS staining within minutes while others must be stained much longer.
5. Samples can be stored at 4 °C for several weeks, but preferentially should be imaged immediately.
6. Keep the same magnification and microscope settings during the same experiment.
7. Image more than ten seedlings per line and experiment. Repeat the experiment at least three times to confirm the results.
8. Analyze identical ROIs (shape and size) in all images used for comparison: Edit → Selection → Restore Selection.
9. Use linear ROIs with an identical length throughout the experiment (Edit → Selection → Restore Selection).

Acknowledgements

We thank Esteban Fernandez for his online posts to scientific questions and David Whittaker for help with the manuscript. This work was supported by the Vienna Science and Technology Fund (WWTF) (Vienna Research Group), Austrian Science Fund (FWF) (P26568-B16; P26591-B16 (to J.K.-V.) and T-728-B16 (to E.F.)), and the European Research Council (ERC) (Starting Grant 639478-AuxinER (to J.K.-V.)).

References

1. Jefferson RA, Burgess SM, Hirsh D (1986) beta-Glucuronidase from Escherichia coli as a gene-fusion marker. Proc Natl Acad Sci U S A 83(22):8447–8451
2. Jefferson RA, Kavanagh TA, Bevan MW (1987) GUS fusions: beta-glucuronidase as a sensitive and versatile gene fusion marker in higher plants. EMBO J 6(13):3901–3907
3. Jefferson RA (1987) Assaying chimeric genes in plants: the GUS gene fusion system. Plant Mol Biol Rep 5:387–405
4. Jefferson RA (1988) Plant reporter genes: the GUS gene fusion system. In: Setlow JK (ed) Genetic engineering, vol 10. Plenum Publishing Corporation, New York
5. Gallagher SR (1992) Quantitation of GUS activity by fluorometry. In: Gallagher SR (ed) GUS protocols: using the GUS gene as a reporter of gene expression. Academic, San Diego, pp 47–59
6. Abramoff MDM, Paulo J, Ram SJ (2004) Image processing with ImageJ. Biophotonics Int 11(7):36–42
7. Barbez E, Kubes M, Rolcik J, Beziat C, Pencik A, Wang B, Rosquete MR, Zhu J, Dobrev PI, Lee Y, Zazimalova E, Petrasek J, Geisler M, Friml J, Kleine-Vehn J (2012) A novel putative auxin carrier family regulates intracellular auxin homeostasis in plants. Nature 485(7396):119–122
8. Barbez E, Kleine-Vehn J (2013) Divide Et Impera—cellular auxin compartmentalization. Curr Opin Plant Biol 16(1):78–84
9. Pozhvanov GA, Medvedev SS (2008) Auxin quantification based on histochemical staining of GUS under the control of auxin-responsive promoter. Russ J Plant Physiol 55(5):706–711

Chapter 9

Imaging *TCSn::GFP*, a Synthetic Cytokinin Reporter, in *Arabidopsis thaliana*

Jingchun Liu and Bruno Müller

Abstract

Cytokinins are classical plant hormones that control numerous developmental processes throughout the plant life cycle. Cytokinin-responsive cells activate transcription via a phospho-relay signaling network. Type-B nuclear RESPONSE REGULATOR (RR) proteins mediate transcriptional activation as the final step in the signaling cascade. They bind to promoters of immediate-early target genes via a conserved Myb-related DNA-binding domain. To monitor transcriptional activation in response to a cytokinin stimulus, we have constructed a synthetic promoter, *TCS* (two-component signaling sensor) that harbors the concatemerized binding motifs for activated type-B RR in an optimized configuration. Here, we describe our protocols for imaging *TCSn::GFP* expression in transgenic *Arabidopsis* plants. The use of the fluorescent reporter GFP allows the visualization of cytokinin-responding cells by fluorescent microscopy without the need for tissue processing steps, or staining reactions. This method is fast and with a low risk of artifacts. However, since cytokinin signaling integrates various environmental information including light, nutrient status, and biotic and abiotic stress, special care needs to be devoted to the control of growth conditions.

Key words Cytokinin signaling, Synthetic reporter, Fluorescent microscopy, Plant development, Plant physiology, *Arabidopsis thaliana*

1 Introduction

Development of a multicellular organism relies on differential gene expression triggered by instructive signals. In plants, chemical hormones such as cytokinins are key signals to govern cell specification and growth of selected cells, as well as to mediate the response to environmental cues ranging from light and stress to nodulation [1, 2]. Cytokinins are a class of adenine N6-substituted organic molecules that trigger transcriptional changes in responding cells. The production of active cytokinins, the nucleobases, involves a number of enzymatic processes that occur in different organs, tissues, and cellular compartments. Cytokinins are then subjected to transport, modifications, and degradation. Each of these processes is regulated, leading to tight control of spatiotemporal distribution of active cytokinins in planta. Cytokinin-responsive cells activate

Jürgen Kleine-Vehn and Michael Sauer (eds.), *Plant Hormones: Methods and Protocols*, Methods in Molecular Biology, vol. 1497, DOI 10.1007/978-1-4939-6469-7_9, © Springer Science+Business Media New York 2017

transcription via a phospho-relay signaling network that involves the following steps: cytokinins bind to the CHASE (cyclases/histidine kinase-associated sensory extracellular) domains of transmembrane hybrid kinases, which autophosphorylate at a conserved histidine. ARABIDOPSIS HISTIDINE TRANSFER PROTEINS (AHP) proteins shuttling between cytosol and nucleus then transfer the phosphoryl group from receptors to the nuclear ARABIDOPSIS RESPONSE REGULATOR (ARR) proteins. Upon phosphorylation, type-B nuclear RR proteins mediate transcriptional activation as the final step in the signaling cascade. They bind to promoters of immediate-early target genes via a Myb-related DNA-binding domain, which is conserved from algae to flowering plants [3], indicating that type-B RR from different plant species recognize common DNA-binding motifs.

To monitor transcriptional activation in response to a cytokinin stimulus, we have constructed a synthetic promoter, *TCS* (*t*wo-*c*omponent signalling *s*ensor), that harbors the concatemerized binding motifs for activated type-B RR in an optimized configuration. Combined with a minimal promoter and a reporter gene, *TCS*-based reporters allow the specific and sensitive detection of activated type-B RR [4, 5]. By now, *TCS*-based reporters in planta have been widely used to visualize the cytokinin response in different developmental contexts, including the female gametophyte [6], gynoecium [7], seed [4, 8], root [9, 10], shoot meristem [11], biotic stress [12], and many others. In combination with short-term treatments or inducible transgenes, the use of the synthetic reporter allows studying changes in the immediate-early responses to cytokinin at cellular resolution, before secondary effects mask the phenotypes. Crosses with permanent mutants can also reveal relevant information. The first version of the *TCS* reporter [4] suffered from silencing, an issue that was corrected in the second version, *TCSn* [5].

In addition, mesophyll protoplast transiently transfected with *TCS::LUCIFERASE (LUC)* reporters can serve as a cellular model for cytokinin-responsive cells, with the LUC activities representing the quantitative cytokinin response [4, 5, 10]. This system is ideal to rapidly screen high numbers of candidate genes and different treatments for potential functions in cytokinin signaling. Readers interested in applying this method are referred to a detailed protocol describing transient transfection assays [13]. Here, we describe our protocols for imaging *TCSn::GFP* expression in various organs derived from transgenic *Arabidopsis* plants and seedlings using confocal microscopy. The choice of a fluorescent reporters such as GFP is advantageous due to the visualization of cytokinin-responding cells without the need for tissue processing steps or staining reactions. Thus, the method is fast and the risk of artifacts is low. However, since cytokinin signaling integrates various environmental information including light, nutrient status, and biotic and abiotic stress, special care needs to be devoted to control the growth conditions (*see* Fig. 1A–H).

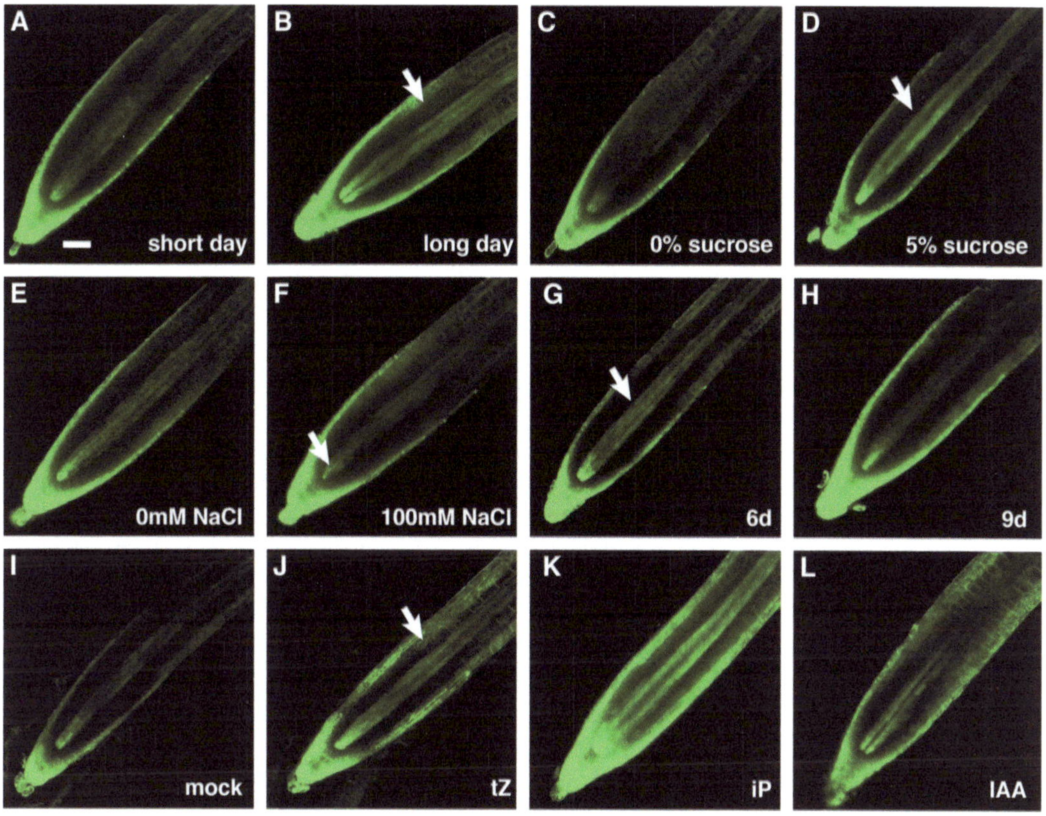

Fig. 1 Parameters influencing *TCSn::GFP* expression in the *Arabidopsis* seedling root apices. Confocal micrographs showing *TCSn::GFP* expression in the root tip of a representative seedling grown at the indicated condition. If not otherwise noted, seedlings were germinated on vertical plates with solid medium as listed in Subheading 2, and grown for 7 days at 12-h light at an intensity 90 μmol/m^2/s at 22 °C during the day and at 18 °C during the night. Of each condition, five seedlings were analysed. (**A**) Short-day conditions (8-h light) and (**B**) long-day conditions (16-h light). (**C**) Seedlings grown on medium without sucrose and (**D**) medium containing high amounts of sucrose, and on medium without (**E**) or with (**F**) 100 mM NaCl to simulate osmotic stress. (**G**) 6-day-old seedlings compared to (**H**) 9-day-old seedlings. (**I–L**) Seedlings were transferred to liquid medium after 7 days and incubated overnight with indicated hormones at a concentration of 1 μM. (**I**) Mock-treated, (**J**) trans-zeatin (tZ), (**K**) isopentenyl-adenine (iP), and (**L**) indole acetic acid (IAA)-treated samples. *Arrows* point to changes in GFP expression. Scale bar 50 μm

2 Materials

Make all solutions using ultrapure water (with resistivity greater than 18 mΩ cm) at room temperature.

2.1 Media, Reagents, and Disposables

1. Liquid growth medium: Combine half-strength MS (Murashige and Skoog) basal salt mixture without vitamins (2.15 g per liter) with 1% (wt/vol) sucrose, and adjust to pH 5.7 using 1 M KOH. Autoclave for 20 min at 15 psi, at 121 °C.
2. Solid growth medium: Add phytagar at a concentration of 0.8% (wt/vol) as a gelling agent to liquid growth medium

(*see* **Note 1**). Autoclave for 20 min at 15 psi, at 121 °C. Mix well, and place in 60 °C oven. When the medium has reached 60 °C, pour 50 ml of medium to each square Petri dish in a laminar flow hood.

3. Seed sterilization solution. Prepare 5 % (vol/vol) sodium hypochlorite and 0.05 % (vol/vol) Tween 20 in water.
4. Standard soil substrate type ED 73.
5. Square plastic pots, 80 mm diameter.
6. Square Petri dishes, 120 × 120 × 15 mm.
7. 6-Well tissue culture plates.
8. Agarose, low-melting temperature, forms gel at <30 °C (Sigma-Aldrich, cat. no. A9045-10G).
9. Mounting medium for shoot apical meristems: Weigh 0.1 g of low-melting-temperature agarose in a 15 ml tube to prepare a 1 % (wt/vol) solution. Add 10 ml of liquid growth medium, mix it thoroughly by vortexing, and place the tube in a water bath set at 70 °C. Incubate for at least 15 min with occasional vortexing until the agarose has dissolved. Alternatively, use a microwave oven to melt the mounting solution. Once the agarose is dissolved and the solution is clear and homogeneous, set the heating block to 40 °C and allow agarose to cool down to this temperature. Keep the agarose solution at 40 °C to use for mounting floral meristems. Dissolved aliquots may be stored for later use at 4 °C for 1–2 months.
10. 1 ml Disposable syringe with 30 G needle.
11. Microscope slides.
12. 5-Well pattern printed microscope slides (Tekdon 6-101, Myakka City, Florida, USA).
13. 24 × 50 mm and 18 × 18 mm (0.13–0.16 mm) cover slips.

2.2 Equipment

1. Forceps (Dumont 0208-5-PS, Actimed SA, Saint-Sulpice, Switzerland) (*see* **Note 2**).
2. Greenhouse and growth chambers with controlled light and temperature conditions.
3. Sterile laminar flow hood.
4. Autoclave.
5. Heated water bath.
6. Stereomicroscope (e.g., Leica MZFLIII).
7. Confocal microscope, equipped with adequate lenses. A Leica SP5 II, quipped with Leica HC PL APO CS2 20× 0.75 IMM, HCX PL APO CS 40× 1.3, and Leica HC PL APO CS2 63× 1.3 GLYC lenses were used in this study.
8. Imaging acquisition software (Leica Application Suite v2.83).

9. 3D-rendering software Imaris 8.1.1 (Bitplane, Zurich, Switzerland).
10. Adobe Photoshop CC, Adobe Illustrator CC.

3 Methods

3.1 Seed Sterilization and Preparation

1. Place seeds from transgenic plants of *Arabidopsis thaliana*, ecotype Columbia (Col-0), stably expressing a *TCSn::GFP* reporter transgene [5] (*see* **Note 3**) into a 1.5 ml microcentrifuge tube and add 1 ml of sterilization solution. Work in the laminar flow hood.
2. Wait for 15 min, let the seeds sink down to the bottom of the tube, and remove the supernatant.
3. Wash the seeds 3–4 times with sterile water, and wait each time till seeds sink to bottom before removing supernatant.
4. To break seed dormancy and stimulate synchronized germination, stratify seeds for 72 h at 4 °C.
5. Resuspend the seeds in 1 ml sterilized 0.1 % agarose solution in a laminar flow hood. The seeds should float in the solution at a density that allows dispensing individual seeds by pipetting.

3.2 Growing Seedlings on MS medium

1. Place about 20 seeds evenly spaced on a line, about 2 cm from the edge of a square 120 × 120 mm Petri dish with appropriate solid growth medium using a sterile pipette tip.
2. Put the plates vertically into a plant growth chamber providing the desired light and temperature conditions, typically 12-h light at an intensity of 75–100 μmol/m^2/s at 22 °C during the day and at 18 °C during the night (*see* **Note 4**).
3. Grow the seedlings for the desired time, usually between 3 and 10 days.
4. Optional: Add 3 ml of liquid growth medium to each well of a 6-well tissue culture plate. Carefully transfer five to seven seedlings with fine forceps to each well. Hormones and other signals at various concentrations may be added to the liquid growth medium to test for effects on *TCSn::GFP* expression. Typically, overnight incubations with signals at different concentrations will be informative. As positive controls, include cytokinins at 1 μM to stimulate *TCSn::GFP* expression, and as negative controls, include a mock-treated sample (*see* Fig. 1I–L).

3.3 Growing Plants on Soil

1. Fill pots with soil to the rim, and compress lightly. Wet the soil with tap water (*see* **Note 5**).
2. Dispense individual seeds directly on soil using a pipette tip, typically one or two seeds per 8 cm plastic pot.

3. Transfer the tray to a greenhouse or growth chamber that provides the desired light, temperature, and humidity conditions (*see* **Note 6**).
4. Cover the tray containing the pots with a transparent lid to maintain high humidity for about 4 days till cotyledons or germinating seedlings are fully expanded.
5. Remove the lid. Water upon need; avoid drought or water stress (*see* **Note 7**).
6. Grow until the desired organs have fully developed (*see* **Note 8**).

3.4 Mounting Organs and Seedlings for Imaging

1. Mounting of seedlings under stereomicroscope: Dispense 200 μl of liquid growth medium on a microscope slide. Using fine dissection forceps, gently grasp the seedlings from the agar plates, transfer them onto a microscopic slide (*see* **Note 9**), cover with a 24 × 50 mm cover slip, and immediately proceed to imaging (*see* **Note 10**).
2. Preparing and mounting female gametophytes under the stereomicroscope.

   Harvest pistils from flower and place it on microscope slide. Slit open the pistil replum on both sides using injection needle and forceps. Add the appropriate amount of liquid medium, about 50 μl, and cover with a 24 × 50 mm cover slip. Proceed immediately to imaging with the confocal microscope (*see* **Note 10**).
3. Dissecting seeds to recover and mount embryos under stereomicroscope.

   Collect a silique (*see* **Note 11**), and place on a 5-well pattern printed microscope slide. Add 10 μl of liquid medium to the silique and also to a neighboring printed well. Slit open the silique with the forceps. Transfer about three to four seeds into the droplet previously placed to the neighboring well. Dissect embryo out of the seed coat (*see* **Note 12**). Cover with a 18 × 18 mm cover slip. Immediately proceed to imaging (*see* **Note 10**).
4. Preparing and mounting shoot apical meristems under stereomicroscope:

   Remove an inflorescence from the plant using forceps. Place on a microscope slide. It is necessary to remove the older, larger flower buds to expose the floral meristems. To do so, cut as close to the base of the meristem as possible, perpendicular to the shoot axis (*see* **Note 13**). If necessary, remove additional older flower buds that block the view to meristem. Add a drop of warm and liquid mounting medium for shoot meristems to a slide. Transfer the dressed meristem to the drop. The meristem should be completely immersed without air bubbles, and with the meristem facing up. Add a 18 × 18 mm cover slip.

Do not push down. By gently repositioning the cover slip, you may also optimize the meristem position. In the meantime, the mounting medium will have cooled and the low-temperature melting agarose solidified, fixing the meristem in the desired position. Immediately proceed to imaging (*see* **Note 10**).

3.5 Confocal Laser Scanning Microscopy Imaging (See Note 14*)*

Seedling root tips may be recorded using a 20× immersion lens to obtain an overview of the root apex as shown in Fig. 1. The shoot meristem shown in Fig. 2C was recorded with 40× oil lens, and female gametophytes or embryos (*see* Fig. 2A, B) were recorded with a 63× immersion lens. Reduce laser power to minimize photodamage that may lead to autofluorescence. The window of the detector for GFP emission was set from 510 to 550 nm. Transmitted light was captured in parallel in transmitted light mode for information about cell outlines and morphology. Tissues were imaged in three dimensions, and interval of sections, also called the axial sampling rate, was chosen to fulfil the Nyquist criterion [14]. When comparing and presenting data from different samples, it is important to include image data representing equivalent regions to avoid bias. Generally, it is preferable to choose parameters that provide fast

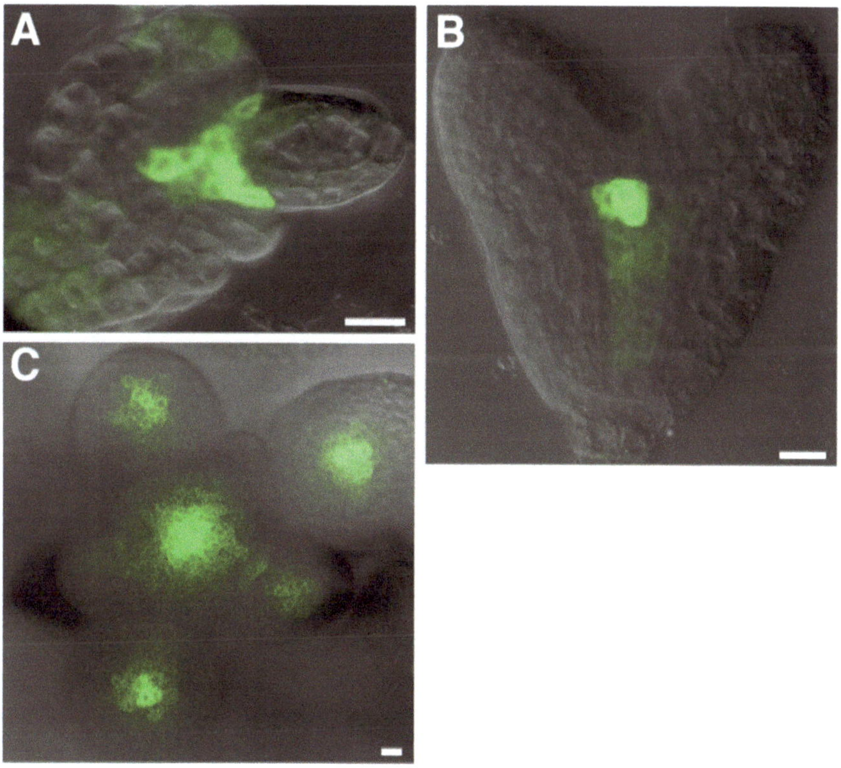

Fig. 2 *TCSn::GFP* expression (**A**) at an early stage of female gametophyte development, (**B**) heart stage, and (**C**) in the floral meristem. The GFP signal is shown in *green*, while transmitted light is shown in *grey* to outline the morphology of the organ. Scale bar 10 μm

scanning and minimal bleaching, but may produce some more noise compared to slow scanning modes that provide higher resolution and less noise but lead to bleaching. Stacks of recorded confocal sections may be processed by Imaris software for measurements, projections, and 3D reconstructions. Final images may then be trimmed and enhanced in Adobe Photoshop. Apply all image processing equally to all samples and the whole area. Composite images may be produced and assembled using Adobe Illustrator.

4 Notes

1. Depending on the biological question, variations to the composition of the solid medium may be introduced. This includes addition of NaCl to simulate salt stress (*see* Fig. **1**E, F), variation in sucrose (*see* Fig. **1**C, D) or nitrate content to vary nutrient conditions, or the addition of signals such as phytohormones.
2. These forceps are very convenient for dissecting and handling small organs such as embryos or shoot meristems, in conjunction with needles. Be careful in handling the forceps, as they are very fine and delicate. The tips will easily bend or break upon touching objects.
3. Depending on the biological question, different mutants or transgenic plants may be crossed to the *TCSn::GFP* reporter line, and used as experimental conditions to be compared against the reporter transgene in wild-type genetic background.
4. Age and light conditions may affect *TCSn::GFP* expression; *see* Fig. **1**A, B, G, H. Control these parameters for consistent results.
5. To restrict the growth of insect larvae of various species, toxin formulations of *Bacillus thuringiensis* may be added to the water; we use SolBac (Andermatt Biocontrol, Grossdietwil, Switzerland).
6. As illustrated in Fig. 1, different environmental conditions (light, nutrition, stress) may affect *TCSn::GFP* expression, which emphasizes the need for diligent control of all parameters.
7. Water consumption depends on growth stage, which usually cannot be accounted for by automatic watering systems that run on fixed schedules. It is recommended to resort to manual watering.
8. Avoid using organs from plants showing senescence, or from stressed plants, unless you are interested in studying the influence of age or stress.
9. Cut off and remove shoot, or root, respectively, depending on which tissue will be imaged to facilitate flat mounting.

10. An advantage of this protocol is the use of fresh and living tissue. Therefore, refrain from accumulating samples but work quickly through the different steps for each sample to minimize the risk of signal loss and artifacts due to dying cells.
11. Embryo stages correlate to silique age. For guidance, we found that the fourth silique, counted from the top, without any sepal or petal remnants typically contains embryos ranging from late globular to transition stage, and the fifth silique contains embryos around the heart stage.
12. This procedure needs practice. Embryos older than heart stage are relatively easy to release from the seed coat, while young embryos are tricky, especially with the suspensor left intact. In our experience it is best to use only one or two well-placed cuts to expose the embryo, and then gently grasp the embryo with the tips of the forceps, relying on the cohesive force of the liquid medium rather than on pressure from the forceps tips to free the embryo from the coat.
13. If cut too far away from the base, it cannot be easily corrected later.
14. For each tissue more sophisticated protocols allowing deeper tissue penetrance, or co-staining with multiple markers, are available. However, these protocols often require special equipment, or additional processing steps.

References

1. Kieber JJ, Schaller GE (2014) Cytokinins. Arabidopsis Book 12:e0168
2. Hwang I, Sheen J, Müller B (2012) Cytokinin signaling networks. Annu Rev Plant Biol 63:353–380
3. Pils B, Heyl A (2009) Unraveling the evolution of cytokinin signaling. Plant Physiol 151: 782–791
4. Müller B, Sheen J (2008) Cytokinin and auxin interaction in root stem-cell specification during early embryogenesis. Nature 453:1094–1097
5. Zürcher E, Tavor-Deslex D, Lituiev D, Enkerli K, Tarr PT, Müller B (2013) A robust and sensitive synthetic sensor to monitor the transcriptional output of the cytokinin signaling network in planta. Plant Physiol 161:1066–1075
6. Bencivenga S, Simonini S, Benková E, Colombo L (2012) The transcription factors BEL1 and SPL are required for cytokinin and auxin signaling during ovule development in Arabidopsis. Plant Cell 24:2886–2897
7. Marsch-Martinez N, Ramos-Cruz D, Irepan Reyes-Olalde J, Lozano-Sotomayor P, Zuniga-Mayo VM, de Folter S (2012) The role of cytokinin during Arabidopsis gynoecia and fruit morphogenesis and patterning. Plant J 72:222–234
8. De Rybel B, Adibi M, Breda AS, Wendrich JR, Smit ME, Novák O, Yamaguchi N, Yoshida S, Van Isterdael G, Palovaara J, Nijsse B, Boekschoten MV, Hooiveld G, Beeckman T, Wagner D, Ljung K, Fleck C, Weijers D (2014) Plant development. Integration of growth and patterning during vascular tissue formation in Arabidopsis. Science 345:1255215
9. Antoniadi I, Plačková L, Simonovik B, Doležal K, Turnbull C, Ljung K, Novák O (2015) Cell-type-specific cytokinin distribution within the Arabidopsis primary root apex. Plant Cell 27:1955–1967
10. Bielach A, Václavíková K, Marhavý P, Duclercq J, Cuesta C, Müller B, Grunewald W, Tarkowski P, Benková E (2012) Spatiotemporal regulation of Arabidopsis lateral root organogenesis by cytokinin. Plant Cell 24:3967–3981
11. Chickarmane VS, Gordon SP, Tarr PT, Heisler MG, Meyerowitz EM (2012) Cytokinin

signaling as a positional cue for patterning the apical-basal axis of the growing Arabidopsis shoot meristem. Proc Natl Acad Sci U S A 109:4002–4007

12. Siddique S, Radakovic ZS, De La Torre CM, Chronis D, Novák O, Ramireddy E, Holbein J, Matera C, Hütten M, Gutbrod P, Anjam MS, Rozanska E, Habash S, Elashry A, Sobczak M, Kakimoto T, Strnad M, Schmülling T, Mitchum MG, Grundler FM (2015) A parasitic nematode releases cytokinin that controls cell division and orchestrates feeding site formation in host plants. Proc Natl Acad Sci U S A 112:12669–12674
13. Yoo SD, Cho YH, Sheen J (2007) Arabidopsis mesophyll protoplasts: a versatile cell system for transient gene expression analysis. Nat Protoc 2:1565–1572
14. Pawley JB (2006) Points, pixels and gray levels: digitizing image data. In: Pawley JB (ed) Handbook of biological confocal microscopy. Springer, New York

Chapter 10

Highlighting Gibberellins Accumulation Sites in *Arabidopsis thaliana* Root Using Fluorescently Labeled Gibberellins

Hilla Schayek, Eilon Shani, and Roy Weinstain

Abstract

The physical location of plant hormones is an important factor in maintaining their proper metabolism, perception, and mediated developmental responses. Thus, unveiling plant hormones dynamics at the molecule's level is essential for a comprehensive, detailed understanding of both their functions and the regulative mechanisms they are subjected to. Here, we describe the use of fluorescently labeled, bioactive gibberellins (GAs) to highlight the dynamic distribution and accumulation sites of bioactive GAs in *Arabidopsis thaliana* roots by confocal microscopy.

Key words Plant hormones, Gibberellins, Fluorescent labeling, Fluorescence microscopy, Chemical biology

1 Introduction

Gibberellins are a class of tetracyclic diterpene carboxylic acids, functioning as plant hormones (phytohormones) and influencing a range of developmental processes including dormancy, germination, root and shoot elongation, and flowering [1, 2]. Over the years, more than 130 GAs have been identified, of which only a handful are bioactive ($GA_{1,3,4,7}$). Plants exert tight regulation over their GA response pathways at multiple levels including biosynthesis, metabolism, and perception [3, 4]. Clearly, to exert their intended roles, GA must be present in correct quantity, at the correct location and in the correct time. To understand how plants synchronize spatial localization, concentration, and timing of GA at their sites of action or metabolism, it is crucial to understand GA dynamics at the molecule's level.

Fluorescence microscopy is a minimally invasive technique that enables high-resolution, real-time tracking of fluorescent molecules in complex biological environments [5, 6]. The specificity and sensitivity of fluorescence microscopy allow the precise location of intracellular components, which can then be monitored over time, as well as

Jürgen Kleine-Vehn and Michael Sauer (eds.), *Plant Hormones: Methods and Protocols*, Methods in Molecular Biology, vol. 1497, DOI 10.1007/978-1-4939-6469-7_10, © Springer Science+Business Media New York 2017

understanding of their associated diffusion coefficients, transport characteristics, and interactions with other biomolecules. As most biological molecules of interest do not inherently fluoresce, their labeling with appropriate fluorescent tags is a perquisite for detection by fluorescence microscopy. Whereas molecular biology techniques are applied to label proteins of interest with fluorescent protein tags [7], synthetic chemistry provides access to fluorescent labeling of small bioactive molecules [8, 9], such as plant hormones, by small-molecule fluorophores [10, 11]. We, and others, have previously demonstrated the substantial utility of synthetic labeling techniques in studying the location, and involvement in dynamic processes, of plant hormones including GA [12], auxin [13], brassinosteroids [14], and strigolactones [15]. Here, we describe how the bioactive, fluorescently labeled GAs can be used to highlight the dynamic distribution and accumulation sites of bioactive gibberellins in *Arabidopsis thaliana* root.

The synthetic procedures utilized to label GA_3 and GA_4 with the small-molecule fluorophore fluorescein, as well as a comprehensive analysis of the labeled GAs, termed GA_3-Fl and GA_4-Fl, bioactivity retention, have been reported before [12] and are not described in this method. Application of GA-Fl to *Arabidopsis* wild-type seedlings, through either solid or liquid medium, leads to uptake of the fluorescent GA into the plant and is followed by movement through the root, resulting in specific accumulation in the root elongating endodermal cells. The accumulation pattern of GA-Fl can be visualized by confocal microscopy and target cells can be identified by their location in the root. Co-staining of the root with propidium iodide, to highlight cell walls, does not interfere with GA-Fl signal detection and assist in observing root morphology. The method described herein can be easily adjusted to visualize GA-Fl distribution patterns under varying developmental and environmental conditions and in plant species beyond *Arabidopsis*. A short summary of the method's advantages and disadvantages is presented in Table 1.

Table 1
Pros and cons for live imaging of fluorescently labeled hormones in the plant

Pros	Cons
Easily transferable to plant species beyond *Arabidopsis*	Exogenous application of the molecule, thus differ from endogenous sites of biosynthesis
Directly monitoring the labeled hormone (opposed to monitoring genetic markers, proteins involved in perception and response)	The hormone is bioactive but structurally modified, thus might differ with respect to transport parameters (charge, crossing membranes, etc.)
No requirement for transgenes	
Live imaging	
High spatial resolution (sub-cellular)	

2 Materials

Carry out all procedures at room temperature, unless otherwise specified.

2.1 GA-Fl Stock Solution (10 mM)

1. Prior to preparation, cover all 0.6/1.5 mL Eppendorf tubes with tin foil (*see* **Note 1**).
2. Dissolve 1.5 mg of solid GA_3-Fl in 180 μL dimethyl sulfoxide (DMSO) to a final concentration of 10 mM, OR 1.5 mg of solid GA_4-Fl in 183 μL DMSO to a final concentration of 10 mM (for calculation, *see* **Note 2**).
3. Vortex to full dissolution. Store at −20 °C (*see* **Note 3**).

2.2 GA-Fl Liquid Work Solution (5 μM)

1. Prepare fresh GA-Fl liquid work solution for each experiment.
2. Wrap 15 mL tube with tin foil.
3. Add ddH_2O, 3 mL and GA-Fl stock solution, 1.5 μL.
4. Vortex well.

2.3 GA-Fl MS Agar Plates (5 μM)

1. Prepare plant growth medium: 0.5 g/L MES (2-(*N*-morpholino) ethanesulfonic acid), 2.3 g/L Murashige and Skoog salt (MS), 10 g/L sucrose, 5 g/L plant agar (final pH 5.7).
2. Prepare the medium in a bottle or a flask (*see* **Note 4**). Stir the mixture using a magnetic stirrer. Once the mixture is homogenic, measure its pH using a pH meter. Adjust to pH of 5.7 by adding the necessary amount of KOH. Pour the first mixture into the new bottle, along with the plant agar.
3. Autoclave the plant growth medium at a temperature of 120 °C for 30 min. Subsequently, mix well by stirring and leave to cool off at room temperature. Once the temperature has dropped to approximately 50 °C (*see* **Note 5**), add the necessary amount of GA-Fl to reach 5 μM. For example, 50 μL of GA-Fl (10 mM stock solution) into 100 mL of MS agar media. Mix well. Once the mixture is homogenic, pour it into 12 × 12 × 1.5 cm plates.
4. Keep the plates open in the hood for the medium to dry off (~20 min, *see* **Note 6**). Store plates upside-down at 4 °C in the dark for up to 1 month.

3 Methods

3.1 Seed Sterilization

Important! The sterilization process must be performed in a *fume hood* to prevent inhalation of volatile materials.

1. Prepare the amount of *Arabidopsis* seeds (*see* **Note** 7) to be sterilized in a 1.5 mL Eppendorf tube (*see* **Note 8**).
2. Add sodium hypochlorite (NaClO), 33 mL, into a 100 mL Erlenmeyer flask.
3. Inside a glass desiccator—place the open Eppendorf tube with the seeds in a stand next to the Erlenmeyer flask containing the NaClO (*see* **Note 9**).
4. Make sure that the fume hood is properly working.
5. Add hydrochloric acid (HCl, concentrated 32%), 1 mL, into the Erlenmeyer flask and *immediately place the lid. CAUTION:* Addition of hydrochloric acid to hypochlorite solution leads to immediate generation of highly toxic chlorine gas (Cl_2).
6. Set a timer for 2 h (*see* **Note 10**).
7. Subsequently, remove the Eppendorf tube from the desiccator (in the fume hood). Make sure that it is properly closed, to prevent contamination.
8. Open the tubes in biological hood for 10 min to allow air exchange.
9. Seeds can be stored for several days (*see* **Note 11**).

3.2 Sowing

1. Sow seeds on a plate containing MS growth medium (*see* Subheading 2.3 for plated preparation and Subheading 3.1 for seed sterilization).
2. Cover the plate with tin foil and store at 4 °C for 48 h for stratification.
3. Place the plates horizontally in an *Arabidopsis* growth chamber with the tin foil off.
4. Allow seedlings to grow for 5 days (depending on the developmental system).

3.3 Seedling Preparation

There are two major administration approaches for applying GA-Fl to roots:

1. *Liquid* (*see* Subheading 2.2)—Immerse the seedlings in the GA-Fl liquid work solution for 0.5–3 h, depending on the developmental question (*see* **Note 12**).
2. *Agar plates* (*see* Subheading 2.3)—Replace the seedlings on GA-Fl agar plates for 2–3 h depending on the developmental question. Make sure that roots are not damaged while replaced (*see* **Note 13**).
3. Prepare a propidium iodide (PI) solution, 10 μg/mL in ddH_2O, in a separate plate.
4. Following GA-Fl application, immerse the seedlings in the PI solution for 2–4 min. Rinse with ddH_2O.

3.4 Specimen Preparation

To prevent dehydration of the seedlings during imaging—first place a drop of ddH_2O on the microscope slide. Use a pipette; the amount of water would vary according to the number of seedlings to be mounted (typically 80–200 μL for 1–3 seedlings). Place 1–5 seedlings on the wet slide. Gently place a cover slip (cover glass) over the mounted seedlings. *Important:* It is crucial not to apply any pressure to the cover slip, as the seedlings can be easily crushed. Mount the slide onto the microscope.

3.5 Imaging

Using confocal microscopy, illuminate the root at a wavelength of 488 nm, according to the absorption spectrum of fluorescein. Fluorescence is expected to be seen mostly in the endodermis layer of the root's elongation zone (*see* **Notes 14** and **15**) (*see* Fig. 1).

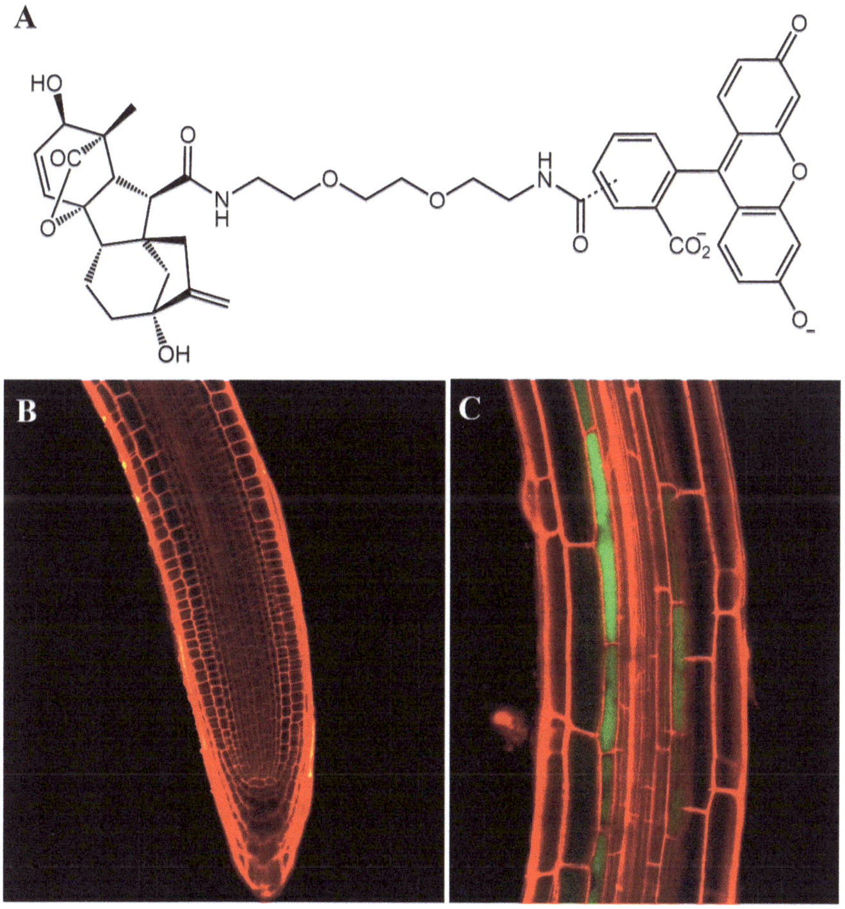

Fig. 1 Imaging fluorescently labeled GA_3 in *Arabidopsis thaliana* root. (**a**) Molecular structure of GA_3-Fl. (**b** and **c**) Confocal images of fluorescently labeled GA_3 (GA_3-Fl) in the elongating endodermal cells of roots. Meristematic zone (**b**), elongating zone (**c**). Root treated with GA_3-Fl (5 μM, 3 h). GA_3-Fl (*green*), propidium iodide (*red*)

4 Notes

1. It is important that GA-Fl solutions remain unexposed to light. Exposure to light will lead to gradual photobleaching of the fluorescein. Thus, tin foil should be used to cover all tubes and plates that contain the GA-Fl solutions, throughout the preparation process and during the experiment itself.
2. According to the formula: $V = \frac{\frac{m}{\mathrm{Mw}}}{C}$

   where m (solid GA_3-Fl or GA_4-Fl) = 1.5 mg
   Mw (GA_3-Fl) = 834 g/mol OR Mw (GA_4-Fl) = 820 g/mol
   In order to reach a concentration of 10 mM, a volume of 180 μL DMSO should be added to GA_3-Fl and 183 μL DMSO to GA_4-Fl.
3. GA-Fl DMSO stock solutions can be kept at −20 °C for at least a year.
4. The precise volume of the flask/bottle is not crucial to the procedure, but at least 2/3 of the volume should remain empty, to prevent spillage while autoclaving.
5. This could be done by carefully touching the bottle to assess its temperature, or more accurately, by using a touch-thermometer.
6. Plates could be kept open in the hood for a longer period of time, ~45 min. This will allow excess water to evaporate.
7. The method is not limited to WT *Arabidopsis* Col-0 seedlings and may be applied to mutants, transgenes, ecotypes, and additional species. However, the protocol listed here is optimized for *Arabidopsis* Col-0.
8. This could be done using a folded paper of a sort, while pouring the seeds onto it and in turn pouring the seeds into the Eppendorf tube.
9. If needed, it is possible for the flask to be tilted, and for the tube stand to sit on top of it, to allow enough room.
10. If there are more than 100 μL seeds in the tube, divide the seeds into several tubes and extend the sterilization to 3 h. Do not exceed 3 h as it might damage the seeds.
11. Sterilized seeds may be stored for several days at room temperature. After about 2 weeks, seeds tend to lose their vitality.
12. GA-Fl accumulation is temperature dependent. Immersing the seedlings in 28 °C dramatically increases GA-Fl uptake. This should be taken into account when deciding on the incubation times.
13. The seedling should be lifted by gently pulling the cotyledon upwards and replacing on top of a new agar plate (not more

than 2 s in the air to prevent desiccation responses). Gently drag the seedling upwards to straighten the root on the new plate.

14. When imaging, use the BF (bright-field) channel in order to examine the root. If dark blots and no definite outlines of cells (in the epidermis it is more prominent) are observed, the seedling was probably harmed while placing the cover glass on the slide. Similarly, patches of PI (red channel) and no definite outlines of cells points on root cells death.
15. Fluorescence may also be detectable around the outlines and on the surface of the root. This happens due to sticking of GA-Fl to the root's surface and can be reduced by increasing washing time or number of washing repetition (Subheading 3.3).

References

1. Achard P, Genschik P (2009) Releasing the brakes of plant growth: how GAs shutdown DELLA proteins. J Exp Bot 60:1085–1092
2. Daviere JM, Achard P (2013) Gibberellin signaling in plants. Development 140:1147–1151
3. Yamaguchi S (2008) Gibberellin metabolism and its regulation. Annu Rev Plant Biol 59:225–251
4. Sun TP (2010) Gibberellin-GID1-DELLA: a pivotal regulatory module for plant growth and development. Plant Physiol 154:567–570
5. Coling D, Kachar B (2001) Theory and application of fluorescence microscopy. Curr Protoc Neurosci Chapter 2, Unit 2 1
6. Combs CA (2010) Fluorescence microscopy: a concise guide to current imaging methods. Curr Protoc Neurosci Chapter 2, Unit2 1
7. Shaner NC, Steinbach PA, Tsien RY (2005) A guide to choosing fluorescent proteins. Nat Methods 2:905–909
8. Sahoo H (2012) Fluorescent labeling techniques in biomolecules: a flashback. RSC Adv 2:7017–7029
9. Jung D, Min K, Jung J, Jang W, Kwon Y (2013) Chemical biology-based approaches on fluorescent labeling of proteins in live cells. Mol Biosyst 9:862–872
10. Lavis LD, Raines RT (2008) Bright ideas for chemical biology. ACS Chem Biol 3:142–155
11. Lavis LD, Raines RT (2014) Bright building blocks for chemical biology. ACS Chem Biol 9:855–866
12. Shani E, Weinstain R, Zhang Y, Castillejo C, Kaiserli E, Chory J, Tsien RY, Estelle M (2013) Gibberellins accumulate in the elongating endodermal cells of Arabidopsis root. Proc Natl Acad Sci U S A 110:4834–4839
13. Hayashi K, Nakamura S, Fukunaga S, Nishimura T, Jenness MK, Murphy AS, Motose H, Nozaki H, Furutani M, Aoyama T (2014) Auxin transport sites are visualized in planta using fluorescent auxin analogs. Proc Natl Acad Sci U S A 111:11557–11562
14. Irani NG, Di Rubbo S, Mylle E, Van den Begin J, Schneider-Pizon J, Hnilikova J, Sisa M, Buyst D, Vilarrasa-Blasi J, Szatmari AM, Van Damme D, Mishev K, Codreanu MC, Kohout L, Strnad M, Cano-Delgado AI, Friml J, Madder A, Russinova E (2012) Fluorescent castasterone reveals BRI1 signaling from the plasma membrane. Nat Chem Biol 8:583–589
15. Tsuchiya Y, Yoshimura M, Sato Y, Kuwata K, Toh S, Holbrook-Smith D, Zhang H, McCourt P, Itami K, Kinoshita T, Hagihara S (2015) Probing strigolactone receptors in Striga hermonthica with fluorescence. Science 349:864–868

Chapter 11

In Silico Methods for Cell Annotation, Quantification of Gene Expression, and Cell Geometry at Single-Cell Resolution Using 3DCellAtlas

Petra Stamm, Soeren Strauss, Thomas D. Montenegro-Johnson, Richard Smith, and George W. Bassel

Abstract

A comprehensive understanding of plant growth and development requires the integration of the spatial and temporal dynamics of gene regulatory networks with changes in cellular geometry during 3D organ growth. 3DCellAtlas is an integrative computational pipeline that semi-automatically identifies cell type and position within radially symmetric plant organs, and simultaneously quantifies 3D cell anisotropy and reporter abundance at single-cell resolution. It is a powerful tool that generates digital single-cell cellular atlases of plant organs and enables 3D cell geometry and reporter abundance (gene/protein/biosensor) from multiple samples to be integrated at single-cell resolution across whole organs. Here we describe how to use 3DCellAtlas to process and analyze radially symmetric organs, and to identify cell types and extract geometric cell data within these 3D cellular datasets. We detail how to use two statistical tools in 3DCellAtlas to compare cellular geometries, and to analyze reporter abundance at single-cell resolution.

Key words 3DCellAtlas, MorphoGraphX, 3D imaging, 3D image analysis, Cell type identification, 3D anisotropy, Digital single-cell analysis

1 Introduction

In order to understand plant growth and development, it is necessary to analyze events across scales ranging from the whole organ level to cellular level, and down to molecular processes. Linking these events in the context of dynamic growing organs requires measuring changes in geometry and gene/protein abundance simultaneously. Generating 3D cellular resolution data sets in combination with reporters enables these spatial and temporal associations to be established. It is these 3D data sets that will allow a comprehensive analysis of plant growth and development [1]. Recent years have seen advances in such 3D imaging techniques, and the number of 3D image data sets is steadily increasing [2–5].

Jürgen Kleine-Vehn and Michael Sauer (eds.), *Plant Hormones: Methods and Protocols*, Methods in Molecular Biology, vol. 1497, DOI 10.1007/978-1-4939-6469-7_11, © Springer Science+Business Media New York 2017

Multiple computational approaches have as well been developed to enable the analysis of such data sets [3, 6–8].

Here we describe a method for the analysis of 3D images of plant organs using 3DCellAtlas, an integrative computational pipeline that has been implemented as an add-on in MorphoGraphX (www.MorphoGraphX.org; [9]), an open-source software developed for the analysis of 3D images that has been used in various studies of plant development (for example [6, 8–10]). The 3DCellAtlas extension [11] provides additional functionality to this software in the analysis of plant growth and development on a whole-organ level. It allows for the semi-automatic identification of cell types in radially symmetric plant organs based on geometry of the cells without the need for transgenic cell type markers or reference atlases, as well as the quantification of 3D cellular anisotropy and reporter abundance at single-cell resolution across whole-plant organs. It enables researchers to pool data from several samples for analysis, providing great statistical power. Thus, using 3DCellAtlas, cell shapes and sizes can be extracted for individual cells in a plant sample; these geometric data can be pooled for a set of samples and compared to another set. 3DCellAtlas is therefore a powerful tool to quantitatively analyze plant development by allowing the integration of 3D organ growth with diverse regulatory network dynamics.

Starting with the acquisition of confocal image stacks of stained plant organs or tissues, we describe how to use MorphoGraphX to process and analyze these 3D images, and how to use the 3DCellAtlas add-on to identify cell types within the plant organ. We further describe the use of two statistical tools in MorphoGraphX, (a) to compare cellular geometries of two sets of samples, and (b) to analyze reporter abundance within the organs at cellular resolution.

2 Materials

2.1 Samples

1. Plant tissue in which cell walls have been stained with propidium iodide and cleared with chloral hydrate [5]. In principle any 3D image set of sufficient quality for 3D segmentation could be used.
2. In these examples, reporter abundance from the GUS enzyme assays is detected using reflectance on a second channel. Again, any reporter can be used in the context of 3D segmented cells.

2.2 Image Acquisition

1. Inverted confocal microscope capable of acquiring high-resolution multi-channel z-stacks, ideally 16-bit images. All images presented here as examples were acquired using a Zeiss LSM 710.
2. As plant tissue is best imaged in liquid (in chloral hydrate clearing solution), dishes with a cover glass bottom are best suited

for imaging (for example Lab-Tek® Chambered Borosilicate Coverglass System, or Greiner Bio-One Cellview™ cell culture dish with glass bottom).

2.3 Image Analysis Software

1. Fiji (http://fiji.sc/Fiji) or equivalent software to convert proprietary microscope format files into single 16-bit multi-page tiffs.
2. MorphoGraphX, which can be downloaded from the developers' website: http://www.MorphoGraphX.org.
   - Recommended to be run on Linux system—Linux Mint 17 (which is based on Ubuntu 14.04), but a version running on Microsoft Windows is available.
   - Cuda-enabled nVidia graphics card with as much RAM as possible to handle large image stacks.
   - A workstation with 32–64 Gb of RAM is recommended.

3 Methods

3.1 Image Acquisition

Acquire confocal z-stacks of the organ of interest. Particular attention should be paid to achieve the highest possible signal-to-noise ratio of cell wall to cell interior for ideal segmentation, and using a thin enough slice interval to attain the best possible resolution along the *z*-axis (*see* **Notes 1–3**). For *Arabidopsis* embryo axes, a suitable image quality is achieved by using a 25× oil objective with a 0.6 zoom factor and an image resolution of 2048 by approximately 600 pixels.

3.2 Importing Images into MorphoGraphX

To open a 3D stack in MorphoGraphX, the image needs to be in *.tiff format (*see* **Note 4**). If multiple channels have been acquired, each channel has to be converted into a separate file via the "split channels" function during the conversion. The *.tiff file can either be dragged onto the open MorphoGraphX window, or it can be accessed via Stack > Stack 1 > Main > Open (Fig. 1a). Functions to process images are accessed via the three tabs on the right of the main window—"Main," "View," and "Process" (Fig. 1b–d).

In the "View" tab, clipping planes 1, 2, and 3 are tools to slice into the 3D image. Clip 1 is initially oriented in the *y*–*z*-plane, clip 2 in the *x*–*z*-plane, and clip 3 in the *x*–*y*-plane. Each clipping plane can be visualized on the image by ticking the box for "Grid," and enabled by ticking "Enable." To control the movement of the clipping planes, make sure that the correct clip is selected in the main tab under "Control-Key Interaction" (Fig. 1b). Holding down the Ctrl key and clicking the left mouse button will rotate the plane, while right-clicking moves the plane, keeping the same orientation. Move the slider below each clip to adjust the thickness of the clipping plane.

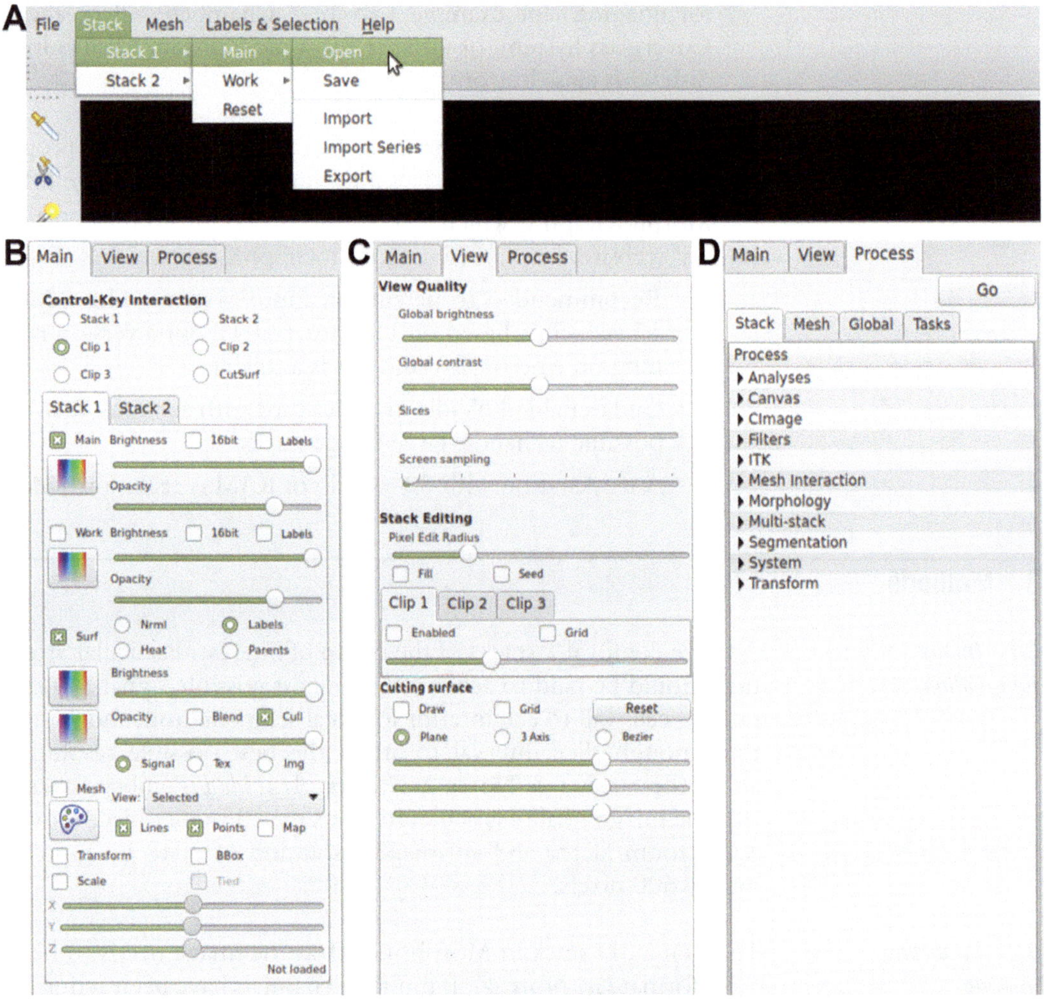

Fig. 1 MorphoGraphX basic functions. (**a**) The mouse pointer highlights the path to open an image in stack 1. (**b**) The main tab with options to choose control key interactions, and to switch between stack 1 and stack 2. (**c**) The view tab which allows control of brightness and contrast and interaction with clipping planes. The orientation of each clipping plane can be visualized by ticking "Grid," and enabled by ticking "Enabled". (**d**) The Process tab which contains all functions to process and analyze stacks and meshes

3.3 3D Segmentation of Cells

Prior to the 3D segmentation of cells, the image needs to be blurred to reduce noise. This is done using "ITK Smoothing Recursive Gaussian Blur," in the "Process" tab, under "Stack" > "ITK" > "Filters" (Fig. 2a). The parameter (Fig. 2b) is set to a radius of 0.5 μm (*see* **Note 5**).

The result of each modification to the main stack will automatically be transferred into the work stack. Stacks can be saved from either position, and can be copied or transferred from one to another via tools in the "Process" tab, under "Stack" > "Multi-stack" (Fig. 2c).

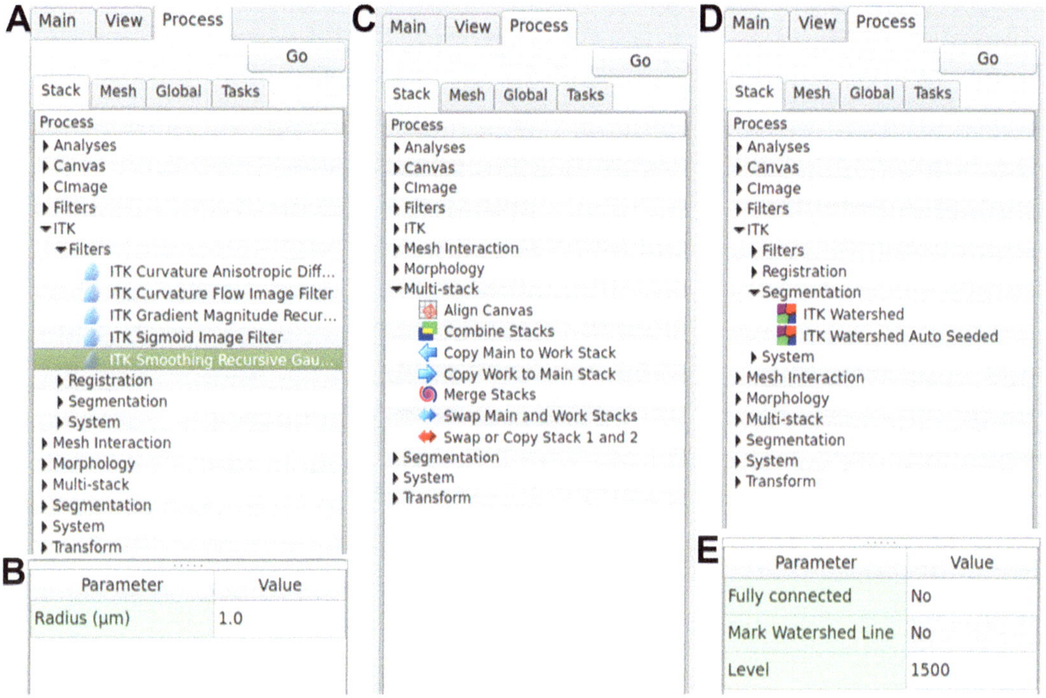

Fig. 2 Close-up of several functions in the Process tab. (**a**) To blur an image as preparation for segmenting it, the "ITK Smoothing Recursive Gaussian Blur" is used. (**b**) The adjustable parameter for this algorithm is the radius of the blurring, in µm. (**c**) Tools for copying and swapping main and work stack can be found in the drop-down menu "Multi-stack". (**d**) Tools for segmenting a stack using ITK segmentation algorithms. (**e**) Adjustable parameters for the "ITK Watershed Auto Seeded" segmentation process are shown

The blurred stack is then segmented, using the "ITK Watershed Auto Seeded" tool, in the "Process" tab, under "Stacks"—"ITK"—"Segmentation" (Fig. 2d). From the adjustable parameters (Fig. 2e) typically only the level is changed from its default to a value between 250 and 700. This level refers to the threshold signal intensity for the detection of "edges" in the segmentation process. This value will have to be adjusted to suit each image stack (depending on signal intensity, and how the image was acquired on the confocal, *see* **Notes 6–9**).

3.4 Editing of 3D Segmented Cells

The segmented stack now needs to be visually inspected, and cells that have mistakenly been assigned more than one label ("oversegmented") can be corrected by editing (note that undersegmented cells cannot be edited, *see* **Note 6**). The tools for this can be found in the tool bar of the MorphoGraphX window (Fig. 3); all of these are used by holding down the Alt key and left-clicking.

The region surrounding the actual image is usually segmented as one big "box," and has to be deleted with the scissors tool.

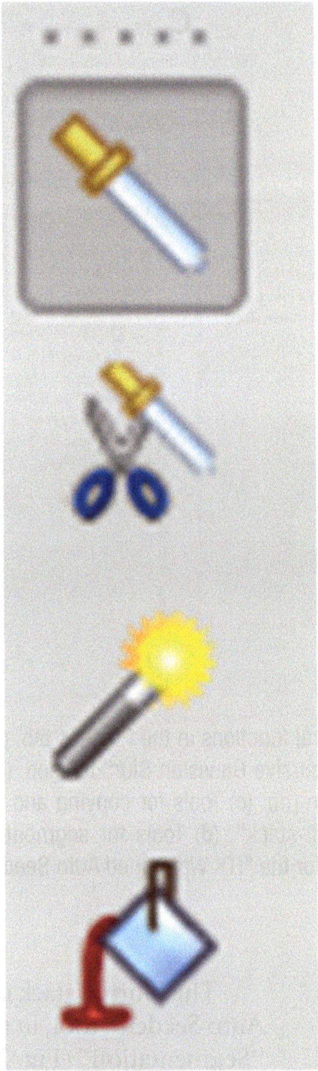

Fig. 3 Tools for editing stacks in MorphoGraphX. The *highlighted* dropper tool is used to pick a label within the segmented stack by holding down the "Alt" key and clicking on a label. The cell number will appear in the bottom left corner of the MorphoGraphX window. The scissors tool can be used to delete labels, while the pixel edit tool will delete pixels within a stack regardless of labels. The paint bucket tool will replace any label with the currently selected one

The dropper icon is used to select a label from the work stack; the number of the label will appear on the bottom left corner of the MorphoGraphX window.

To replace another label with the selected one, the paint bucket tool is selected. Clicking on the label to be changed will replace the original label with the active selected one.

To edit cells on the inside of the stack, any of the three clipping planes can be used to move through the image. Progressive changes during editing of the segmented image stack need to be saved frequently, as there is no "undo" function in MorphoGraphX.

3.5 Creating the 3D Cell Mesh

Once the segmented image is fully edited, a cell mesh describing the surface of each individual cell is created. The tool for this process is called "Marching Cubes 3D," and can be found in the "Process" tab, under "Mesh" > "Creation" (Fig. 4a).

Under the parameters (Fig. 4b), "Cube size" determines the distance between each point in the mesh; this will have to be adjusted to suit the size of the cells. A cube size of 2 μm is typically used (*see* **Note 10**).

"Min Voxels" determines the minimum size in voxels of a segmented cell to be assigned a surface.

"Smooth passes" is the number of times the mesh is smoothed during its creation. Repeated smoothing will shrink segmented cell meshes. This procedure must be used sparingly, if at all (*see* **Note 11**).

"Label" refers to the cell ID the mesh is being created on. If a number is chosen, the marching cubes will only run on this particular label; if it is "0" it runs on the whole sample.

3.6 Creating the Surface Mesh

A second mesh, which describes the surface of the whole organ to be analyzed, needs to be created; this is to capture the global organ shape rather than that of individual cells. For this, the original image is blurred with the "ITK Smoothing Recursive Gaussian Blur," at a radius of 5 μm (*see* Fig. 2a, b), and the surface mesh is created using the "Marching Cubes Surface" tool, in the "Process" tab, under "Mesh" > "Creation" (Fig. 4a). The cube size is typically set to 5 μm for this mesh. The threshold refers to the signal intensity (Fig. 4c). This will have to be adjusted to suit each image to get a smooth description of the surface.

3.7 Trimming the Open End of the Surface Mesh

The end of the surface mesh, where the sample was truncated during imaging, needs to be cut off for 3DCellAtlas. Surface meshes of hypocotyls have two "open ends," while embryo axes and roots where the tip is included need to be cut at one end only. To trim the ends of a surface mesh, use any of the clipping planes. Move the clipping plane to include the end to be cut off and select it by running "Select Clip Region" in the "Process" tab, under "Mesh" > "Selection." The selected region will be highlighted in red (Fig. 5). Pressing the Delete button on the keyboard will delete the selected region. The final surface mesh with the end(s) cut off is then saved.

3.8 Creating the Bezier Curve

To create a Bezier curve that describes the geometric center of the radial sample, in the "View" tab, under Cutting Surface, tick "Draw," "Grid," and "Bezier," and then click "Reset" (Fig. 6a). This will generate a 2D Bezier plane in the *x*–*y* plane with 5 × 5

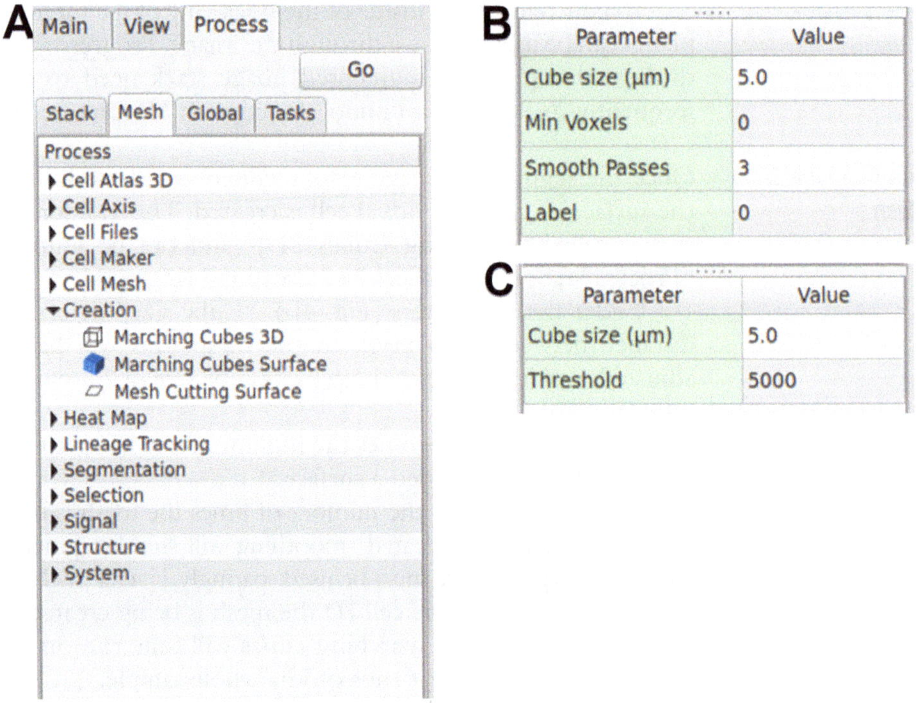

Fig. 4 Tools and parameters for creating meshes. (**a**) Algorithms used to create meshes. A mesh describing the surface of individual cells of a segmented stack is created using the "Marching Cubes 3D" tool, while a mesh describing the overall surface of the whole organ is generated with the "Marching Cubes Surface" algorithm. (**b**) Adjustable parameters for the "Marching Cubes 3D" tool. (**c**) Adjustable parameters to create a surface mesh with the "Marching Cubes Surface" tool

handles. To collapse these into a single line, go to "Mesh" > "3D Cell Atlas" > "Tools" > "Collapse Bezier Points" in the "Process" tab. This will transform the Bezier plane into a single line with five handle points. These five handle control points need to be adjusted to fit the Bezier line to the center. The "Select points in mesh" (Fig. 6b) tool is used to select handles, by holding down the Alt key and clicking on, or dragging over, a control handle (Fig. 7). Selected handles will turn red. These can be moved by simultaneously holding down the Alt and Shift key while clicking and dragging them to the required position (*see* **Note 12**). The positions of the Bezier points are stored in MorphoGraphX View (*.mgxv) files, and need to be saved from the top main menu via "File" > "Save as."

3.9 Preparing the Sample for 3DCellAtlas

The 3DCellAtlas tool will create a coordinate system through the plant organ using a Bezier line through its centrer and a surface mesh on its outside. To run 3DCellAtlas, open the Bezier view file (*see* Subheading 3.8), load the segmented cell mesh into "Mesh 1" (*see* Subheading 3.5), and the organ surface mesh into "Mesh 2"

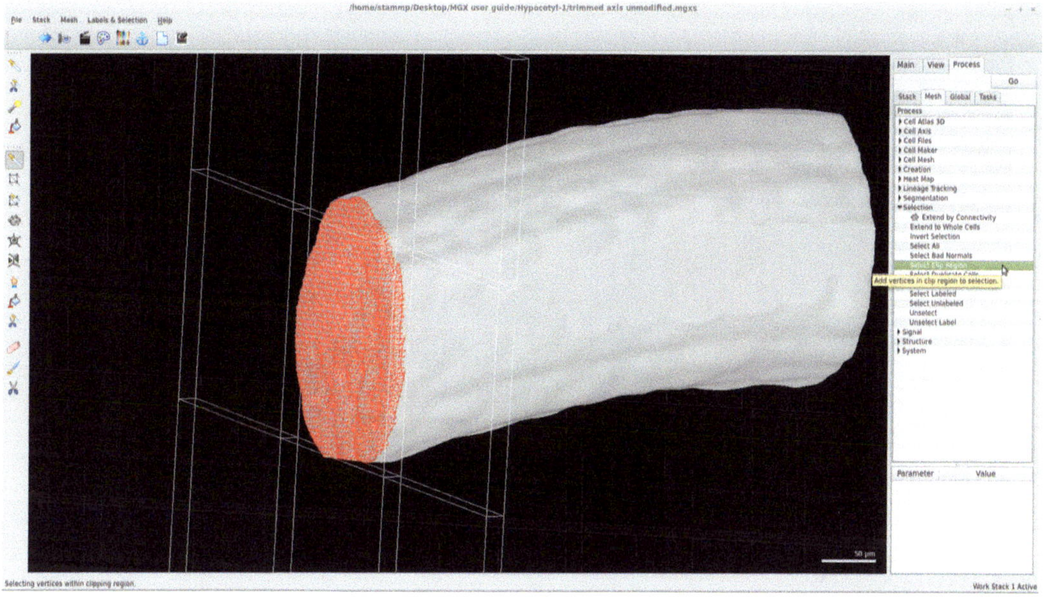

Fig. 5 Selecting and deleting parts of meshes. Screenshot of the MorphoGraphX window showing a surface mesh of an *Arabidopsis* hypocotyl stack. The grid of clip 1 is visible, which has been used to select one end of the surface mesh. The mouse pointer highlights the function to select the part of the mesh within the clipping plane. Once selected, the vertices of the mesh will be highlighted in *red*. Highlighted regions can then be deleted by pressing the "Delete" key on the keyboard

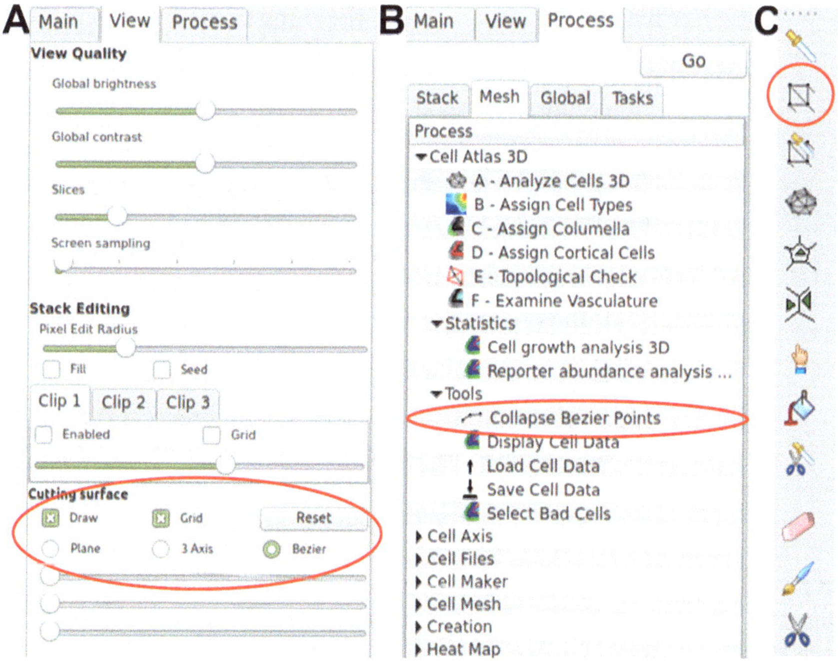

Fig. 6 Creation of and interaction with a Bezier curve. (**a**) To create a Bezier plane, "Draw," "Grid," and "Bezier" need to be ticked in the View tab before pressing "Reset." (**b**) The resulting Bezier plane is collapsed into a Bezier curve with five handle points using the "Collapse Bezier Points" tool in the Process tab, under Mesh > Cell Atlas 3D > Tools. (**c**) Tools for editing meshes and Beziers. The tool for selecting and moving Bezier handles is *circled*

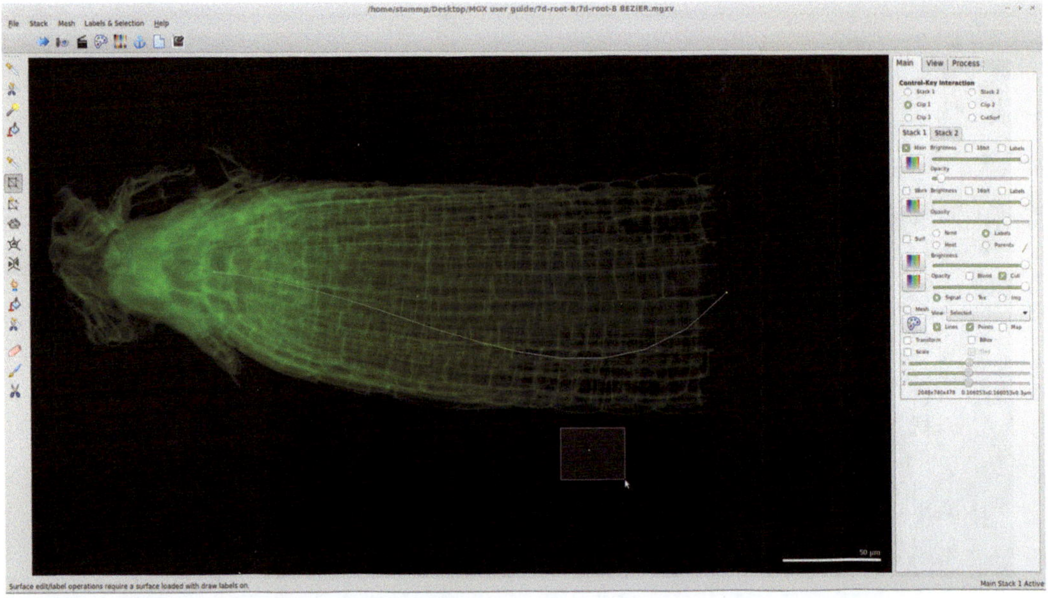

Fig. 7 Adjusting the Bezier curve to align with the axis of the imaged plant organ. MorphoGraphX window with a Bezier curve through the center of an *Arabidopsis* root. The mouse pointer has been dragged over one of the handles while holding down the Alt key to select it

(*see* Subheading 3.7). This can be saved again as the current *.mgxv file. In this way opening the *.mgxv file will open the cell mesh, organ surface mesh, and the Bezier curve, all together in the correct locations.

Highlight a cortical cell at one end of the sample in the segmented mesh using the "Select Connected Area" tool (Fig. 8a, circled) with a left click while holding down the Alt key. Note that the "Stack 1" tab needs to be selected in the "Main tab" (thus, Mesh 1 will be active, instead of Mesh 2). This cortical cell will identify cell position "1" for the cortical referencing system.

If the organ has two different cell topologies, for example roots where the root cap splits, or embryo axes which have both radicle and hypocotyl cellular arrangements, an additional cell on the outside of the sample needs to be highlighted (Fig. 8b). This second outside cell should be selected at the position where the cellular topology changes. This is not required for organs with a single-cell topology, for example hypocotyls (Fig. 8c).

3.10 Step 1 of 3DCellAtlas: Analyze Cells 3D

In the "Process" tab, under "Mesh" > "Cell Atlas 3D," "A—Analyze Cells 3D" will calculate all geometric data of the mesh, and determine the positions of all cells relative to the Bezier (centre) and the surface (Fig. 9a). Hereby, the coordinates and lengths along the three principal directions are calculated for each cell. The three directions are longitudinal (along the Bezier line), radial (relative location

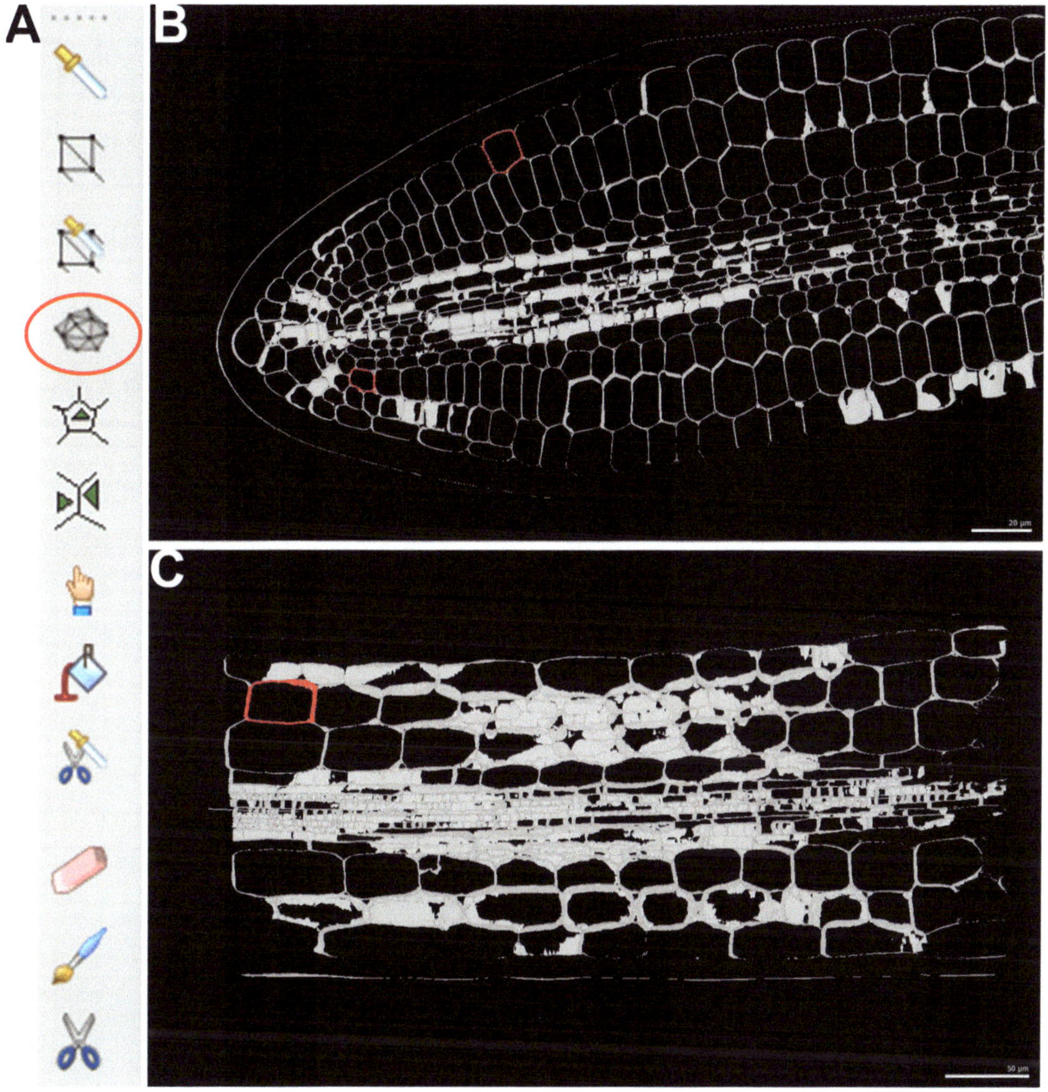

Fig. 8 Setting up samples to be analyzed using 3DCellAtlas. (**a**) Tools to interact with meshes in MorphoGraphX. The *highlighted tool* is used to select cells within a mesh. (**b** and **c**) Screenshots of MorphoGraphX windows with the setup required for running the 3D Cell Atlas tools: the segmented mesh of the organ in stack 1, the surface mesh in stack 2, and the Bezier curve describing the center of the axis. (**b**) *Arabidopsis* embryo with two cells *highlighted*, the first cell in a cortical cell file, and a cell at the position of the topological switch. (**c**) Arabidopsis seedling hypocotyl with the first cortical cell *highlighted*

between the Bezier line and the surface mesh), and circumferential (angular position within the radially symmetric cross section).

In the parameters (Fig. 9b), a minimum volume to be identified as a cell can be set. Thus, cells with a smaller volume than this set value will be ignored. Also, other "malformed" cells will be ignored and marked as "bad"; those can be visualized using the process "Select Bad Cells."

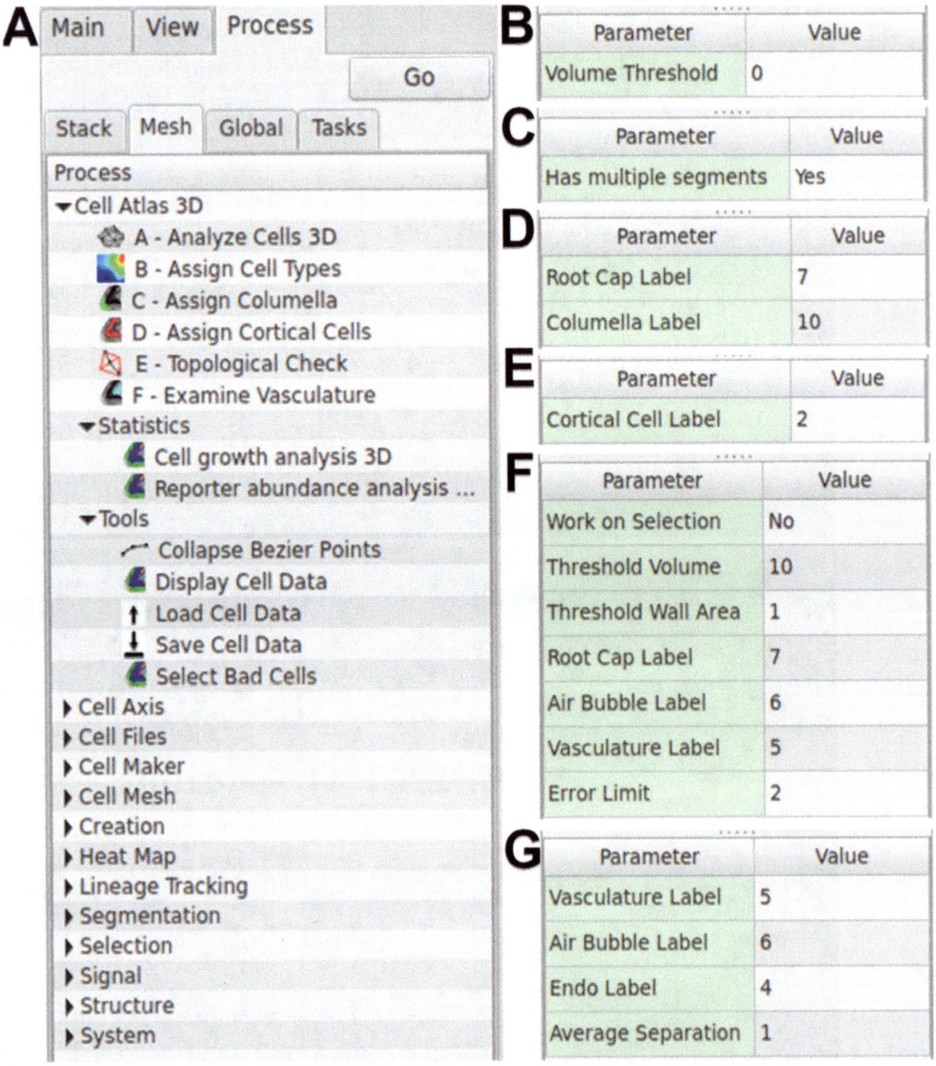

Fig. 9 Cell Atlas 3D tools and algorithms. (**a**) Close-up of the Process tab with the drop-down menus for the Cell Atlas 3D tools. (**b**–**g**) Adjustable parameter for each function. (**b**) Parameters for "A—Analyze Cells 3D." The Volume Threshold can be set to exclude cells below a certain size from the analysis. (**c**) Parameters for "B—Assign Cell Types." "Has multiple segments" refers to segments within the organ to be analyzed that have different topologies. (**d**) Parameters for "C—Assign Columella." The values for Root Cap and Columella label can be changed, but must be numbers that have not been used during the assigning of cell types. (**e**) Parameters for "D—Assign Cortical Cells." The value has to be the number assigned to cortical cells. (**f**) Parameters for "E—Topological Check." "Work on Selection" determines which cells will be analyzed; set to "No" to examine all cells in the sample. A "Threshold Volume" defines the minimum size of cells to be examined, while "Threshold Wall Area" determines the minimum shared cell wall area between neighboring cells to be considered as a connection. Labels for root cap, air spaces, and vasculature need to match the numbers assigned to these cell types. The "Error Limit" is the minimum number of wrong connections (impossible edges) between cell types for a given cell to be identified as mislabeled. (**g**) Parameters for "F—Examine Vasculature." Here, labels of vasculature, air spaces, and endodermis need to be matched up with the numbers assigned for these cell types

3.11 Step 2 of 3DCellAtlas: Assign Cell Types

Next, run "B—Assign Cell Types" (Fig. 9a). For organs with a single-cell topology, for example hypocotyls, the parameter "Has multiple segments" needs to be "No." If the organ has two different topologies, for example embryo axes and roots, the parameter needs to be "Yes" (Fig. 9c). Running the process will prompt a GUI to open (Fig. 10). The drop-down next to "Part of organ," with the options "segment 1" and "segment 2," refers to the parts of the organ with different topologies, divided at the second selected cell prior to analysis. This option will be disabled if "B—Assign Cell Types" was run with the parameter "No" (Fig. 9c). By changing the selection of the segment (1 or 2), the heatmap will show all cells located within each respective region. This facilitates the selection of cell clusters across organs with multiple topologies. Without this feature, cell clusters of diverse cell types would overlap.

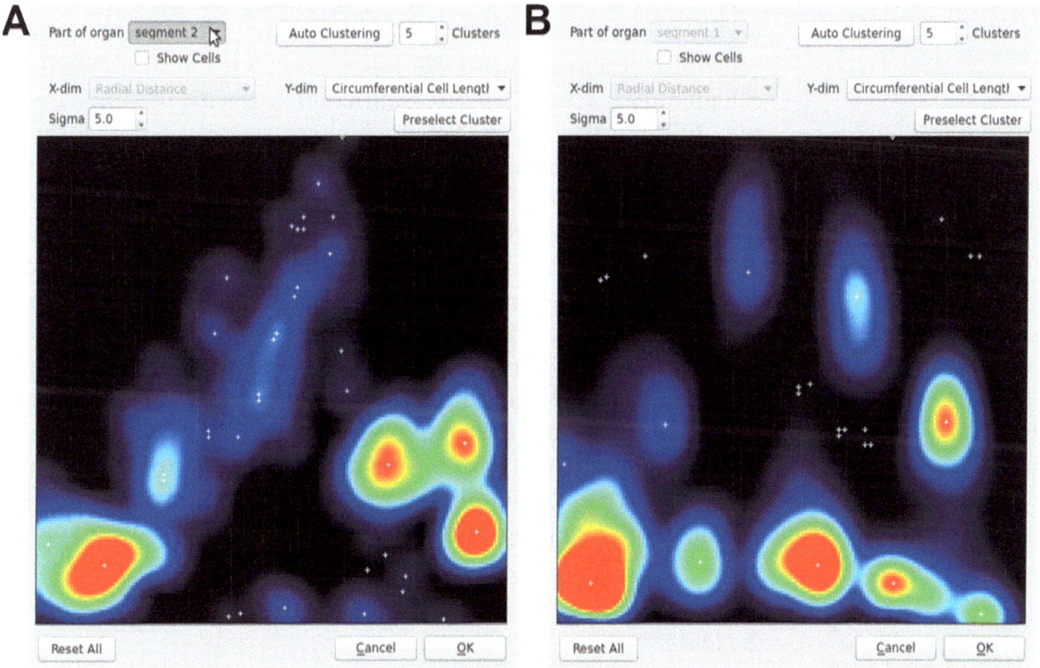

Fig. 10 Screenshots of the GUI opened upon running "Assign Cell Types." (**a**) If an organ with two topologies is analyzed, these two regions will be separately annotated, and appear as segment 1 and segment 2 in a drop-down menu next to "Part of organ." (**b**) This option is disabled for organs with a single topology. Cell data are plotted in heatmaps, displaying the radial distance of cells from the central Bezier curve on the *x*-axis, and a chosen geometric property of cells along the *y*-axis. The *y*-dimension "Y-Dim" can be changed on the drop-down menu to display either of the three principal cell lengths, to find the best option for ideal spread of cell clusters. Tick the box next to "Show Cell" to display each cell on the heatmap. "Sigma" is the variance parameter for a 2D Gaussian representing each cell's parameters. A larger sigma will lead to a smoother heatmap

3.12 Step 3 of 3DCellAtlas: Selecting Cell Clusters

The 2D heatmap GUI generates cell clusters by plotting the radial distance of cells from the central Bezier curve on the *x*-axis, and a chosen geometric property of cells along the *y*-axis. The heatmap (red—high value, blue—low value) is created by using the radial distance between the Bezier and the surface mesh in the x-dimension ("X-Dim"), and a user-defined cell length in the y-dimension "Y-Dim" and summing Gaussians around the coordinates of each cell. "Y-Dim" can be changed on the drop-down menu to display the cell length along the three different directions (circumferential, radial, or longitudinal, *see* Subheading 3.10). The best option for ideal spread of cell clusters can be chosen to define cell types accurately. Tick the box next to "Show Cell" to display the data point of each cell within the heatmap.

The value of "Sigma" is the variance parameter for a 2D Gaussian representing each cell's parameters. A larger sigma will create larger overlap of the "heat" of single cells, which leads to a smoother heatmap.

In this 2D heatmap GUI, left mouse clicks are used to place seeds upon cell clusters, and to assign these cell cluster (representing distinct cell types) unique numbers. It is important to keep track of labels assigned to cell types, and to make sure that these cell identity labels match across different segments in samples with two different topologies. Small white crosses refer to the location of local maxima in the heatmap; placed seeds are assigned to their nearest local maximum.

Clicking onto a cluster will place a seed onto it, automatically assigning this the number "1." To change the cell identity label a seed gives to a cluster, hover the mouse over the seed, and press a number key on the keyboard. A maximum of ten different cell identities can be assigned (numbers 0 through 9). To move placed seeds, drag and drop them to the required position. To remove seeds from the GUI, click and drag them out of the visible area.

We recommend a cell identity numbering system where outermost cells are given number 1 and numbers increase towards the middle with each progressive cell layer:

1. Epidermis
2. Outer cortex
3. Inner cortex
4. Endodermis
5. Vasculature
6. Air spaces
7. Lateral root cap

Additional layers including the columella are assigned higher numbers (*see* Subheading 3.13).

The "Auto Clustering" button automatically identifies the selected number of clusters by placing seeds on the highest peaks present on the heatmap.

By pressing "Preselect cluster" after having placed a seed onto one cell cluster, all cells assigned to this cluster will be given this particular label, and will be removed from the heatmap. This will also re-scale the heatmap with the remaining cells. Preselecting clusters can be undone using the button "Reset All."

Cell type labels for all cells are stored as parent labels, and following GUI selection, cells are false colored based on their parent label. If cells were mis-annotated, it is possible to return to 2D Heatmap GUI in "B—Assign Cell Types" and iteratively adjust seed locations to improve accuracy. Typical distribution of cell clusters and placement of seeds for an *Arabidopsis* hypocotyl, root, and embryo axis are shown in Fig. 11.

3.13 Step 4 of 3DCellAtlas: Assign Columella

For roots and embryos, cells in front of the highlighted first cortical cell are not included in the cell type analysis described above. To assign columella and lateral root cap at this end of the sample, run "C—Assign Columella." This will identify cells as root cap when their relative wall area touching other cells is below a certain

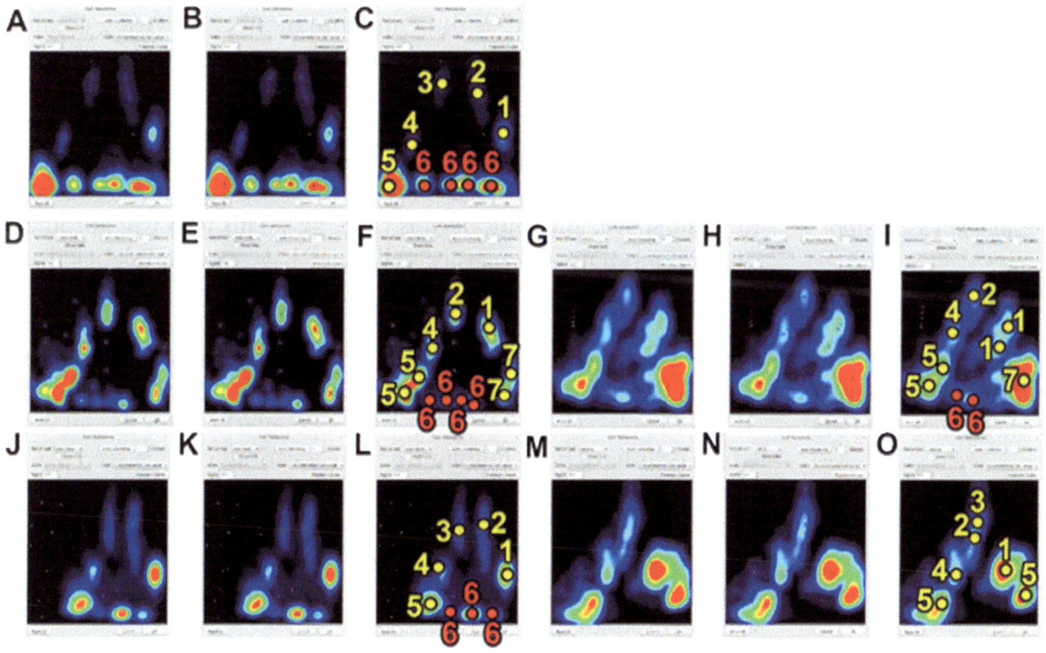

Fig. 11 Heatmaps and locations where annotated seeds are placed when identifying cell clusters in 3DCellAtlas. (**a**–**c**) *Arabidopsis* hypocotyl. (**d**–**i**) *Arabidopsis* root showing segment 1 in (**d**–**f**) and segment 2 in (**g**–**i**). (**j**–**o**) *Arabidopsis* embryo showing segment 1 in (**j**–**l**) and segment 2 in (**m**–**o**). *Yellow labels* indicate true cell clusters and *red labels* those for segmented air spaces

threshold. In the parameters (Fig. 9d), the cell identity value needs to be changed to the number assigned to these cell types, and not a number used previously for other cell types.

3.14 Step 5 of 3DCellAtlas: Assign Cortical Cells

Next, run "D—Assign Cortical Cells." With the first cortical cell selected, this will assign the cortical cell referencing system of cell positions to all cells in the mesh. Make sure that the value in the parameters (Fig. 9e) contains the correct label for cortical cells.

3.15 Step 6 of 3DCellAtlas: Topological Check

Mis-annotated cells can be detected and highlighted by running "E—Topological Check." Hereby a neighborhood graph is used to determine unlikely cell neighborhoods, which will be labeled as "wrong connections." Note that this process requires a sufficient accuracy of the input data to work correctly, as it depends on a majority of correct neighborhoods.

In the parameters (Fig. 9f), "Work on Selection" determines which cells will be analyzed; set to "No" to examine all cells in the sample.

"Threshold Volume" refers to the minimum size of cells to be examined, while "Threshold Wall Area" defines the minimum shared cell wall area between neighboring cells to be considered as a connection.

Make sure that the labels for root cap, air space, and vasculature match the earlier selection.

The "Error Limit" is the minimum number of wrong connections (impossible edges) between cell types for a given cell to be identified as mislabeled.

3.16 Step 7 of 3DCellAtlas: Examine Vasculature

Following the topological check, run "F—Examine Vasculature." Make sure that the correct parent labels are assigned to vasculature, air space, and endodermis (Fig. 9g). This tool will automatically re-annotate vasculature cells that have been mislabeled (*see* **Note 13**).

3.17 Step 8 of 3DCellAtlas: Saving and Loading Cell Data

Once all parent labels have been correctly identified, save all cell data. This is done using "Save Cell Data," under "Cell Atlas 3D" > "Tools" (Fig. 9a). This stores geometric data for all cells, which cells have been highlighted, the locations of each seed placed in the 2D Heatmap GUI, cell position according to the referencing system (in the column "associated cortical cell"), and parent labels associated with each original unique cell ID in the mesh. Using "Load Cell Data," all this information can be reloaded onto the corresponding mesh.

3.18 Step 9 of 3DCellAtlas: Displaying Cell Data

The "Display Cell Data" tool can be used to visualize and validate measured cell data (Fig. 12). These will be displayed in the form of a false-colored heatmap on the analyzed organ. The drop-down menu allows the selection of the parameter to visualize.

3.19 Step 10 of 3DCellAtlas: Statistical Analysis of 3D Cell Anisotropy

To analyze differences in 3D cell shape, data saved as described in Subheading 3.18 in csv files for multiple samples and/or time points are placed into separate directories for "control" and "treatment" data. Additionally, create a directory in which the output of these statistical analysis is to be placed (*see* **Note 14**).

In the "Process" tab, under "Mesh" > "Cell Atlas 3D" > "Statistics," run "Cell growth analysis 3D" (Fig. 9a) (*see* **Note 15**).

If any of the values for "Control folder," "Treatment folder," and "Output folder" are left blank (Fig. 13), the program will prompt the user to guide it to each missing directory.

Under "Output Type," either "Control" or "Treatment" can be chosen. This refers to the samples that will be used to generate heatmaps with the result of the analysis.

"Sliding Average" is an optional smoothing function, which averages the value for each cell based on its neighbors. The window for this averaging needs to be an odd number, and can be set under "Window" (*see* **Note 16**).

The output of this analysis is a number of text files for each and every control or treatment sample analyzed, depending on which has been chosen under "Output Type."

Per sample, two text files are created. The first contains the data for the mean ratio (treatment over control) of all geometric measurements for each cell ("[filename].csv"). The second file contains the value of elongation factors (E-factors) ("[filename]_eFactor.csv"). These text files may be imported into MorphoGraphX as heatmaps to false color their respective segmented cellular organ mesh files.

An additional file named "Data_CI.csv" is also generated as part of the process output. This file contains the ratio of each

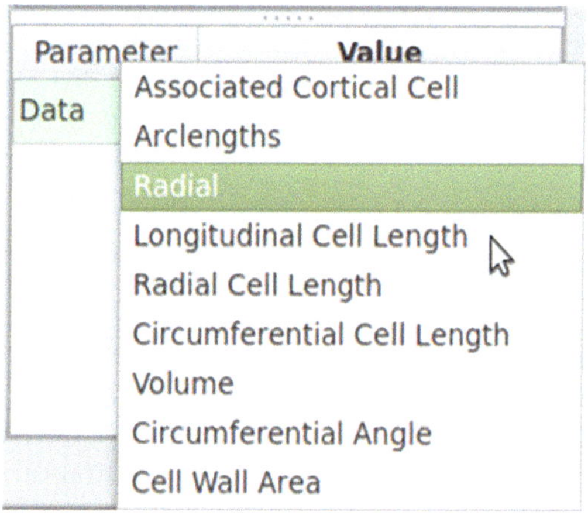

Fig. 12 Drop-down menu for the "Display Cell Data" tool. Any of the geometric data can be chosen, and will be displayed as false colors on the mesh

Parameter	Value
Control folder	
Treatment folder	
Output folder	
Output Type	Treatment
Sliding Avg	No
Window	3

Fig. 13 Parameters for the "Growth Analysis 3D" tool. "Control folder" and "Treatment folder" are directories that contain the raw data of samples to be analyzed and compared. "Output folder" is the directory in which the results of this analysis will be saved. Under "Output Type," either "Control" or "Treatment" can be chosen, referring to the samples that will be used to generate heatmaps with the result of the analysis. "Sliding Average" is an optional smoothing function which averages the value for each cell based on its neighbors. The window for this averaging needs to be an odd number, and can be set under "Window"

metric for each cell type, at each position, in addition to their respective 95 % confidence intervals (based on the *t*-distribution) at each data point. Plots can be generated to visualize these data by importing the Data_CI.csv file into a spreadsheet program such as Excel. Note that the confidence intervals will not be smoothed, if smoothing has been enabled; the "sliding average" smoothing function will only be applied to the metric itself.

3.20 Step 11 of 3DCellAtlas: Statistical Analysis of 3D Reporter Abundance

To statistically analyze the abundance of a reporter within each cell of the organ in question, heatmaps for reporter abundance need to be created for each individual sample. For this, the segmented mesh of a sample is loaded, together with the z-stack of the reporter channel of that same sample in the main stack 1.

In the "Process" tab, under "Mesh" > "Heatmap," run "Heatmap" (Fig. 14). This will open a window in which Heat Map Type and Heat Map Visualisation can be chosen from drop-down menus (Fig. 15). For reporter abundance, Volume and Interior signal are chosen to quantify reporter concentration.

Ticking the box next to "Signal average" will calculate the reporter abundance relative to the volume of each cell, so as to reflect reporter concentration for each cell.

Under "Report to Spreadsheet," type in a name for the output file to be created. This will be a csv file, which contains each cell ID of the sample and its respective reporter abundance.

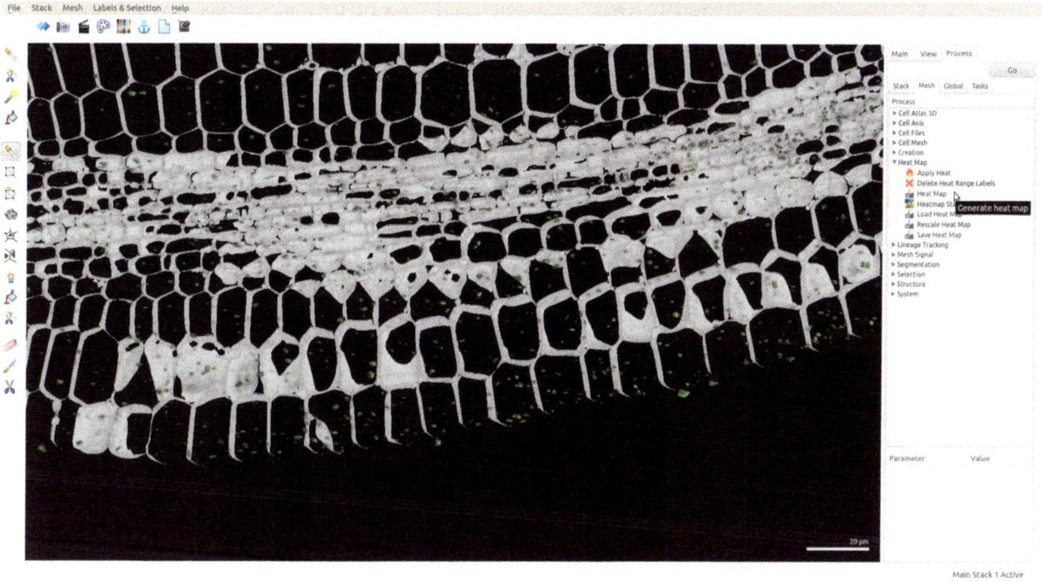

Fig. 14 Reporter abundance analysis in MorphoGraphX. Screenshot of a MorphoGraphX window with a sample mesh and reporter z-stack file loaded. The mouse pointer highlights the "Heatmap" function to be used to calculate reporter abundance in each cell

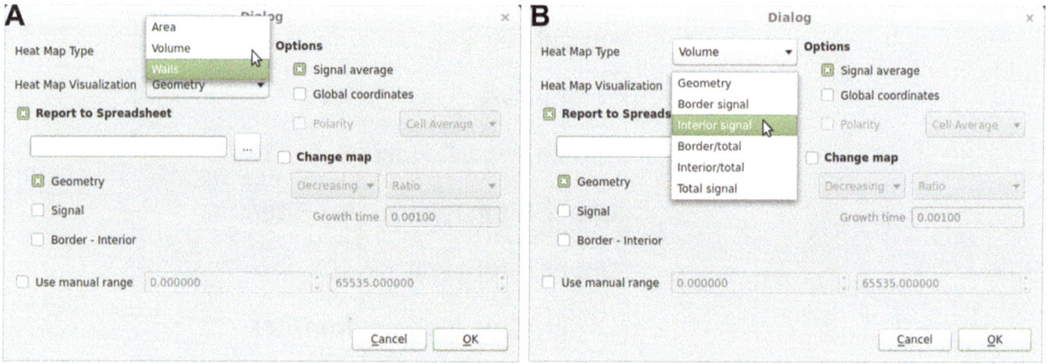

Fig. 15 Windows displaying options for generating heatmaps. (**a**) The type of heatmap can be chosen from Area, Volume, or Walls. (**b**) The drop-down menu for heatmap visualization offers several options; for reporter abundance, "Interior signal" is chosen

Save cell data and reporter abundance files for all samples to be analyzed in a directory. The "3D Reporter Abundance Analysis" process will recognize pairs of files with the same file name, with the reporter abundance file name containing an additional string that can be chosen by the user. In this way, the cell type and position information can be matched to the corresponding reporter abundance file.

In the "Process" tab, under "Mesh" > "Cell Atlas 3D" > Statistics, run "Reporter abundance analysis." This will prompt the user to direct the program to the required directory.

In the parameters (Fig. 16), if the value for "Input folder" is left blank, the program will prompt the user to guide it to the correct directory.

To visualize data on a mesh that is not part of the samples to be analyzed, a separate sample (in the form of its "cell_data.csv" file) can be set as "Output File." "Merge With File" will then have to be set as "Yes."

Raw data can be optionally filtered in the process by setting "Upper Filter Type" and/or "Lower Filter Type" to one of the options. "No Filter" will include all data points, while "Percentage" or "Value" will cut off data points beyond the given limit. The

Parameter	Value
Input folder	
Reporter file string	_reporter
Merge With File	Yes
Output File	
Sliding Avg	No
Sliding Avg Window	3
Upper Filter Type	No Filter
Upper Filter Limit	100
Lower Filter Type	No Filter
Lower Filter Limit	0

Fig. 16 Parameters for the 3D reporter abundance analysis tool. "Input folder" is the directory that contains cell data and corresponding reporter abundance data of all samples to be analyzed. To visualize the results on a separate sample mesh, the "cell_data.csv" file of this mesh can be set as "Output File." This requires the "Merge With File" option to be set to "Yes." "Sliding Average" is an optional smoothing function which averages the value for each cell based on its neighbors. The window for this averaging needs to be an odd number, and can be set under "Sliding Avg Window." Raw data can be optionally filtered in the process by setting "Upper Filter Type" and/or "Lower Filter Type" to one of the options. "No Filter" will include all data points, while "Percentage" or "Value" will cut off data points beyond the given limit. The limits for each filter can then be set to a value defined under "Upper Filter Limit" and "Lower Filter Limit"

limits for each filter can then be set to a value defined under "Upper Filter Limit" and "Lower Filter Limit."

If the mesh to be used for visualization of these data is loaded into MorphoGraphX, the results will automatically be loaded onto it.

The data output file, labeled "output_CI.csv," will automatically be created in the same directory. It contains the mean of the reporter abundance for each cell type, at each position, in addition to the confidence intervals at each data point. From there, plots can be generated to visualize these data as described in the growth analysis (*see* **Note 17**).

3.21 Visualization of Data on a Segmented Organ Mesh

Any cell data obtained through 3DCellAtlas, be it individual geometric data of an individual sample, growth data from the comparison of two sets of samples, or reporter abundance data from a set of samples, can be displayed on a segmented cellular organ mesh of your choice. Cells will thus be false colored according to the value in the data file. For this, a *.csv file with the respective data is loaded onto a segmented mesh as a heatmap in the "Process" tab, under "Mesh" > "Heat Map" > "Load Heat Map" (Fig. 17). It is important that the data are correctly matched to the cell IDs of the segmented mesh of choice with regard to cell type and position

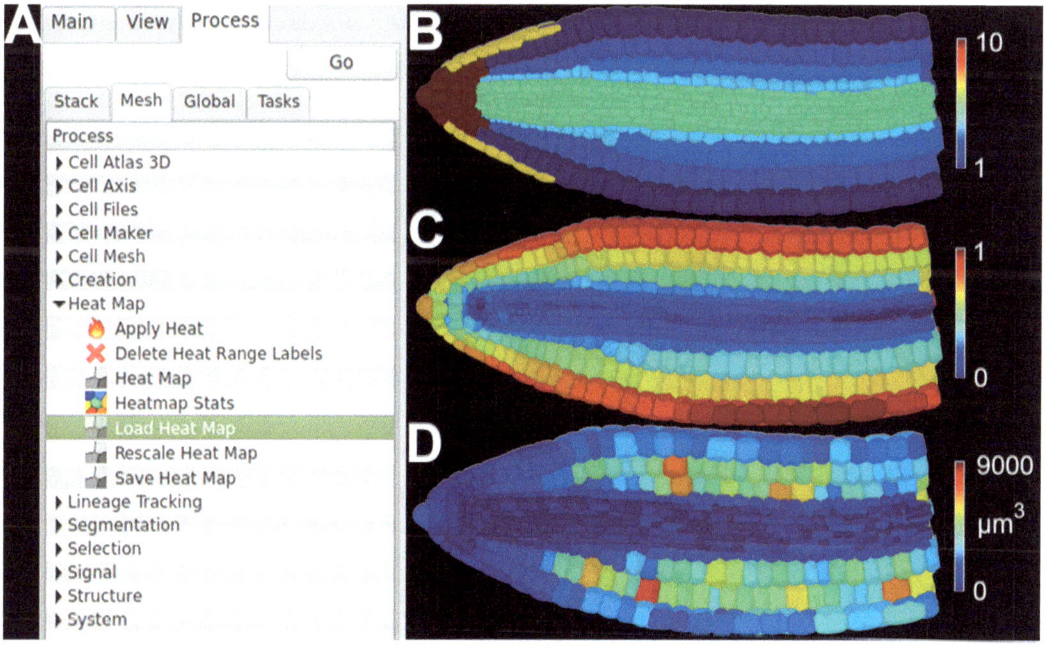

Fig. 17 Visualizing data as heatmaps on a segmented mesh. (**a**) Close-up of the Process tab with the drop-down menu for the heatmap tools. *Highlighted* is "Load Heat Map," which is used to upload a data file (in *.csv format) to false-colored cells in a segmented cellular organ mesh. (**b**–**d**) Examples of false-colored meshes of an *Arabidopsis* embryo axis, halved to show cells on the inside. Cells have been false colored to show (**b**) cell types, (**c**) the relative radial distance of each cell from the Bezier line, and (**d**) cell volume

(*see* **Note 18**). "Rescale Heat Map" allows the user to change the range of the scale displayed. To save a screenshot, the camera tool is used; image type and resolution can be adjusted according the user's preferences.

4 Notes

1. Image stacks need to be stored as 16-bit images; if 8-bit images are acquired, they will be too dark to see in MorphoGraphX. This can be overcome, however, by auto-adjusting the transfer function. In the "Main" tab, click on the color map button for the stack that is open, and click on "Auto Adjust" in the window thatpopsup.Inthe"Process"tab,use"Stack" > "Filters" > "Brighten Darken" with an amount of 16 to convert 12-bit images, or 256 for 8-bit images, then reset the transfer function, and save the 16-bit stack.
2. The pinhole should be as small as practical (depending on signal intensity) to narrow the optical slice thickness. This will reduce noise to a minimum, but will also decrease the amount of fluorescence signal being captured.
3. For best quality and resolution, choose a z-stack interval of the same size as the *x–y* resolution in each slice (i.e., if the resolution is 0.7 μm per pixel in the *x–y* direction, a slice interval of 0.7 μm is recommended).
4. An extensive user manual and FAQs can be found on the developers' website: http://www.MorphoGraphX.org.
5. The radius for blurring the image depends on the magnification used during its acquisition; a 0.5 μm radius is typically used for images acquired with 20 or 25× objectives. A lower radius will work better for smaller cells, for example images acquired with 40× or higher objectives.
6. A segmentation threshold that is too high will cause thinner cell walls with low signal intensity to be missed and cells to be merged, while a threshold that is too low over-segments cells into numerous segments. It is worth noting that oversegmented cells can be corrected, while merged cells cannot, so it is advisable to choose a threshold with minimum to no bleeding of adjacent cells, which could lead to some cells being oversegmented.
7. If a lot of cells within the image are merged after segmenting the image ("undersegmented" or "bleeding"), the signal-to-noise ratio in the image might be too low.
   - Likely, the threshold used for segmenting the image has been too high. Use a lower threshold for segmentation. It is also possibly to improve image quality by normalizing

the image before blurring (in the "Process" tab, under "St ack > Filters > Normalization"). This will even out differences in signal across the sample by narrowing the distribution of signal intensities.

- Bleeding of cells could also be due to the sample not being clear enough. If that is the case, continue clearing the sample, with changes of chloral hydrate clearing solution, to reduce noise inside of cells, and re-image.
- Similarly, if the staining has not worked well, the intensity from the cell walls could be too low to be sufficiently distinguishable from noise.

8. If cells are excessively oversegmented after segmentation, repeat the segmentation using a higher threshold.
9. A common error message in MorphoGraphX upon segmenting an image is the following: "Exception occurred during SingleMethodExecute [...]: Number of objects greater than maximum of output pixel type." This indicates that the ITK segmentation has found too many regions to label in 16-bit, which could be due to the image not having been blurred, or the radius of the Gaussian blur not having been big enough.
10. A mesh size of 2 μm is typically used. This value describes the distance between points (vertices) within the mesh. A higher value will generate a more coarse mesh; this risks misrepresentation of cell sizes, as corners may be cut off. A smaller value will generate a smoother mesh with more points, which will increase file size, but is likely to represent the cell sizes and shapes more accurately.
11. The smoothing operation for meshes in MorphoGraphX uses the Laplacian smoothing; it repositions vertices to an average position along a mesh. As a side effect, this will slightly shrink the mesh, and thus misrepresent cell sizes. This operation should therefore be used with caution [12].
12. While adjusting the handle points of the Bezier to align with the axis of the imaged plant organ, it is helpful to reduce the opacity of the image, and make use of clipping planes to determine the exact position of the Bezier curve within the 3D image.
13. After cell types have been identified and corrected via the topological check function, it is still possible that some cells are not assigned the correct type. It is important to note that the accurate determination of cell types strongly depends on the quality of the segmentation. Cell types are saved as parent labels. In the main tab, next to "Surface," the display can be changed to show "Nrml" (no color), "Labels" (individual cell IDs), "Heat" (values based on a heatmap), or "Parents" (cell types). To edit parent labels, check that "parents" is ticked in the main

tab, select a parent label using the dropper tool amongst the mesh tools (*see* Fig. 6c), and change the parent label of another cell to the selected one using the paint bucket tool.

14. It is important that all files to be analyzed are in csv file format. If one or all files are not comma delimited, an error message will occur, and the analysis will not run.
15. During this process, data from multiple samples will be pooled by cell type, and then associated cortical cell, into bins representing unique cells. Data will be log transformed, and the mean for each cell size metric will be calculated. Total growth is then determined as a ratio of the means of treatment over control. For more detail of the exact calculations, *see* ref. 11.
16. If a sliding window average of three is chosen, the value for each cell will be changed to the average of three associated cortical cell positions, one cell on either side. Cells in the first and last position within samples, possessing only a single neighbor, will be reduced to the average of two cells.
17. Reporter concentration values are assigned to the appropriate cell type and position in this process, and data from multiple samples will be pooled. The statistical analysis determines the mean reporter concentration for each cell type and position as described in **Note 15** for growth analysis. The 95 % confidence intervals for reporter abundance data are computed using bootstrapping, with a minimum sample size of seven cells. Thus, should there be less than seven data points available for one particular cell type at a given position, the confidence interval will be returned as a value of −1.
18. Each individual cell of the segmented mesh is assigned a cell type and position; the values of each metric to be displayed for each of these combinations have to be matched according to these cell type-cell position combinations. This can be done quickly using database software, for example Microsoft Access, to match the cell data file of the segmented cellular mesh to be used, and the data file to be displayed.

Acknowledgments

G.W.B. was funded by Biotechnology and Biological Sciences Research Council (BBSRC) Grant BB/L010232/1 and a Birmingham Research Fellowship. P.S. was funded by BBSRC grant BB/J017604/1. T.D.M.J. is supported by a Royal Commission for the Exhibition of 1851 Fellowship. R.S.S. was funded by Swiss National Science Foundation Interdisciplinary Project Grant number CR32I3_143833 and the Max Planck Society.

References

1. Roeder AH et al (2011) Computational morphodynamics of plants: integrating development over space and time. Nat Rev Mol Cell Biol 12(4):265–273
2. Fernandez R et al (2010) Imaging plant growth in 4D: robust tissue reconstruction and lineaging at cell resolution. Nat Methods 7(7):547–553
3. Kierzkowski D et al (2012) Elastic domains regulate growth and organogenesis in the plant shoot apical meristem. Science 335(6072):1096–1099
4. Roeder AH et al (2012) A computational image analysis glossary for biologists. Development 139(17):3071–3080
5. Truernit E et al (2008) High-resolution wholc-mount imaging of three-dimensional tissue organization and gene expression enables the study of Phloem development and structure in Arabidopsis. Plant Cell 20(6):1494–1503
6. Bassel GW et al (2014) Mechanical constraints imposed by 3D cellular geometry and arrangement modulate growth patterns in the Arabidopsis embryo. Proc Natl Acad Sci U S A 111(23):8685–8690
7. Cunha AL, Roeder AH, Meyerowitz EM (2010) Segmenting the sepal and shoot apical meristem of Arabidopsis thaliana. Conf Proc IEEE Eng Med Biol Soc 2010:5338–5342
8. Yoshida S et al (2014) Genetic control of plant development by overriding a geometric division rule. Dev Cell 29(1):75–87
9. Barbier de Reuille P et al (2015) MorphoGraphX: a platform for quantifying morphogenesis in 4D. eLife 4:e05864
10. Shapiro BE et al (2015) Analysis of cell division patterns in the Arabidopsis shoot apical meristem. Proc Natl Acad Sci U S A 112(15): 4815–4820
11. Montenegro-Johnson TD et al (2015) Digital single-cell analysis of plant organ development using 3DCellAtlas. Plant Cell 27(4):1018–1033
12. Bassel GW (2015) Accuracy in quantitative 3D image analysis. Plant Cell 27(4):950–953

Chapter 12

Analyzing Cell Wall Elasticity After Hormone Treatment: An Example Using Tobacco BY-2 Cells and Auxin

Siobhan A. Braybrook

Abstract

Atomic force microscopy, and related nano-indentation techniques, is a valuable tool for analyzing the elastic properties of plant cell walls as they relate to changes in cell wall chemistry, changes in development, and response to hormones. Within this chapter I will describe a method for analyzing the effect of the phytohormone auxin on the cell wall elasticity of tobacco BY-2 cells. This general method may be easily altered for different experimental systems and hormones of interest.

Key words Atomic force microscopy, Auxin, Cell growth, Cell wall, Elasticity

1 Introduction

Cell wall elasticity has been correlated with important developmental transitions and growth in several plant systems [1–5]. There are several methods currently used to measure cell wall elasticity at the cell and tissue level (reviewed in [6]); one of the most prevalent modern methods of measuring cell wall elasticity *in planta* at the cellular level is atomic force microscopy (AFM) [1, 3–5, 7, 8]. Non-AFM methods include other nano-indentation devices [9–11] and osmotic swelling and shrinking of tissues [2, 9]. Previously, AFM-based indentation has been used to demonstrate that the phytohormone auxin induces an increase in cell wall elasticity (decrease in the elastic modulus) of cells near the shoot apex in *Arabidopsis thaliana* [1]. In order to simplify the experimental system for this chapter, I will examine cell wall elasticity in growing tobacco bright yellow 2 (BY-2) culture cells upon treatment with auxin. Using the simple method presented here, I have confirmed that auxin treatment increases cell length in BY-2 cells and decreases cell wall elastic modulus. This is consistent with data showing that BY-2 cells respond positively, with respect to growth, when treated with a cell wall-acidifying agent or the wall-loosening agent expansin [12].

Jürgen Kleine-Vehn and Michael Sauer (eds.), *Plant Hormones: Methods and Protocols*, Methods in Molecular Biology, vol. 1497, DOI 10.1007/978-1-4939-6469-7_12, © Springer Science+Business Media New York 2017

This method proposes a two-step process: first, the experimenter should determine the concentration of hormone to use given their experimental question; second, the experimenter should choose a time point for AFM-based elasticity analysis that further suits their question and system. Here, a concentration of 10 μM IAA and a treatment time of 1 h were chosen after initial experiments; a more complex set of treatment times and concentrations could easily be constructed using this method as a basis.

2 Materials

2.1 Plant Material

Tobacco BY-2 cells, non-transgenic, were obtained from the Leibniz Institute DSMZ-German Collection of Microorganisms and Cell Cultures (DSMZ, Cat # PC-1181). Cells were cultured in MSBY media (see below), in the dark at 25 °C, with shaking at 100 rpm. Cells were sub-cultured weekly into fresh media at a 1/10 dilution. All experiments were performed 1–2 days after sub-culturing, when cells were growing in their exponential phase.

2.2 Solutions

All solutions should be prepared fresh, using ultrapure water whose pH has been verified as neutral (*see* **Note 1**).

0.55 M Mannitol, for plasmolysis.

10 mM Indole acetic acid (IAA) stock solution: 1.97 mg of IAA (Sigma, Cat # I5148) is dissolved in 1 mL of ethanol. After dissolving, ultrapure water is added to 10 mL. Store in the dark at −20 °C.

MSBY media: Full MS media supplemented with 1 mg/L thiamine, 100 mg/L myo-inositol, 0.2 mg/L 2,4-D, 3% sucrose, pH 5.7 (*see* **Note 2**).

2.3 Equipment and Consumables

Atomic force microscope: The experiments described were performed on a JPK Nanowizard 3 (JPK Instruments, DE).

Cantilevers (without tips) of spring constant ~45 N/m (Windsor Scientific, UK, Cat #TL-NCL-10). Silica beads of ~5 μm radius were mounted as per [7] (Microspheres-Nanospheres, Corpuscular, USA, Cat# C-SIO-5.0). Note that tip size and cantilever stiffness should be chosen carefully with respect to the question and system.

Small petri dishes coated with adhesive agent (*see* **Note 3**).

Pipette and tips.

A dissecting microscope with camera for observation and length measurements.

250 mL Flasks for sub-culturing and growth of cells.

A sterile bench.

3 Methods

3.1 Assessing an Appropriate Hormone Concentration and Time of Treatment (See Note 4)

1. Use the literature to determine a range of concentration of hormone to test. You are looking for a hormone concentration that is not too high so as to be unrealistic, but also shows consistent and significant changes in the phenotype you are interested in. Here 1, 10, and 50 μM IAA were tested and 10 μM was chosen (*see* Fig. 1).
2. At the time of sub-culturing, make four extra sub-culture flasks for this initial experiment. To two flasks of 30 mL MSBY media add 30 μL of 10% ethanol (Mock), and to the other two flasks add 30 μL of 10 mM IAA stock (10 μM IAA treatment). Subculture 1/10 volume of week-old culture into each flask. Grow the cells for 24 h in culture conditions.
3. After 24 h, sterilely remove a small aliquot of cells and place on a glass glide with a cover slip. Return the cells to culture conditions. Using a dissecting microscope, take several images of the slide area. Make sure that you have the correct calibration for length measurements. Figure 1 shows the effect of 10 μM IAA on BY-2 cell length after 24 h of treatment.
4. Using ImageJ (open source, ImageJ.net), measure the length of cells within the images. Select cells at random, covering a wide area of each image and thus the whole slide area.
5. Repeat these measurements for further time points of interest (*see* **Note 5**). In the experiments here, cell lengths were mea-

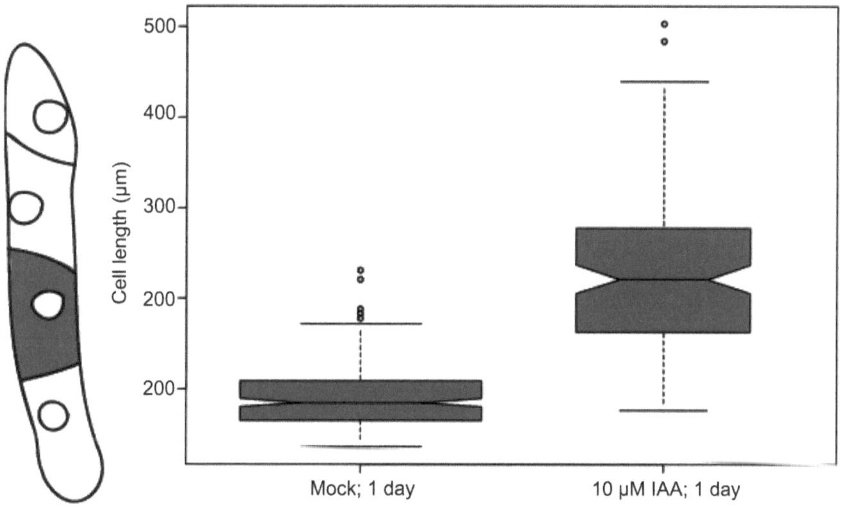

Fig. 1 Cell length of BY-2 cells after culture with the auxin IAA. Cell length of randomly chosen individual cells with a BY-2 cell chain (diagram on *left*, *shading* not meant to denote a fixed cell position) was measured after 24 h of growth with mock (diluted ethanol) or 10 μM IAA treatment. Cell length was measured from two replicate flasks per treatment and from 100 cells per replicate. A notched box plot was generated in Matlab

sured at 1, 2, and 3 days. As cell length was increased at all time points, it was decided to assess the effect of IAA on wall elasticity before 24 h (when a statistically significant phenotype was observable, Fig. 1a).

3.2 Preparation of Cells for AFM

1. Using cells 1–2 days post-sub-culture, take a sterile aliquot of 3 mL. Apply the aliquot of cells to the coated petri dish. Allow the cells to sit for ~5 min, then tip the plate to remove the media and un-stuck cells, rinse three times with MSBY media (−2,4D), and then fill the dish halfway with fresh MSBY (−2,4-D) media. Check that the cells are remaining stuck to the dish during washes. The media will be replaced with 0.55 M mannitol just before the initial AFM experiments begin (*see* **Note 6**). Figure 2a shows an outline of the experiment.

3.3 AFM-Based Cell Wall Elasticity Measurements

1. Prepare a cantilever with a 5 μM bead and calibrate as in [7]. Different tips may be used for different purposes and experimental questions, as detailed in [7].
2. Place the petri dish of cells (in 0.55 M mannitol) under the AFM head.
3. Locate an area of the dish where several cells are well adhered to the surface and flat (*see* **Note 7**). Take a picture of the cells so that you can locate them again after treatment.
4. Position the AFM tip over a cell-cell junction and perform an AFM scan. Here I have used a force of 500 nN which yields an indentation depth of roughly 250–500 nm. Details of AFM procedure can be found in [7] and a schematic of how the indentation relates to BY-2 cell dimension in Fig. 2b. Here, a

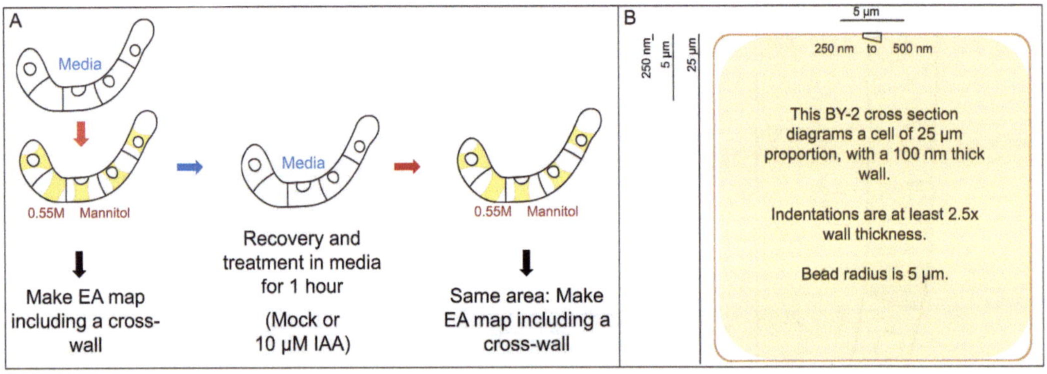

Fig. 2 Diagram of experimental procedures. (**a**) The experimental workflow for auxin treatment of BY-2 cells for AFM indentation. Cells are plasmolyzed in mannitol for an initial AFM scan; they are then recovered and treated in media for 1 h, then plasmolyzed again, and re-scanned with the AFM. (**b**) Diagrammatic representation of a BY-2 cell in cross section with dimensions and scale of AFM indentation. BY-2 cells are usually ~25 μm in diameter and have a wall thickness of ~100 nm [13]. Indentations here were performed with a 5 μm diameter bead to a depth of 250–500 nm

map of 32 × 32 indentations was made over a 30 μm^2 area. Cell-cell junctions are essential as the cross wall will be used for further analysis; in plasmolyzed cells, cross walls will be evident as stiffer regions within the scan. You can use the height and stiffness maps produced during the AFM scan to help ensure that you are in the right area.

5. Proceed to make cross-wall scans for several cells within your field of view. It is likely that you will lose some cells between the subsequent steps, so starting with more than you need is advisable. Take a picture of each cell you scan so that you can re-locate it after treatment. Try to keep your time of collection under 1 h to keep your cells happy.
6. After you have collected enough maps, remove the AFM head and carefully use a pipette to remove the mannitol. Try not to move the plate so that the position of cells under the AFM does not shift too much. Wash the cells once with MSBY media (−2,4-D), and then fill the dish with 5 mL of MSBY (−2,4-D). For cells to be mock treated, add 5 μL of 10% ethanol. For cells to be auxin treated, add 5 μL of 10 mM IAA stock (to get 10 μM treatment conditions). Let the cells sit for 1 h (*see* **Note 8**).
7. Exchange the treatment media with 0.55 M mannitol (including one wash step). Place the plasmolyzed cells under the AFM head.
8. Using your field of view image from **step 3**, locate the same area in your dish.
9. Using your cell images from **steps 4** and **5**, re-find the same cells and perform AFM maps over the same cell-cell junctions using identical settings.

3.4 AFM Data Analysis

1. Indentation force maps may be analyzed in several ways. Usually, AFM manufacturers supply analysis software. Here, I have used JPK SPM Data Processing (JPK Instruments, DE. Version "spm 4.3-10"). During analysis, you will need to choose whether you will fit models for linear stiffness or a Hertzian indentation (yielding a Young's modulus, here "EA") [7]. In brief, stiffness fits may be useful when indentation depth is constant and there is adhesion in the sample; in these cases the approach indentation should be analyzed. In cases where more complex indentation behavior is observed, and where the entire indentation curve is desired for analysis, fitting a Hertzian model is preferable. Here, I have fit a Hertzian model to the indentation curve in order to obtain an apparent Young's modulus (EA) for each map.
2. Select the data from along the cross wall in each map. Use this data to obtain values of EA along cross walls for each cell junction before and after treatment. Figure 3a shows the EA value

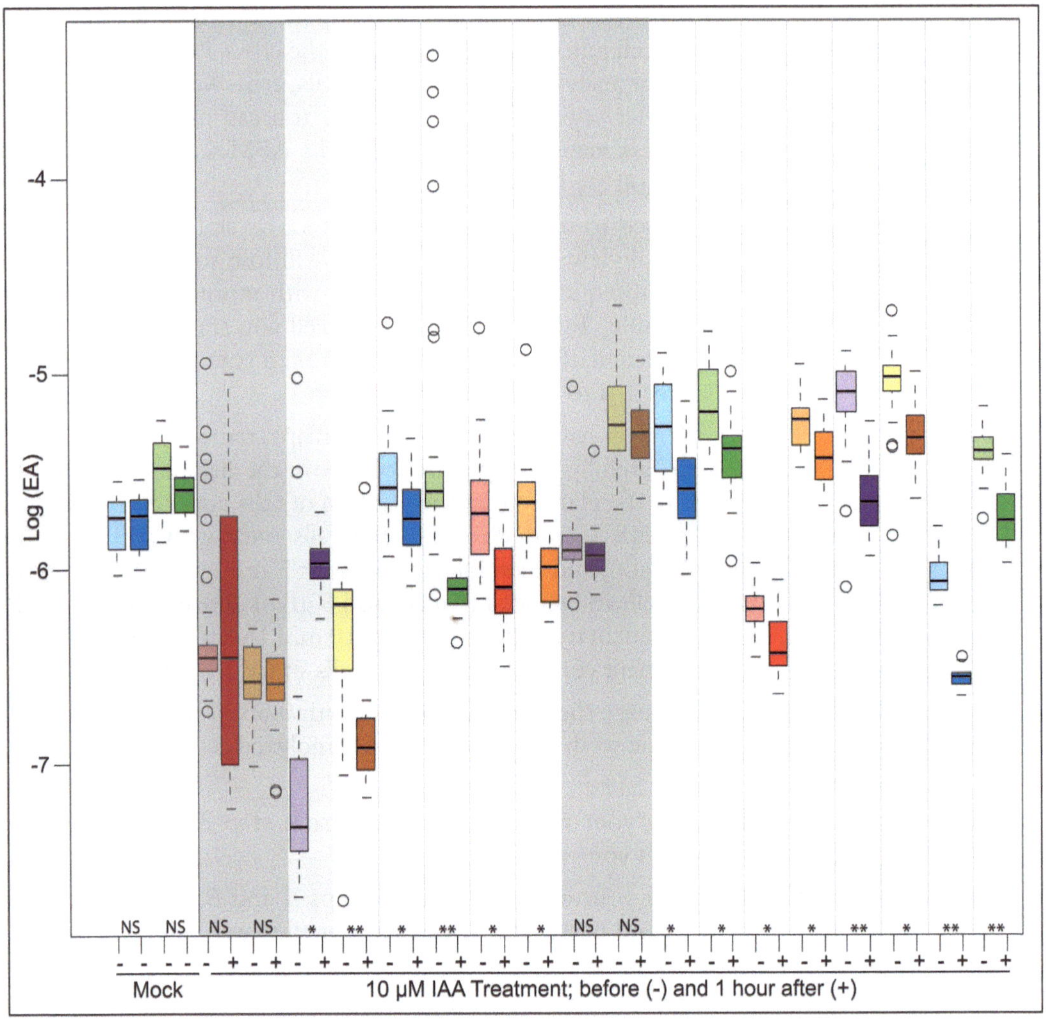

Fig. 3 Young's modulus (EA) of BY-2 cell walls before and after treatment. The log of EA (in Pascals) was plotted as box plots for each equivalent cross wall before (–) and after (+) treatment. IAA-treated cells which showed no significant changes are shaded in *dark grey*. One replicate showing an increase in EA (stiffening) after IAA treatment is shaded in *light grey*. *Boxes* are colored by individual cell, with the *lighter shade* being before treatment and the *darker* after treatment. It is evident that there is variation in EA between cells and also in the repose of individual cells to auxin, although the majority of cells responded by decreasing EA (*t*-test; $P<0.05$ (*), $P<0.01$ (**))

for each cell in this experiment, before and after treatment. Here it can be seen that there is a large variance between samples, even without treatment. For this reason, it may be advisable to keep biological samples separate for analysis and to focus on relative trends. For example, in 13 of the cells treated with IAA a significant decrease in EA was obtained after 1 h of treatment (*t*-test; $P<0.05$). There was one sample where the EA increased after treatment (light grey over-shading), and four where no difference was observed (dark grey over-shading). As BY-2 cells are not always perfectly synchronized in growth, this may represent a change in biological response to the treatment.

3. For this example, data was collected, analyzed, and graphed in MatLab.

4 Notes

1. Water pH should be neutral as pH can affect cell wall elasticity and cell growth.
2. As 2,4-D is an auxin, for treatments cells were washed three times with media minus 2,4-D before first measurement. Application of mock (1/1000 dilution of 10% ethanol) or 10 μM IAA was performed in media without 2,4-D.
3. It is essential that cells do not move during the experimental procedure. After much testing, it appears as though tissue section adhesives (e.g., Biobond Tissue Adhesive, Agar Scientific, UK) work well. Roughly 10 μL of Biobond adhesive can be placed in the bottom of a small petri dish. A small piece of parafilm is then used to spread the adhesive over the whole plate bottom. The plate is placed in a 50 °C oven to dry for at least 10 min.
4. When assessing the concentration of hormone to use, it is important to calibrate your method to your system and question. Here, I present a method for BY-2 cells in culture. You might be interested in how auxin (or another hormone) affects some other growth process in another system (perhaps gibberellins and hypocotyl elongation). It is essential that you assess several concentrations for the phenotype you wish to examine, and also make assessments in time, thus allowing you to wisely choose the best concentration, and treatment time, for your AFM experiment.
5. If at all possible, measurement of your phenotype of interest should be made with as a high a temporal resolution as possible. For example, in BY-2 cells one could use time-lapse imaging as in [12] if one had the correct equipment. This would allow for very accurate choosing of time points for ANF analysis.
6. For the method presented here, it is essential that cells are plasmolyzed as the indentation depth is more than wall thickness and thus turgor pressure would affect our measurements. It is important to know how quickly your cells plasmolyze for these experiments. I recommend testing this beforehand using a membrane dye (such as FM 4-64) or transgenic membrane marker, and monitoring plasmolysis under a confocal microscope. In our

hands, BY-2 cells plasmolyze in 0.55 M mannitol within 5 min. There are methods which do not require plasmolysis; please refer to [3, 7, 8].

7. If cells are not flat, there is a strong possibility that they will be knocked loose during the AFM scanning. You want to ensure that the cells you begin with will have staying power for the duration of the experiment.
8. The time of treatment will vary depending on your question. Here, it was desirable to see if changes in wall elasticity preceded changes in length and the first experiments revealed that 24 h of treatment showed length changes; thus a 1-h treatment should precede this. You might wish to perform a more detailed time series for the AFM scans. To do this, simply use this basic protocol as a starting point but add in more time steps. It may be efficient to stagger treatments and reads.

Acknowledgements

I would like to thank Cris Kuhlemeier, Theres Imhof, and Alexis Peaucelle for help with initial experiments with BY-2 cells and AFM. I would also like to thank my research team at SLCU for their patience with me while writing. This work was started in the lab of C. Kuhlemeier, supported by an NSF-IRFP (Award #0853105; to S.A.B.), and completed at The Sainsbury Laboratory, University of Cambridge, supported by a Gatsby Career Development Fellowship (GAT3396/PR4) and the BBSRC (BB. L002884.1) to S.A.B.

References

1. Braybrook SA, Peaucelle A (2013) Mechano-chemical aspects of organ formation in *Arabidopsis thaliana*: the relationship between auxin and pectin. PLoS One 8(3):e57813
2. Kierzkowski D et al (2012) Elastic domains regulate growth and organogenesis in the plant shoot apical meristem. Science 335(6072): 1096–1099
3. Milani P et al (2011) In vivo analysis of local wall stiffness at the shoot apical meristem in Arabidopsis using atomic force microscopy. Plant J 67(6):1116–1123
4. Peaucelle A et al (2011) Pectin-induced changes in cell wall mechanics underlie organ initiation in Arabidopsis. Curr Biol 21(20): 1720–1726
5. Uyttewaal M et al (2012) Mechanical stress acts via Katanin to amplify differences in growth rate between adjacent cells in Arabidopsis. Cell 149(2):439–451
6. Milani P, Braybrook SA, Boudaoud A (2013) Shrinking the hammer: micromechanical approaches to morphogenesis. J Exp Bot 64(15):4651–4662
7. Braybrook SA (2015) Chapter 13—Measuring the elasticity of plant cells with atomic force microscopy. In: Ewa KP (ed) Method cell biology. Academic, New York
8. Sampathkumar A et al (2014) Subcellular and supracellular mechanical stress prescribes cytoskeleton behavior in Arabidopsis cotyledon pavement cells. eLife 3
9. Weber A et al (2015) Measuring the mechanical properties of plant cells by combining micro-indentation with osmotic treatments. J Exp Bot 66(11):3229–3241

10. Routier-Kierzkowska A-L et al (2012) Cellular Force Microscopy for in vivo measurements of plant tissue mechanics. Plant Physiol 158(4): 1514–1522
11. Hayot CM et al (2012) Viscoelastic properties of cell walls of single living plant cells determined by dynamic nanoindentation. J Exp Bot 63(7):2525–2540
12. Link BM, Cosgrove DJ (1998) Acid-growth response and α-Expansins in suspension cultures of Bright Yellow 2 tobacco. Plant Physiol 118(3):907–916
13. Sabba RP, Durso NA, Vaughn KC (1999) Structural and immunocytochemical characterization of the walls of Dichlobenil-habituated BY-2 tobacco cells. Int J Plant Sci 160(2):275–290

Chapter 13

FRET-FLIM for Visualizing and Quantifying Protein Interactions in Live Plant Cells

Alejandra Freire Rios, Tatyana Radoeva, Bert De Rybel, Dolf Weijers, and Jan Willem Borst

Abstract

Proteins are the workhorses that control most biological processes in living cells. Although proteins can accomplish their functions independently, the vast majority of functions require proteins to interact with other proteins or biomacromolecules. Protein interactions can be investigated through biochemical assays such as co-immunoprecipitation (co-IP) or Western blot analysis, but such assays lack spatial information. Here we describe a well-developed imaging method, Förster resonance energy transfer (FRET) analyzed by fluorescence lifetime imaging microscopy (FLIM), that can be used to visualize protein interactions with both spatial and temporal resolution in live cells. We demonstrate its use in plant developmental research by visualizing in vivo dimerization of AUXIN RESPONSE FACTOR (ARF) proteins, mediating auxin responses.

Key words FRET, FLIM, Protein interactions, Fluorescent proteins

1 Introduction

In the mid-1990s, commercial confocal microscopes were introduced, which has revolutionized the investigation of biomolecules in living cells [1]. Visualizing biological processes has made great progress due to significant advances in instrument and detector design as well as the introduction of genetically encoded fluorescent proteins in vivo. Although confocal microscopy provides very-high-quality multicolored images, spatial resolution is limited by the diffraction of light [2]. Since 2006, several fluorescence imaging techniques have been developed known as super-resolution imaging technology, which provide fine-detailed and high-contrast images to create images down up to 20 nm spatial resolution [3, 4]. However, these imaging techniques are still unable to visualize protein interactions in vivo [5, 6]. Förster resonance energy transfer (FRET) is a phenomenon of radiation-less energy transfer from a fluorescent donor molecule to an acceptor molecule (either fluorescent or not) through

Jürgen Kleine-Vehn and Michael Sauer (eds.), *Plant Hormones: Methods and Protocols*, Methods in Molecular Biology, vol. 1497, DOI 10.1007/978-1-4939-6469-7_13, © Springer Science+Business Media New York 2017

dipole-dipole coupling. This process only takes place when the donor and acceptor molecules are in very close proximity (typically 2–10 nm; [7]). The energy transfer rate is inversely proportional to the sixth power of the distance between donor and acceptor, making FRET very sensitive to small changes in distance and useful as a molecular ruler ([8]; *see* Fig. 1). In order to accomplish FRET, the fluorescent donor and acceptor molecules have to fulfill several prerequisites: (1) a spectral overlap between donor emission and acceptor absorption spectra, (2) close proximity between donor and acceptor (<10 nm), and (3) finite dipole orientation factor [7]. FRET can be quantified using different imaging methods of which fluorescence lifetime imaging microscopy (FLIM) is the most robust and straightforward approach [9]. A fluorescence lifetime can be defined as the average time a fluorophore remains in the excited

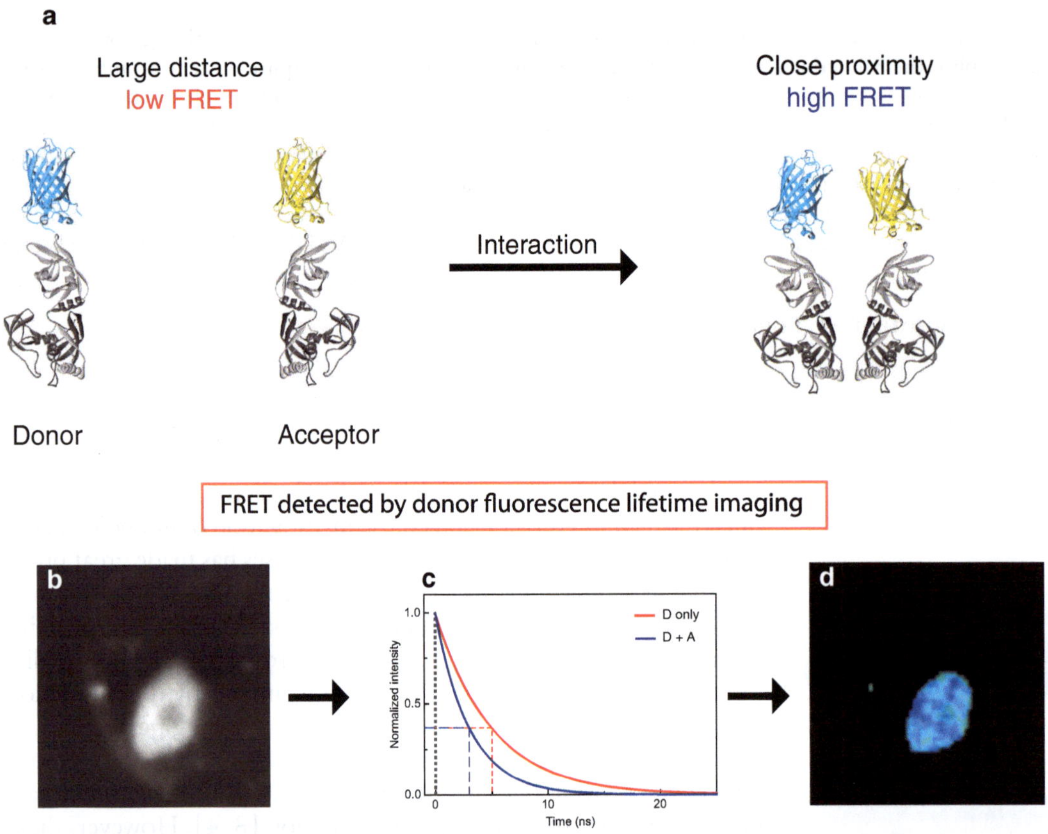

Fig. 1 Illustration of FRET principle and FRET analysis. (**a**) Two DNA-binding domains of ARF5 are coupled to CFP (donor) and YFP (acceptor). Large distance between CFP and YFP results in low FRET whereas high FRET signals are obtained if ARF domains are in close proximity and correct orientation. FRET can be quantified by fluorescence lifetime analysis. A fluorescence intensity image (**b**) shows the nuclear expression of the donor ARF5-sCFP3A in a protoplast. Per pixel the time-resolved photon distribution is plotted (**c**) of which the fluorescence lifetime is calculated and shown as a false color-coded fluorescence lifetime image (**d**). Interaction of a donor with an acceptor results in a decrease of the fluorescence lifetime (**c**)

state before returning to the ground state. The fluorescence lifetime is an intrinsic property of a fluorophore and is independent of concentration, absorption by the sample, sample thickness, photobleaching, and laser excitation intensity [10]. However, the fluorescence lifetime is very sensitive to the microenvironment of the fluorophore such as pH, ion or oxygen concentration, molecular binding, or the proximity of energy acceptors [11], which makes this method unique for FRET analysis. FRET-FLIM measurements are based on determination of the fluorescence lifetime of the donor molecule. The close proximity of acceptor molecules creates an additional relaxation path of the donor-excited fluorophores resulting in a decreased donor fluorescence lifetime ([9]; *see* Fig. 1). The amount of donor fluorescence lifetime reduction is directly correlated with the FRET efficiency (E) via $E = (1 - \tau_{DA}/\tau_D)$ where τ_{DA} is the fluorescence lifetime of the donor in the presence of acceptor and τ_D is the fluorescence lifetime of the donor alone. In order to obtain a comprehensive overview and critical essentials concerning the fundamentals of FRET and the methods of measurements, we refer to a detailed description [12].

In this chapter, a detailed description of FRET-FLIM technique and analysis is presented by showing recent data on the dimerization of the transcription factor family AUXIN RESPONSE FACTORs (ARFs) [13] in living plant cells. ARFs act dependent on the signaling molecule auxin, which is a small organic molecule important for cell division, elongation, and differentiation [14]. Specificity in auxin responses is generated through the DNA-binding domain (DBD) of ARFs. Crystal structures of the ARF5 and ARF1 DBDs were solved showing homodimerization of DBDs generating cooperative DNA binding, which is critical for in vivo ARF5 function [13]. ARF5 dimerization was also demonstrated in *Arabidopsis* protoplasts using a FRET-FLIM analysis [13]. In this assay ARF5-sCFP3A and ARF5-sYFP2 fusion proteins were expressed and donor fluorescence lifetime analysis revealed specificity in homodimerization of ARF5 proteins. Full-length ARF5 homodimerization revealed a decrease of donor fluorescence lifetime resulting in a FRET efficiency of 10% (*see* Subheading 3) [13], whereas mutations within the dimerization domain of the DBD of ARF5 resulted in FRET efficiencies around background level. These FRET data excluded the possibility of ARF DBD dimerization being an artifact of the crystallization process but confirmed that these interactions take place in vivo. Furthermore, site-directed mutations and subsequent phenotypic analysis shed light on the importance of dimerization for the ARF function in vivo.

Here we present a stepwise protocol to perform FRET-FLIM analysis and show data on ARF5 homodimerization as an example. Protein expression in plant protoplasts, FRET-FLIM analysis, quantification of FRET efficiencies, use of proper controls, and different methods to display protein interactions are discussed.

2 Materials

2.1 Cloning of Transient Expression Vectors

1. Plant expression vector: LIC vectors based on pMON999 (Monsanto, USA) [15].
2. The cDNA of ARF5 was cloned into LIC vectors harboring either SCFP3A fluorescent protein (sCFP3A) or yellow fluorescent protein (sYFP2) (*see* **Note 1**).
3. Constructs were generated using ligase-independent cloning procedure (*see* **Note 2**).

2.2 Protoplast Isolation

Protocol adapted from Refs. [16] and [17]:

1. Plant material: Rosette leaves of *Arabidopsis thaliana* plants (ecotype Columbia) grown for ±3 weeks under long-day conditions (16-h light/8-h dark) at ±22 °C.
2. Mannitol solution: 0.4 M Mannitol, 20 mM KCl, and 20 mM 2-(*N*-morpholino) ethanesulfonic acid (MES) pH 5.7 in Milli-Q water.
3. Calcium chloride solution: 1 M $CaCl_2$ in Milli-Q water.
4. Enzyme solution: 1% (w/v) cellulose "Onozuka" R10 (Serva Electrophoresis, GmbH, Germany) and 0.2% (w/v) Macerozyme R10 from Rhizopus sp. (Serva Electrophoresis, GmbH, Germany) dissolved in mannitol solution; subsequent addition of $CaCl_2$ to a final concentration of 10 mM.

2.3 Transfection Protocol Adapted from [16] and [17]

Protocol adapted from Refs. [16] and [17]:

1. PEG/Ca^{2+} solution: 40% (w/v) Polyethyleneglycol 4000 (PEG) 0.2 M mannitol, and 100 mM $Ca(NO_3)_2$ in Milli-Q water (*see* **Note 3**).
2. W5 solution: 154 mM NaCl, 125 mM $CaCl_2$, 5 mM KCl, and 2 mM MES pH 5.7.
3. W5/glucose solution: W5 solution containing 1 mM glucose.
4. MMg solution: 0.2 M Mannitol, 15 mM $MgCl_2$, and 4 mM MES pH 5.7.
5. Plastic round-bottom tubes.
6. Microscope 8-well slides

2.4 Confocal Imaging and FLIM

1. Leica SP5X (*see* **Note 4**).
2. B&H SPC 730/830 module (Becker & Hickl, Germany).

3 Methods

3.1 Protoplast Isolation

1. All steps are carried out at room temperature unless otherwise stated.

2. Rosette leaves of ±3-week-old *A. thaliana* plants are fixed on Time tape adhered to the upper epidermis [18].
3. 3 M Magic tape is fixed to the lower epidermis to make a "Tape-*Arabidopsis* Sandwich" ([18]; *see* **Note 5**).
4. A clean Petri dish (9 cm diameter) is covered with three stripes of tape covered with leaves sufficient for six independent transfections.
5. Add 25 mL of enzyme solution, and swirl to dampen all plant material.
6. The Petri dish is transferred on a platform shaker (shaking ± 60 rpm) and incubated at 27 °C for 20 min (**Note 6**).
7. Protoplasts are released from the leaf matrix by carefully pipetting the enzyme solution on the taped leaves.
8. Transfer the protoplasts to a clean plastic round-bottom tube.
9. Protoplasts are collected by centrifugation for 3 min at 50 × *g* using a tabletop centrifuge; wash once with 5 mL of W5 solution.

3.2 Protoplast Transfection

1. All steps are carried out at room temperature.
2. Pipette plasmid DNAs in round-bottom tubes; for single transfections 10–20 μg of the respective DNA is required and for double transfections, 10–15 μg of each construct. Single transfection is used to obtain protoplasts expressing the donor construct only, whereas the transfections of two plasmids yield protoplasts for the interaction studies.
3. Collect the isolated and washed protoplasts (*see* Subheading 3.1) by centrifugation for 3 min at 50 × *g* in a tabletop centrifuge and remove the supernatant carefully with a pipette.
4. Resuspend the protoplasts in 1.2 mL of MMg (*see* **Note 6**) and transfer 200 μL aliquots to round-bottom tubes that contain plasmid DNA already (*see* **Note 7**).
5. Add 220 μL of PEG/Ca^{2+} solution to the protoplasts, mix well but carefully, and incubate for 5 min (*see* **Note 8**).
6. Stop the transfection by addition of 800 μL of W5 solution to the protoplast solution and collect the protoplasts by centrifugation at 50 × *g* for 90 s in a tabletop centrifuge.
7. Remove the supernatant with a pipette and wash with 5 mL of W5 solution.
8. Collect protoplasts by centrifugation at 50 × *g* for 3 min, remove the supernatant with a pipette, and resuspend the protoplasts in 1 mL of W5 solution containing 1 mM glucose.
9. Protoplasts can stay in round-bottom tubes incubated at 22 °C in the dark for 16 h (*see* **Note 9**).

3.3 Image Acquisition

1. In this chapter the FRET-FLIM experimental procedure is optimized for a Leica SP8. Single-photon excitation pulses of 40 Mhz (*see* **Note 10**) are generated by a diode-pulsed laser resulting in excitation pulses of ±10 ps. (*see* **Note 11**).
2. Image acquisition is performed using a 60×/1.2 water-immersion objective.
3. Set objective for optimal glass thickness (*see* **Note 12**).
4. Cyan fluorescence emission was selected from 450–495 nm using spectral detection (*see* **Note 13**).
5. Single photons are captured using internal hybrid detector (*see* **Note 14**).
6. Fluorescence images of 128 × 128 pixel size are set for both the Leica acquisition software and the B&H SPC 830 FLIM module (*see* **Note 15**).
7. The analog-digital converter (ADC) of TCSPC module is set at 256 channels (*see* **Note 16**).
8. FLIM acquisition is performed using line scanning at 400 Hz (*see* **Note 17**) using an average count rate around 10^4 photons per second for an acquisition time of 90 s (*see* **Note 18**).
9. Single transfected protoplasts expressing C-terminally sCFP3A-tagged ARF5 result in the donor fluorescence lifetime required as reference. To reveal dimer formation of ARF5, protoplasts expressing ARF5 sCFP3A- and sYFP2-tagged proteins are investigated (*see* **Note 18**).
10. The protoplasts are transferred into an 8-well chamber for imaging.

3.4 Analysis of FLIM Data with SPCImage 5.2

1. SPCImage is a software package, which is included with the B&H acquisition card. This software determines the instrumental response function from the rise of the decay curve and performs data analysis using an exponential model function (*see* **Note 20**).
2. After importing the raw data, a fluorescence intensity image will appear. A blue crosshair is visible that can be used to point anywhere in the image for displaying the fluorescence decay of that pixel.
3. Within the histogram showing the fluorescence decay, the limits in between the fluorescence lifetime analysis should be performed, and can be defined. Typical values are around 2 ns (before the rising edge) for the left and about 20 ns for the right border.
4. A threshold of fluorescence intensity can also be set. Pixels will be excluded from the analysis if the numbers of photons in that pixel is below that value.

5. In case the number of photons is too low for fluorescence lifetime analysis, static pixel binning can be applied. This specific binning option is a procedure where the selected pixel is analyzed including the photon statistics of neighboring pixels for calculation of the fluorescence lifetime. The binning factor can be calculated according to the following formula: binning factor $= (2n + 1)^2$, where n is the number of pixels. For the FLIM images shown in Fig. 2, a binning factor of 1 has been applied (*see* **Note 21**).
6. Before fitting the raw data, the number of components for the underlying exponential fit function has to be defined (here one or two). All other parameters should remain unfixed independently of the fit model.
7. After the fit procedure, the fluorescence lifetimes are calculated per pixel and displayed as a false-colored image (*see* Fig. 1d and Fig 2a, c). During the fitting process, the chi-square value between model function and data is minimized [19]. For a good fit, the chi-square value should be around one and the displayed residuals in the box below the fitted curve should scatter around zero.
8. To reveal if protein interactions take place one compares donor fluorescence lifetime in the absence and presence of acceptor. Single-transfected protoplasts (donor only) are evaluated using one-component analysis (*see* **Note 22**).
9. Double-transfected protoplasts (donor and acceptor) are analyzed based on a two-component model. The fluorescence lifetimes and amplitudes obtained can be assigned to two donor populations—one transferring energy to an adjacent acceptor molecule resulting in a reduced fluorescence lifetime, and the second showing no FRET and hence exhibiting fluorescence decay kinetics as donor alone (*see* **Note 23**).
10. After FLIM analysis, a fluorescence intensity image and a false-colored donor fluorescence lifetime image (*see* Fig 2a, c) are shown. The color distribution of the FLIM image is automatically set. To compare donor-alone versus donor-acceptor samples it is possible to set the color limits of fluorescence lifetime (*see* Fig. 2c, d).
11. The nucleus of the fluorescence intensity image can be selected using the "region of interest" (ROI) tool. The corresponding fluorescence lifetime distribution will directly be adapted.
12. After performing the calculations, the mean fluorescence lifetime, the distribution of the single-pixel values, as well as a false-color-coded lifetime image will be displayed for the selected ROI.

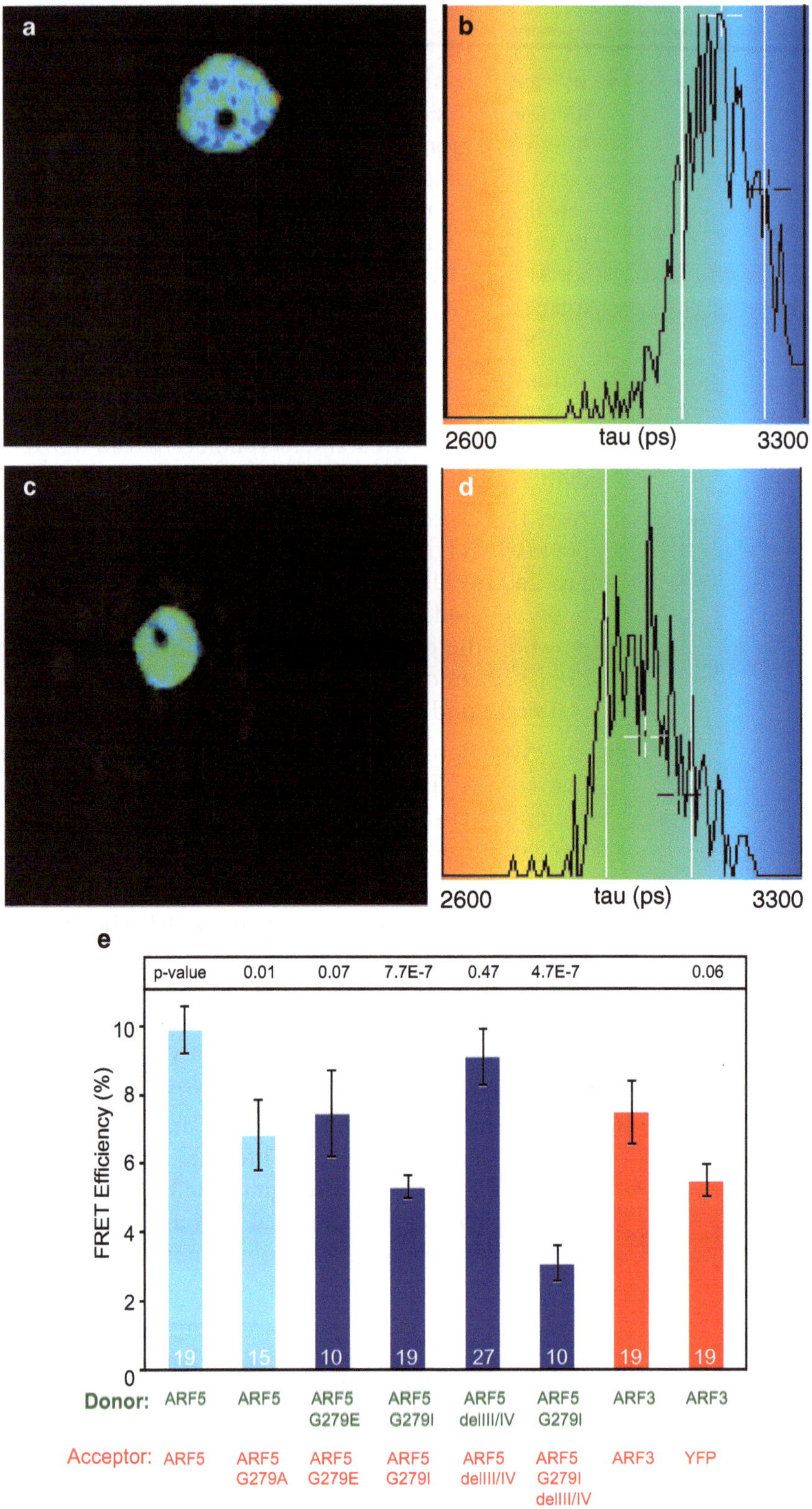

Fig. 2 FRET-FLIM analysis of ARF5 homodimerization in plant protoplasts. Donor fluorescence lifetime images of ARF5-sCFP3A alone (**a**) and ARF5-sCFP3A in the presence of ARF5-sYFP2 proteins (**c**). Histograms of fluorescence lifetime values of the nucleus (**b** and **d**) show a change of color (from *blue* to *green*) when donor fluorescence

13. In order to determine if a particular protein couple is interacting, three independent transfections of 10–15 samples of each construct are measured and average fluorescence lifetimes are determined for all cells. In this example we show a histogram of the FRET efficiencies of different combinations tested (*see* Fig 2e).

4 Notes

1. Over the years many variants of the cyan family were developed. For fluorescence lifetime analysis mTurquoise2 is the most optimal fluorophore, as it shows the longest fluorescence lifetime (*see* more information in [20]).
2. An extended protocol for ligation-independent cloning can be found in Wendrich and coworkers [21].
3. Heat twice for 6 s at 300 W in a microwave to dissolve PEG.
4. FLIM can be performed on any confocal microscope (or multiphoton microscope). The only requirement is the photon registration of the detector, which is coupled to a time-correlated single-photon counting card.
5. The Time tape supports the top side of the leaf during manipulation, while tearing off the 3 M Magic tape allows easy removal of the lower epidermal layer and exposes mesophyll cells to cell wall-digesting enzymes when the leaf is later incubated in an enzyme solution. The protoplasts released into solution are collected and washed for further use.
6. Incubation can be performed at lower temperatures but this will reduce the enzyme activity and duration should be adapted.
7. MMg solution is stressful for protoplasts; therefore the transfection procedure should take place quickly.
8. Use tips with enlarged openings for pipetting to reduce shear forces.
9. In general, imaging can be performed about 16 h after transfection but this depends on the protein expression. It is recommended to check the expression levels at several time points to determine the duration for optimal expression.

Fig. 2 (continued) lifetime is reduced. (**e**) Quantification of in vivo ARF dimerization measured by FRET-FLIM in protoplasts. Protein interaction is expressed as average FRET efficiency (±SEM), number of protoplasts is indicated in each bar, and the *p*-value for significance of difference between mutants and wild type is shown (two-sided Student's *t* test). Mutations within the dimerization domain of the ARF5 DBDs abolished the interactions, resulting in FRET efficiencies around background level. The donor is indicated in *green*, and acceptor in *red*. The histogram is reproduced from Ref. [13]

10. Pulsed diode lasers have the possibility to vary the repetition rate of the laser, which is set depending on the fluorescence lifetime. Rule of thumb is the fluorescence lifetime multiplied by 5 to know what time window is appropriate.
11. Fluorophores absorbing in the range of 470–670 nm can be excited using a super-continuum laser, which have typical excitation pulses of 1 ps.
12. This can be achieved by selecting 488 nm laser, setting AOBS in reflection mode, and selecting regular PMT.
13. In case filters are used, narrow band-pass filters should be used to avoid contamination of acceptor signal in donor channel.
14. External hybrid (HyD) or MCP-PMT (Hamamatsu Photonics, Japan) detectors can also be used for time-correlated single-photon counting experiments. HyD detectors show high quantum efficiencies, low dark noise, and large dynamic range and have a typical time transient of 100 ps, which offers opportunities like fluorescence lifetime imaging, but can also be used for single-molecule type of experiments like fluorescence correlation spectroscopy. Hamamatsu R3809U MCP photomultiplier suffers in quantum efficiency but exhibits best time resolution (30 ps).
15. FLIM analysis is optimal at sufficient photon distribution per pixel. Imaging at higher pixel resolution requires a prolonged data acquisition time to ensure the detection of a statistically relevant number of photons.
16. Time resolution of the ADC can be set ranging from 16 to 4096 channels. In case of low photon counts, the ADC can be set to 64 channels.
17. Scanning should not be slower than 200 Hz and bidirectional scanning is not appropriate.
18. TCSPC principle only holds when single photons are counted. Higher count rates can cause so-called pile-up effects, which affects fluorescence lifetime estimation.
19. In order to obtain information about the expression levels of the sCFP3A- and sYFP2-tagged proteins a sequential confocal image of donor and acceptor should be taken. Before a FLIM measurement, confocal images of donor and acceptor should be taken to select cells that show similar expression levels.
20. Other analysis software like TIMP developed by [22] or FLIMfit [23] can be used.
21. In these studies, the effect of binning has been checked between binning 0 (no pixel binning) and 1. No significant difference of the average fluorescence lifetime distribution was observed. However, image acquisition was taking five times

longer in case binning 0 was used for obtaining sufficient number of photons for data analysis.

22. Donor samples are also evaluated using two-component analysis to check for false positives.
23. Using two-component analysis one can choose to fix the long lifetime to value found for donor-only sample.

Acknowledgements

This study was supported by the Netherlands Organization for Scientific Research (NWO; NWO-NSFC grant number 846.11.001 and ECHO grant 711.011.002 to D.W.). B.D.R. was funded by the Netherlands Organization for Scientific Research (NWO; VIDI 864.13.001) and by The Research Foundation—Flanders (FWO; Odysseus II G0D0515N and Post-doc grant 12D1815N). The authors would like to thank Dr. S. Lindhoud for providing us with Fig. 1. FRET-FLIM experiments were performed on a multimode confocal microscope supported by an NWO Middelgroot Investment Grant (721.011.004; J.W.B.).

References

1. Amos WB, White JG (2003) How the confocal laser scanning microscope entered biological research. Biol Cell 95:335–342
2. Cox G, Sheppard CJ (2004) Practical limits of resolution in confocal and non-linear microscopy. Microsc Res Tech 63:18–22
3. Betzig E, Patterson GH, Sougrat R, Lindwasser OW, Olenych S, Bonifacino JS, Davidson MW, Lippincott-Schwartz J, Hess HF (2006) Imaging intracellular fluorescent proteins at nanometer resolution. Science 313:1642–1645
4. Hess ST, Girirajan TP, Mason MD (2006) Ultra-high resolution imaging by fluorescence photoactivation localization microscopy. Biophys J 91:4258–4272
5. Bastiaens PI, Pepperkok R (2000) Observing proteins in their natural habitat: the living cell. Trends Biochem Sci 25:631–637
6. Bücherl CA, Bader A, Westphal AH, Laptenok SP, Borst JW (2014) FRET-FLIM applications in plant systems. Protoplasma 251:383–394
7. Clegg RM (1996) Fluorescence resonance energy transfer. In: Herman XFWB (ed) Fluorescence imaging spectroscopy and microscopy. Wiley, New York, pp 179–252
8. Stryer L (1978) Fluorescence energy-transfer as a spectroscopic ruler. Annu Rev Biochem 47:819–846
9. Borst J, Visser AJ (2010) Fluorescence lifetime imaging microscopy in life sciences. Meas Sci Technol 21
10. Lakowicz JR (2006) Principles of fluorescence spectroscopy. Springer, New York, NY
11. Valeur B (2002) Molecular fluorescence. Principles and applications. Wiley-VCH, Weinheim
12. Clegg RM (2009) Förster Resonance Energy Transfer – FRET what is it, why do it, and how it's done. In: Gadella T (ed) FRET and FLIM techniques. Elsevier Science, Amsterdam, pp 1–57
13. Boer DR, Freire-Rios A, van den Berg WA, Saaki T, Manfield IW, Kepinski S, Lopez-Vidrieo I, Franco-Zorrilla JM, de Vries SC, Solano R et al (2014) Structural basis for DNA binding specificity by the auxin-dependent ARF transcription factors. Cell 156:577–589
14. Ljung K (2013) Auxin metabolism and homeostasis during plant development. Development 140:943–950
15. De Rybel B, van den Berg W, Lokerse A, Liao CY, van Mourik H, Moller B, Peris CL, Weijers D (2011) A versatile set of ligation-independent cloning vectors for functional studies in plants. Plant Physiol 156:1292–1299

16. Sheen J (2001) Signal transduction in maize and Arabidopsis mesophyll protoplasts. Plant Physiol 127:1466–1475
17. Bücherl C, Aker J, de Vries S, Borst JW (2010) Probing protein-protein Interactions with FRET-FLIM. Methods Mol Biol 655:389–399
18. Wu FH, Shen SC, Lee LY, Lee SH, Chan MT, Lin CS (2009) Tape-Arabidopsis Sandwich - a simpler Arabidopsis protoplast isolation method. Plant Methods 5:16
19. Becker W, Bergmann A, Hink MA, Konig K, Benndorf K, Biskup C (2004) Fluorescence lifetime imaging by time-correlated single-photon counting. Microsc Res Tech 63:58–66
20. Goedhart J, von Stetten D, Noirclerc-Savoye M, Lelimousin M, Joosen L, Hink MA, van Weeren L, Gadella TW Jr, Royant A (2012) Structure-guided evolution of cyan fluorescent proteins towards a quantum yield of 93%. Nat Commun 3:751
21. Wendrich JR, Liao CY, van den Berg WA, De Rybel B, Weijers D (2015) Ligation-independent cloning for plant research. Methods Mol Biol 1284:421–431
22. Laptenok SP, Snellenburg JJ, Bucherl CA, Konrad KR, Borst JW (2014) Global analysis of FRET-FLIM data in live plant cells. Methods Mol Biol 1076:481–502
23. Warren SC, Margineanu A, Alibhai D, Kelly DJ, Talbot C, Alexandrov Y, Munro I, Katan M, Dunsby C, French PM (2013) Rapid global fitting of large fluorescence lifetime imaging microscopy datasets. PLoS One 8:e70687

Chapter 14

In Vivo Identification of Plant Protein Complexes Using IP-MS/MS

Jos R. Wendrich, Sjef Boeren, Barbara K. Möller, Dolf Weijers, and Bert De Rybel

Abstract

Individual proteins often function as part of a protein complex. The identification of interacting proteins is therefore vital to understand the biological role and function of the studied protein. Here we describe a method for the in vivo identification of nuclear, cytoplasmic, and membrane-associated protein complexes from plant tissues using a strategy of immunoprecipitation followed by tandem mass spectrometry. By performing quantitative mass spectrometry measurements on biological triplicates, relative abundance of proteins in GFP-tagged complexes compared to background controls can be statistically evaluated to identify high-confidence interactors. We detail the entire workflow of this approach.

Key words Protein–protein interaction, Arabidopsis, Immunoprecipitation, Tandem mass spectrometry, Protein complex identification, GFP

1 Introduction

To understand a biological process, one can focus on the function or activity of the individual proteins that are involved. However, these proteins often perform their function as part of intricate complexes, comprising several and/or different proteins bound and functioning together. Therefore, being able to identify which proteins interact with one another is a very important step towards a thorough understanding of their biological function.

Several methods are currently used to test the interaction of two (or more) selected proteins, including both in vitro and in vivo techniques like yeast-two/three-hybrid (Y2H, Y3H; [1]), co-immunoprecipitation (Co-IP) followed by Western blotting [2], and Förster resonance energy transfer measured by fluorescence lifetime imaging (FRET-FLIM) (also described in this issue by Freire Rios et al.; [3]). The drawback of these types of techniques is that they are limited to test the interaction of the selected proteins, and often require prior knowledge regarding the interactions or

Jürgen Kleine-Vehn and Michael Sauer (eds.), *Plant Hormones: Methods and Protocols*, Methods in Molecular Biology, vol. 1497, DOI 10.1007/978-1-4939-6469-7_14, © Springer Science+Business Media New York 2017

require all tested proteins to be labeled, limiting the amount of interactions that can be examined and creating a bias in the measurements. Regardless of this, techniques like FRET-FLIM are powerful as a means to characterize individual interactions, as they allow for testing of dynamic interactions at the subcellular level [3].

Here, we describe an optimized protocol for immunoprecipitation (IP) followed by (label-free) tandem mass spectrometry (MS/MS) that allows the identification of proteins interacting with a fluorescently labeled bait in a non-biased and (semi-) quantitative fashion. This methodology has been successfully used to identify complexes of multiple types of proteins, including transcription factors and other nuclear proteins [4, 5], membrane-associated proteins [6], trans-membrane proteins [6], and cytosolic proteins [7]. When considering IP-MS/MS for the identification of interacting proteins to a protein of interest, it is important to keep in mind that this is not a saturated identification method. Due to the finite amount of peptides that can be measured by the MS per run, not all interacting proteins may be visible. Also some (previously known) interacting proteins may be absent from the results list, due to very low abundance in the sample. It is, therefore, of vital importance to keep contaminations to a minimum, as these will fill the MS with unwanted peptides.

This step-by-step protocol for IP-MS/MS, using anti-GFP-coated magnetic beads, describes the workflow that is used, from collection of plant material up to the final data analysis (*see* Fig. 1), and emphasizes important points for obtaining high purity and efficiency.

2 Materials

2.1 Plant Material

Homozygous transgenic lines should be generated, expressing the protein of interest tagged with a GFP (or derivative) in the desired tissue. Generally, 3 g of powdered material is used per replicate in the IP. As a rule of thumb, for three replicates, 1 mL of seed is needed when studying 6-day-old whole seedlings. For the control sample, use either a non-transgenic line or a line expressing only GFP.

2.2 Immuno-precipitation

1. Liquid nitrogen.
2. Mortar and pestle.
3. Standard 50 mL Falcon tubes.
4. Standard 14 mL Falcon tubes.
5. Extraction buffer without detergent (~60 mL is needed per sample; EB−):

   Tris–HCl, pH7.5 1 M (5 mL), NaCl, 5 M (3 mL), protease inhibitor complete (two tablets), H_2O to 100 mL.

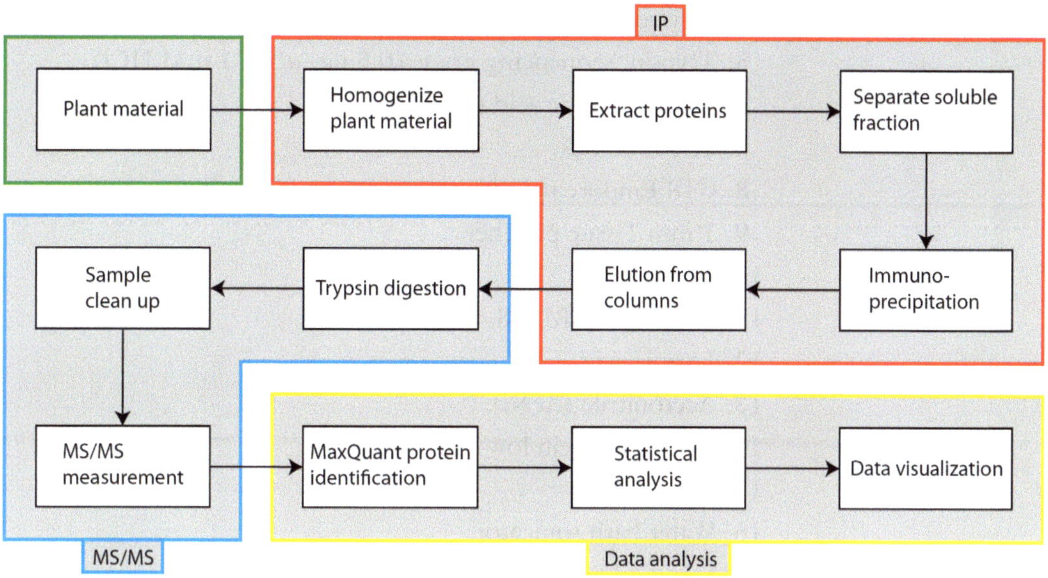

Fig. 1 IP-MS/MS workflow. The protocol starts with collecting plant material, then homogenizing it, and extracting the proteins. The soluble fraction is then separated and immunoprecipitation is performed. Beads and proteins are subsequently eluted and trypsin digestion of the proteins is done. After cleanup, label-free tandem MS measurements are performed followed by MaxQuant protein identification, statistical analysis, and data visualization

6. Extraction buffer with 1% detergent (~10 mL needed per sample; EB+):
   Nonidet P-40; 0.1 mL, EB−; 9.9 mL.
7. Needle sonicator.
8. 50 mL Centrifuge tubes.
9. Beckman-Coultier Avanti J26 XP—JA 25.50 rotor.
10. Standard tabletop balance, 0.05 g accurate.
11. 40 μm Cell strainer.
12. Anti-GFP-coated magnetic beads (μMACS, Miltenyi).
13. Magnetic μ columns (μMACS, Miltenyi).
14. MultiStand (μMACS, Miltenyi).
15. μMACS Separator (Miltenyi).
16. ABC buffer: 50 mM NH_4HCO_3 in H_2O.
17. Thermoblock.
18. 0.5 mL Protein low-bind tubes.

2.3 Sample Preparation Mass Spectroscopy

1. ABC buffer: 50 mM NH_4HCO_3 in H_2O.
2. Dithiothreitol (DTT).
3. Iodoacetamide (IAA).

4. l-Cysteine.
5. Trypsin, sequencing grade (0.5 μg/μL in 1 mM HCl).
6. Trifluoroacetic acid (TFA).
7. Thermoblock.
8. C18 Empore membrane.
9. 1 mm Tissue puncher.
10. Methanol.
11. LiChroprep RP-18.
12. Formic acid.
13. Acetonitrile (AcNi).
14. 0.5 mL Protein low-bind tubes.
15. SpeedVac.
16. Water bath sonicator.
17. Standard tabletop centrifuge.

3 Methods

3.1 Immunoprecipitation

Day 1:

1. Thoroughly grind plant material (both transgenic line and the control) in liquid nitrogen with mortar and pestle. Transfer the ground material to a precooled 50 mL tube and place it in liquid nitrogen.
2. For each of the replicates, weigh the desired amount of plant powder on a scale into a precooled tube (*see* **Note 1**).
3. Store at −80 °C until next day (*see* **Note 2**).

Day 2:

1. Precool mortars on ice and prepare extraction buffers.
2. Add weighed material to mortars and add as little EB+ as possible to solubilize the plant material (*see* **Note 3**) and grind again very thoroughly with mortar and pestle.
3. Transfer the protein extract to a 14 mL tube (if you have more than 3 mL protein extract, divide over several 14 mL tubes) and sonicate three times for 15 s on ice with at least 15 s pause in between (*see* **Note 4**).
4. Keep the samples on ice for 30 min to thoroughly extract the proteins (*see* **Note 5**).
5. Dilute the protein extract to 0.2% NP40 by adding EB− and transfer the protein extract to centrifuge tubes (*see* **Note 6**).

6. Centrifuge for 15 min at 4 °C at 18,000 rpm/39.000 g (*see* **Note 7**). Transfer the supernatant to another centrifuge tube and centrifuge again for 15 min at 4 °C at 18,000 rpm/39.000 g.
7. Transfer the supernatant through a 40 μm cell strainer into a new 50 mL tube (*see* **Note 8**).
8. Add 50–100 μL anti-GFP μBeads (*see* **Note 9**). Mix well by swirling the tube. Rotate for 2 h at 4 °C (*see* **Note 10**).
9. Place μColumn in the magnetic field of the μMACS Separator.
10. Add 200 μL clean EB+ to the column.
11. Prepare EB with 0.1 % NP40.
12. Add 500 μL EB with 0.1 % NP40 to the column.
13. Apply cell lysate onto the column and let the lysate run through (*see* **Note 11**).
14. Prepare ABC buffer and heat 1 mL to 95 °C in a heat block for later use.
15. Rinse column four times with 200 μL EB with 0.1 % NP40.
16. Rinse column twice with 500 μL ABC buffer to remove all detergent (*see* **Note 12**).
17. Remove the column from the magnet, immediately place a 0.5 mL low-bind Eppendorf tube below the column, and add 50 μL preheated ABC buffer to the column to transfer the beads into the Eppendorf tube (*see* **Note 13**).
18. Transfer column into a new low-bind tube and add another 50 μL preheated ABC buffer to the column (second elution with ABC, *see* **Note 14**) to check whether all beads were transferred in **step 17**. Combine with the beads in **step 17** when beads are visible.
19. Store at −20 °C.

3.2 Sample Preparation Mass Spectrometry

Day 3:

1. Add 1 μL 500 mM DTT in ABC buffer to the beads containing eluate and incubate for 2 h at 60 °C (*see* **Note 15**).
2. Add 1 μL 750 mM IAA in ABC buffer to the samples. Mix well and incubate for 2 h at room temperature in the dark (*see* **Note 16**).
3. Add 1 μL 200 mM l-cysteine in ABC buffer to the samples (*see* **Note 17**). Mix well.
4. Add 1 μL sequence-grade trypsin, mix well, and incubate for 16 h at 20 °C (*see* **Note 18**).

Day 4:

1. Using 10% TFA, adjust the pH of the samples to approximately 3; check the pH with 0.1 μL of sample on a piece of pH paper (*see* **Note 19**).

2. Use a tissue puncher to cut a small (~1 mm) piece of a C18 Empore membrane and insert it into a polypropylene 200 μL pipet tip [8].
3. Add 200 μL methanol to the tip with membrane (*see* **Note 20**).
4. Prepare a 50% (v/v) slurry of LiChroprep RP-18 column material in methanol and apply 4 μL of the slurry to the tip/μColumn in the methanol.
5. Elute the μColumn and wash once with 100 μL methanol (*see* **Note 21**).
6. Equilibrate the μColumn once with 100 μL 0.1% formic acid (*see* **Note 22**).
7. Spin down the beads at full speed for 15 min in a standard tabletop centrifuge.
8. Add sample to the μColumn and elute through (*see* **Note 23**).
9. Wash the μColumn once with 100 μL 0.1% formic acid.
10. Transfer the μColumn to a 0.5 mL protein low-bind tube.
11. Add 50 μL AcNi / 0.1% formic acid (1:1) and manually elute the μColumn directly into the 0.5 mL protein low-bind tube (*see* **Note 24**).
12. For LCMS analysis, reduce the AcNi content by putting the samples with open cap in a SpeedVac at 30–45 °C for ~2 h. The final volume should be well below 10 μL. Adjust the sample volume to 50 μL with 0.1% formic acid (*see* **Note 25**).
13. The sample is now ready to be loaded onto an LCMS and can be stored at −20 °C.

3.3 Tandem Mass Spectrometry

1. Separate peptides by reversed-phase nano liquid chromatography connected to the MS via an electrospray interface.
2. Measure MS spectra online by accurate mass spectrometry (MS) as well as collision-induced dissociation fragmented MS-MS spectra according to the settings in Tables 1 and 2 (*see* **Notes 26** and **27**).

3.4 MaxQuant Protein Identification and Statistical Analysis

1. Analyze all LC-MS/MS runs obtained with all MS/MS spectra with the database search algorithm Andromeda from MaxQuant 1.3.0.5 [9, 10] in the "Label Free Quantification" mode with settings described in Table 3 (*see* **Note 28**).
2. Normalized quantitative information (Label Free Quantification [LFQ] intensity) is obtained for all protein groups identified when two or more peptides could be properly quantified (*see* **Note 29**).
3. Group the replicates together. An optional final filtering can be performed in the Perseus filtering and statistics software to improve the statistics by deleting those proteins that are identified in less than half of the replicates (*see* Table 4 and **Note 30**).

Table 1
Chromatography settings

Nano LC	Proxeon EASY nLC II
Setup	Vented column
Pre-concentration column	0.10 × 32 mm Magic C18AQ 200A 5 μm beads prepared in-house
Analytical column	0.10 × 250 mm Magic C18AQ 200A 3 μm beads prepared in-house
Column temperature	Room temperature
Autosampler temperature	7 °C
Sample loading volume and flow	30 μL at 270 bar (generally 7–8 μL/min) with eluent A
Autosampler wash	2 cycles with 27 μL of 50 % (v/v) acetonitrile/50 % (v/v) 0.1 % formic acid in water, 100 μL of 0.1 % formic acid in water
Eluent A	5 mL/L acetic acid in LCMS-grade water
Eluent B	5 mL/L acetic acid in LCMS-grade acetonitrile
Measurement gradient (time in min (% eluent B))	0 (8), 50 (33), 53 (50), 58 (50)
Cleaning gradient (time in min (% eluent B))	0 (20), 10 (80), 15 (80). Run at least one cleaning gradient after each measurement gradient
Injection volume	18 μL of sample (measurement gradient) or 15 μL of 50 % (v/v) acetonitrile/50 % (v/v) 0.1 % formic acid in water
High-voltage connection	Apply an electrospray potential of 3.5 kV directly to the eluent via the stainless steel needle fitted into the vented waste line of a P777 Upchurch microcross that was positioned between the pre-concentration and analytical column
Pre-column re-equilibration	12 μL Eluent A at 270 bar
Analytical column re-equilibration	3 μL Eluent A at 270 bar

4. Perform T-tests between groups in Perseus. This not only results in the *p*-value (column: -Log t-test *p*-value), but also immediately yields the ratio of the group's average LFQ values (column: t-test Difference) when applied to two groups. Significance at a variable false discovery rate (FDR) is judged by both *p*-value and ratio, of which the weight is set by a variable S_0 value (*see* **Note 31**). Both FDR and S_0 should be optimized to yield no or only a few proteins significantly different on the control side of the volcano plot obtained (*see* Fig. 2).
5. In a successful experiment, the GFP and the bait (tagged) protein are in the top 10 when the list is sorted by ratio (*see* **Note 32**). The high ratio and low *p*-value should make the difference between sample and control clearly significant. Depending

Table 2
Mass spectrometry settings

MS	LTQ-Orbitrap XL
Measurement software	Xcalibur 2.1
Tuning and calibration	From *m/z* 150–2000 by direct infusion of LTQ ESI calibration solution for positive mode containing caffeine, peptide MRFA, and Ultramark 1621 according to instructions of the manufacturer
nESI	Capillary temperature = 200 °C, capillary voltage =35 V, tube lens voltage = 100 V, source voltage = 3.5 kV (*see* high-voltage connection in Table 1)
Tune file parameters	FTMS Full Microscans = 1, Ion Trap MSn Microscans = 1, FTMS Full Max Ion Time = 500 ms, Ion Trap MSn Max Ion Time = 100 ms, FTMS Full AGC Target = 1,000,000, Ion Trap MSn AGC Target = 10,000
MS detector acquire time (min)	58
Start delay (min)	0
Segments	1
Scan events	2
Event 1	Analyzer = FTMS, Mass range = Normal, Scan Range = 380-1400 m/z, Resolution = 60.000, Scan type = Full, Polarity = Positive, Data type = Profile
Event 2 (data dependent)	Analyzer = Ion Trap, Mass range = Normal, Scan Rate = Normal, Data type = Centroid, Dependent scan On
Lock mass	Disabled (FTMS is calibrated every day)
Global dynamic exclusion	Repeat Count = 1, Repeat duration = 0 s, Exclusion list size = 500, Exclusion duration = 45 s, Exclusion mass width Relative to reference mass = 25 ppm (low, and high) Early expiration = disabled
Segment	Preview mode for FTMS master scans = Enabled, Charge state screening and Rejection = Enabled, Monoisotopic precursor selection = Enabled, Charge state +1 and 4 and up = Rejected, Unassigned charge state = Rejected
Scan event	Minimum MS signal Threshold for MS2 trigger = 5000, Mass determined from scan event = 1, Nth most intense ion = Enabled, Analyze top N peaks = 4, Activation type = CID, Activation Default charge state = 3, Activation Isolation width (m/z) = 2.0, Normalized collision energy = 30, Activation Q = 0.250, Activation time (ms) = 15
Mass lists and global mass lists	NOT used

on the starting tissue that is used, some highly abundant proteins like glyceraldehyde-3-phosphate dehydrogenase and Rubisco can always be observed, but these should give ratios around one (=0 on a logarithmic scale).

Table 3
Database search and quantification software settings

Software	MaxQuant 1.3.0.5
Group-specific parameters	Variable modifications = Oxidation (M), Acetyl (Protein N-term), Deamidation (NQ), Separate variable modifications for first search = disabled, Multiplicity = 1, Enzyme = Trypsin/P, First search ppm = 20, Main search ppm = 6, Max number of modifications per peptide = 5, Max. missed cleavages = 2, Max. charge = 7, individual peptide mass tolerances = Enabled, Type = Standard, Separate enzyme for first search = disabled
MS/MS sequences	ITMS MS/MS tolerance = 0.5 Da, Include contaminants = Enabled FASTA files = Arabidopsis thaliana Uniprot reference database, Reverse, Special AAs = KR, Separate file for first search = a contaminants database with 60 proteins including GFP/YFP/CFP/eYFP and BSA (P02769, bovin serum albumin precursor), Trypsin (P00760, bovin), Trypsin (P00761, porcin), Keratin K22E (P35908, human), Keratin K1C9 (P35527, human), Keratin K2C1 (P04264, human) and Keratin K1CI (P35527, human) Fixed modifications = Carbamidomethyl (C)
Identification and quantification	All default with deamidation (NQ) added to the protein quantification peptide list Experimental design file should be made and selected
Misc.	Match between runs = Enabled (2 min), Label-free quantification = Enabled (LFQ min ratio count = 2), iBAQ and Log fit = Enabled

Table 4
Filtering and statistics on identified protein groups

Software	Perseus
Selectively load data from the result file	Expression: all"LFQ_name" rows, Categorical annotation: Only identified by site, Reverse, Contaminant, Textual annotation: Protein IDs, Majority protein IDs, Protein names, Gene names, Fasta headers, (remove "Proteins"); Numerical annotation: id, Proteins, Peptides, Unique Peptides, iBAQ and any other info you may need like: sequence coverage (%), Mol. Weight [kDa], Sequence length and/or PEP score
Extra filtering	Filter out "Reverse" and "Only identified by site" proteins. "Only identified by site" marks those proteins that have been identified by modified peptides only Filter out proteins identified by only one peptide or no unique peptide
Logarithmic transformation	Transform the LFQ and iBAQ to its logarithmic values
Grouping of samples	Group biological replicate measurements
Optional extra filtering 2	Filter out those protein groups that have less than two LFQ values in at least one group
Replace NaNs	Replace/imputate missing values by a constant that is slightly lower than the lowest (Log) value measured. This is done to make sensible ratio calculation possible
T-test	Perform a first *T*-test with FDR = 0.01 and $S_0 = 1$

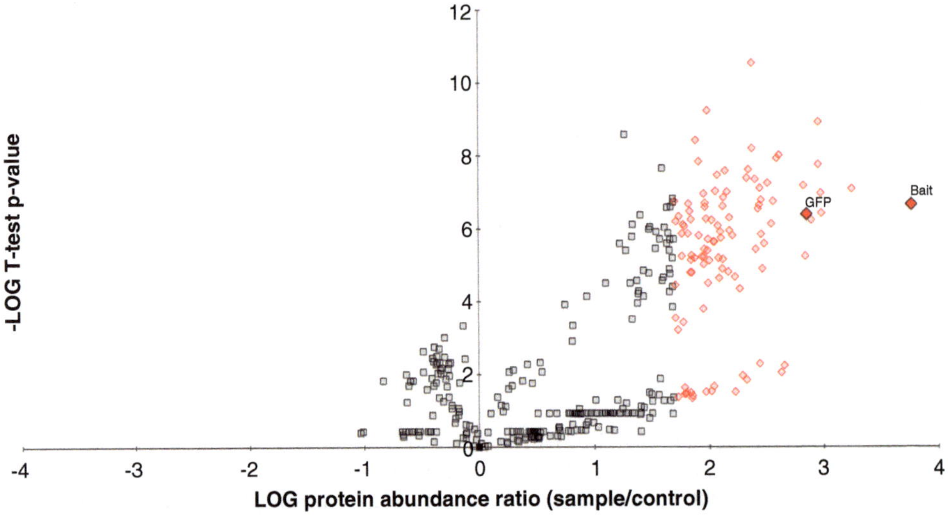

Fig. 2 Volcano plot example. Volcano plot showing all identified proteins after filtering and statistical analysis, with their corresponding protein abundance ratios over the *T*-test p-value. Depicted in *red* are proteins that are significantly different from the control, with the bait and GFP among the top hit list

4 Notes

1. Usually about 1–3 g of ground plant material is used per replicate/sample (for low-expressed proteins use 3 g of tissue and use only tissue where your protein is expressed to limit the amount of background proteins in the sample).
2. This is recommended because it ensures that the mortars are not too cold, which in some cases can lead to complications with fluidity of the sample.
3. Usually ~8–9 mL of EB+ is needed for 3 g material. Using less of the buffer in this step will yield more concentrated sample later in the protocol.
4. Sonication is used to break up cell membranes and release cellular components in the liquid. For the described sonicator, use 3/4 power and middle tune. Optimization might be required depending on the sonicator. It is very important to keep the samples on ice the whole time, especially during sonication, to prevent complex deformation through heat.
5. If desired, a small aliquot can be taken as INPUT sample for Western blotting.
6. Too high detergent concentrations may increase background proteins and disrupt MS measurements.
7. Make sure that the tubes are balanced properly (±0.05 g).
8. This ensures that any leftover cell debris is filtered out, as these could clog the IP μColumns. If desired, a small aliquot can be taken as supernatant sample for Western blotting.

9. Use 50 μL when 1 g of material is used and 100 μL when 3 g of material is used. Make sure to resuspend the beads before pipetting with a cut tip.
10. In some cases it can be advisable to speed up the whole protocol. This incubation step can then be decreased to 1 h.
11. The columns are "flow-stop" and do not run dry. If desired, an aliquot of the flow-through can be taken as sample for Western blotting.
12. Any detergent present in the final MS sample may interfere with the chromatography and the MS measurement, as this may cause peak broadening.
13. Make sure that the "end" of the column is always in contact with the eluate. This way more beads will be eluted.
14. If desired, a small aliquot can be taken as IP sample for Western blotting.
15. DTT reduces disulfide bonds in the proteins, leaving them more exposed for the trypsin digestion later in the protocol.
16. IAA alkylates the proteins, blocking disulfide formation.
17. l-Cysteine stops the alkylation.
18. Do not leave this longer than 16 h, as this may result in chymotrypsinic cleavages.
19. Lowering the pH inactivates trypsin, stopping digestion. Avoid a pH of lower than 2.
20. If some air is trapped between the liquid and the frit/column material, flick the μColumn to remove.
21. The prepared μColumns can be eluted by hand using a 1 mL syringe with a rubber-lined tip or using a vacuum pump. Whatever method used, do not let the columns run dry.
22. Methanol concentration should be brought below 5 % to allow binding of peptides.
23. Peptides in the sample will bind the column material and detergent and other contaminants will be washed out.
24. Peptides are released from the column material in acetonitrile concentrations above 5 %; 50 % is used to limit yield losses in this step.
25. If samples are completely dry, resuspend in 50 μL 0.1 % formic acid and sonicate several times for 10 s, vortexing in between.
26. A high chromatographic resolution as well as high MS resolution, accuracy, and sensitivity are required to successfully measure transcription factor interactors.
27. Similar results can be obtained using instruments from suppliers other than mentioned in the tables when they have comparable or improved specifications like some more modern instruments do.

28. MaxQuant internally re-calibrates MS spectra resulting in MS deviations which generally lie below 2 ppm. The advantage of using the "Match between runs" option is that the software will search for all peptides identified in all MS runs based on retention time, m/z measured, and isotopic distribution. This way, quantitative information can be obtained for all peptides identified in all runs even when the peptide may only have been selected for fragmentation in a single run.
29. By default a FDR of 1 % is used.
30. This will restrict the results table to only those proteins with a high reliability.
31. Significant FDR can be set manually. S_0 values <1 give more weight to the *p*-value and S_0 values >1 give more weight to the ratio, when determining the significance. Generally an S_0 of 1 is chosen to give both *p*-value and ratio an equal weight.
32. Sometimes, GFP is found somewhat lower in the list, probably due to its compact folding, making it more difficult to digest with trypsin.

References

1. Ferro E, Trabalzini L (2013) The yeast two-hybrid and related methods as powerful tools to study plant cell signalling. Plant Mol Biol 83:287–301
2. Albrecht C, Boutrot F, Segonzac C, Schwessinger B, Gimenez-Ibanez S, Chinchilla D, Rathjen JP, de Vries SC, Zipfel C (2012) Brassinosteroids inhibit pathogen-associated molecular pattern-triggered immune signaling independent of the receptor kinase BAK1. Proc Natl Acad Sci U S A 109:303–308
3. Bücherl CA, Bader A, Westphal AH, Laptenok SP, Borst JW (2014) FRET-FLIM applications in plant systems. Protoplasma 251:383–394
4. De Rybel B, Möller B, Yoshida S, Grabowicz I, Barbier de Reuille P, Boeren S, Smith RS, Borst JW, Weijers D (2013) A bHLH complex controls embryonic vascular tissue establishment and indeterminate growth in Arabidopsis. Dev Cell 24:426–437
5. Saiga S, Möller B, Watanabe-Taneda A, Abe M, Weijers D, Komeda Y (2012) Control of embryonic meristem initiation in Arabidopsis by PHD-finger protein complexes. Development 139:1391–1398
6. Zwiewka M, Feraru E, Möller B, Hwang I, Feraru MI, Kleine-Vehn J, Weijers D, Friml J (2011) The AP-3 adaptor complex is required for vacuolar function in Arabidopsis. Cell Res 21:1711–1722
7. Smaczniak C, Li N, Boeren S, America T, van Dongen W, Goerdayal SS, de Vries S, Angenent GC, Kaufmann K (2012) Proteomics-based identification of low-abundance signaling and regulatory protein complexes in native plant tissues. Nat Protoc 7:2144–2158
8. Rappsilber J, Mann M, Ishihama Y (2007) Protocol for micro-purification, enrichment, pre-fractionation and storage of peptides for proteomics using StageTips. Nat Protoc 2:1896–1906
9. Cox J, Mann M (2008) MaxQuant enables high peptide identification rates, individualized p.p.b.-range mass accuracies and proteome-wide protein quantification. Nat Biotechnol 26:1367–1372
10. Cox J, Neuhauser N, Michalski A, Scheltema RA, Olsen JV, Mann M (2011) Andromeda: a peptide search engine integrated into the MaxQuant environment. J Proteome Res 10:1794–1805

Chapter 15

Assaying Auxin Receptor Activity Using SPR Assays with F-Box Proteins and Aux/IAA Degrons

Mussa Quareshy, Veselina Uzunova, Justyna M. Prusinska, and Richard M. Napier

Abstract

The identification of TIR1 as an auxin receptor combined with advanced biophysical instrumentation has led to the development of real-time activity assays for auxins. Traditionally, molecules have been assessed for auxinic activity using bioassays, and agrochemical compound discovery continues to be based on "spray and pray" technologies. Here, we describe the methodology behind an SPR-based assay that uses TIR1 and related F-box proteins with surface plasmon resonance spectrometry for rapid compound screening. In addition, methods for collecting kinetic binding data and data processing are given so that they may support programs for rational design of novel auxin ligands.

Key words Surface plasmon resonance (SPR), Biacore, F-box proteins, Auxin binding, Compound screen, Kinetics

1 Introduction

The study of hormone-active proteins requires robust assays for binding. Most binding assays are developed for receptor candidates, but both enzymes and transport proteins also carry binding sites specific for their substrates. Historically, the basis of most analyses was radiolabeled ligand-binding assays. More recently a range of biophysical techniques has offered new, label-free tools for binding analysis and while no single technique can address all experimental systems, surface plasmon resonance (SPR) has found widespread utility. It yields definitive kinetic data for binding interactions and may also be used for thermodynamics and high-throughput compound screening. SPR requires that one partner is immobilized on a chip surface which can be a limiting factor, but in most cases the variety of available chip surface chemistries allows binding reactions to be measured by immobilizing either receptor or ligand. Ideally kinetics are recorded in both orientations.

Jürgen Kleine-Vehn and Michael Sauer (eds.), *Plant Hormones: Methods and Protocols*, Methods in Molecular Biology, vol. 1497, DOI 10.1007/978-1-4939-6469-7_15, © Springer Science+Business Media New York 2017

The most revealing kinetic parameter sought from binding assays is the affinity of the site for its ligands. Let affinity be K_D, the dissociation constant for the interaction under the stated conditions. The units are molar and K_D is the concentration of ligand at which the receptor-binding sites are at half occupancy. Radiolabel assays in various formats generally yield the affinity. For example, most binding analyses on auxin-binding protein 1 (ABP1) used radiolabel displacement assays [1], competing radiolabel off the binding site with increasing concentrations of unlabeled auxin to give K_D from a Scatchard analysis. However, label-free assays like SPR are able to provide additional valuable information. How much information depends somewhat on the availability of pure protein. For example, SPR typically requires micrograms of protein and can provide binding on- and off-rates as well as K_D, with the off-rate indicating the durability of the interaction-a useful feature in ligand screening. SPR can also be used for determining the thermodynamics of binding, measuring changes in free energy, enthalpy and entropy, as well as affinity to reveal aspects of the mechanism of binding. We will explain how this experiment is run, but with more protein available (milligrams) isothermal titration calorimetry (ITC) will give more definitive thermodynamic values given that the ITC reaction is entirely in solution.

Many candidate auxin receptors have been listed [1, 2] and two have been characterized extensively, ABP1, and transport inhibitor resistant 1 (TIR1) which mediates auxin-inducible changes in transcription [3, 4]. Both are soluble proteins and structures have been determined for each, with binding sites empty and with auxin bound [5, 6]. Structural data are invaluable in terms of mechanistic insights, but crystallography offers only a snapshot of a binding interaction taken under extreme conditions. We describe here an SPR assay developed for TIR1 (and its homolog AFB5) [7] in which binding is measured in real time. The assay not only gives us kinetic data on auxin binding, but it also provides a platform for novel auxin and anti-auxin discovery.

A schematic of the cascade of interactions involved in TIR1-mediated auxin perception in vivo is shown in Fig. 1 (*see* [8] for more details). The SPR assay mimics this by using the interacting peptide from the Aux/IAA degron as the "bait" molecule on the SPR chip surface. Purified TIR1 protein is then passed over the "bait" with or without auxin. In the presence of an active auxin TIR1 will bind and the SPR technique interprets this increase in mass on the chip surface to give a binding sensorgram (Fig. 2).

2 Materials

1. Biacore 2000
2. Maintenance Chip and SA Sensor Chip (streptavidin coated for coating with biotinylated molecules) (GE Healthcare BR 1006-51, 1000-32).

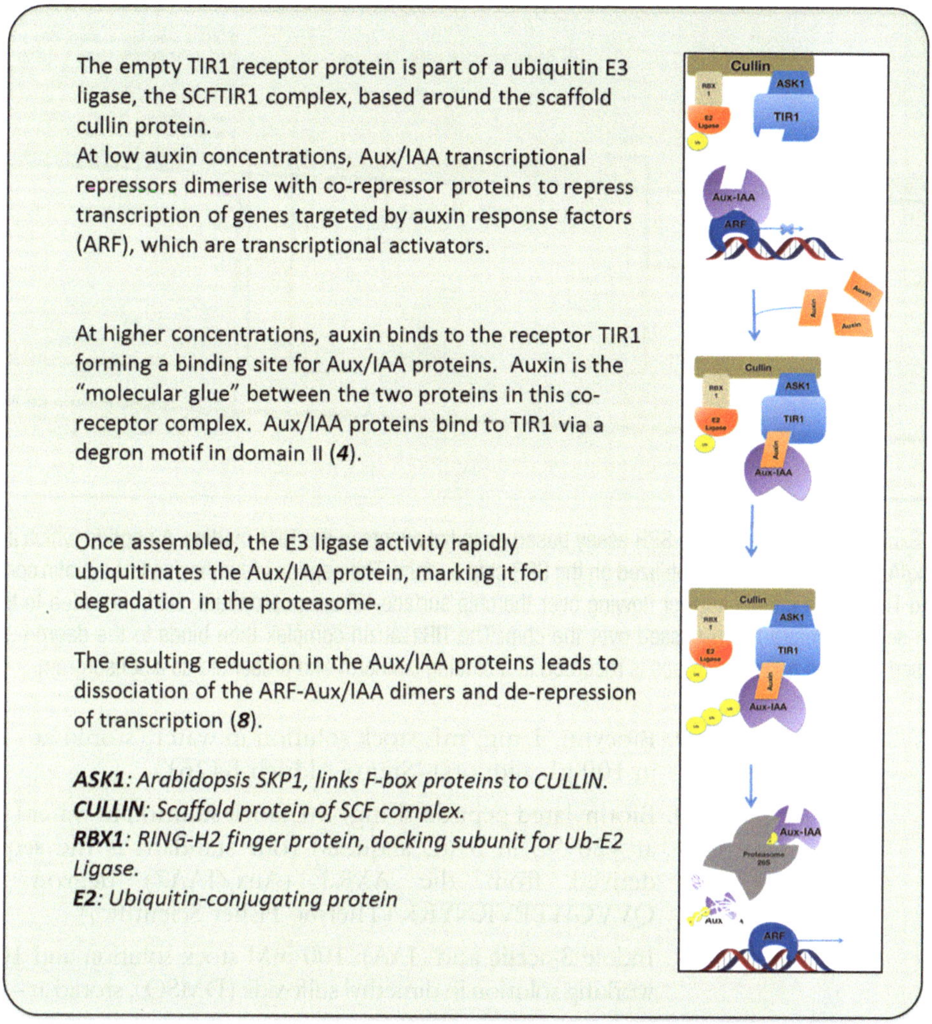

Fig. 1 Schematic representation of the cascade of events during auxin perception by TIR1

3. Plastic vials (7 mm, 0.8 mL) with rubber caps, type 3 (GE Healthcare BR 1002-12, 1005-02).
4. Desorb kit consisting of BIAdesorb solutions 1 and 2 (GE Healthcare BR 1008-23).
5. Chip cleaning solution: 1 M NaCl, 50 mM NaOH.
6. Surfactant P20 10 % v/v (GE Healthcare BR 1000-54).
7. SPR rinse solution, filtered (0.2 μm) and degassed: 0.05 % v/v Surfactant P20.
8. SPR buffer, filtered and degassed: 10 mM HEPES pH 7.4, 150 mM NaCl, 3 mM EDTA, 50 μM phytic acid, 1 mM TCEP, 0.05 % v/v Surfactant P20 (*see* **Note 1**).

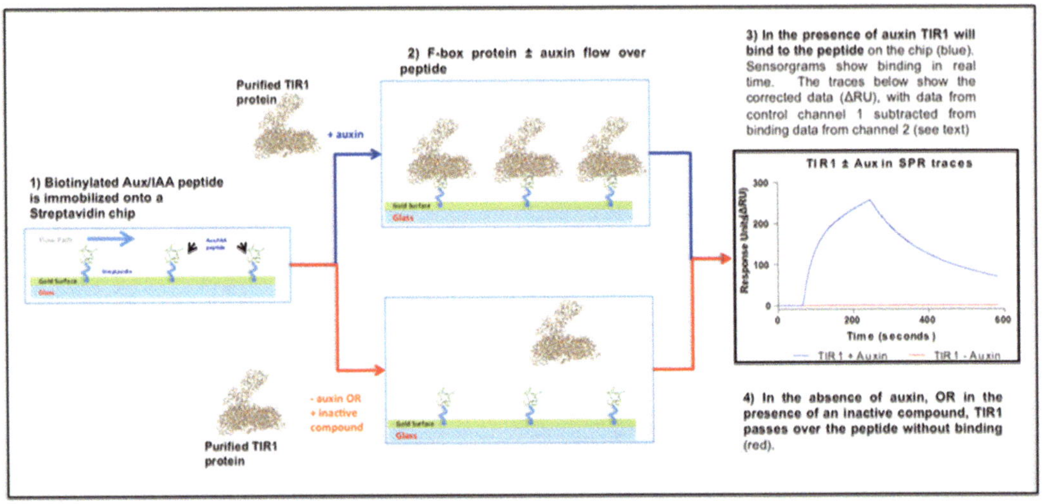

Fig. 2 Experimental setup for the SPR assay based on initial events in the TIR1 binding. A peptide which includes the Aux/IAA degron motif is immobilized on the SPR chip surface. The peptide is synthesized as a biotin conjugate. Purified TIR1 is the mobile partner flowing over the chip surface. Where appropriate, auxin is added to the TIR1 protein solution before being passed over the chip. The TIR1-auxin complex then binds to the degron and this increased mass on the chip surface is recorded in a binding isotherm and presented as a sensorgram

9. Biocytin, 1 mg/mL stock solution in water (stored at −20 °C in 100 μL aliquots) (Sigma-Aldrich B4261).
10. Biotinylated peptide, 1 mg/mL stock solution in water (stored at −20 °C in 5 μL aliquots) (our standard is the sequence derived from the AXR3 (Aux/IAA7) degron, biot-QVVGWPPVRNYRK (Thermo Fisher Scientific)).
11. Indole-3-acetic acid (IAA): 100 mM stock solution and 10 mM working solution in dimethyl sulfoxide (DMSO), stored at −20 °C.
12. TIR1/ASK1 purified protein complex in SPR buffer.

3 Methods

The methods discussed below relate to Biacore 2000 and 3000 instruments, but may be adapted for other SPR platforms. The control software opens as shown below:

From the menu bar Command, Run and Tools are used most commonly. The software guides users through all the routines and gives instructions for sample placement, volume requirements, etc.

We recommend that new Biacore users take training on the instrument before their first experiment. However, the instructions given by the software are comprehensive. For more details see the Biacore 2000 Instrument Handbook.

3.1 Chip Preparation

Prior to coating a new sensor chip (or coating additional channels on a part-used chip), always run a Desorb routine to clean the fluid lines. Desorb is run using a Maintenance Chip.

1. Place the buffer inlet tubes into a bottle of SPR rinse solution.
2. Command -> Undock; wait until the undocking routine completes (the countdown time is shown, see below). Place a maintenance chip in the holder. Command -> Dock; wait for docking to complete.

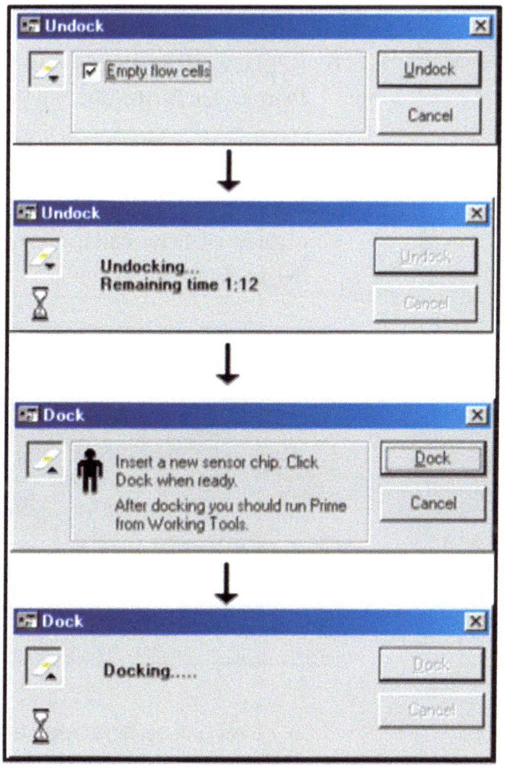

3. You will be asked if you want to Prime the system. Accept the Prime option. Read the notes in the dialog box.
4. Tools -> Working Tools -> Desorb. Read the warnings displayed. You need 3 mL of Desorb solutions 1 and 2 placed in the nominated rack positions. -> Start.

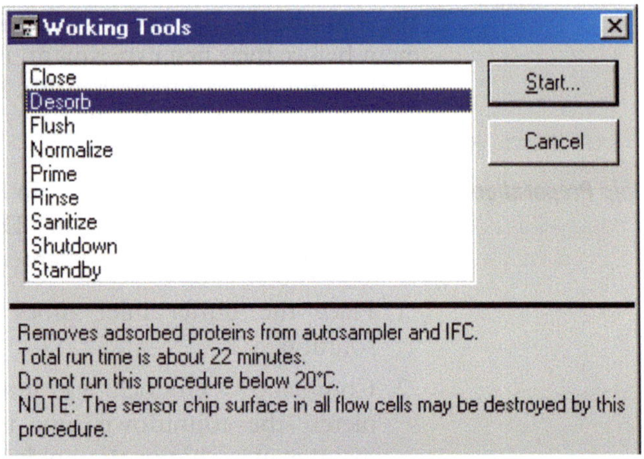

5. As in 2, undock to remove the maintenance chip and dock a new SA chip.
6. Replace SPR rinse solution with SPR buffer. As above, select Prime (*see* **Note 2**).
7. Perform a chip conditioning routine using Manual mode and Chip cleaning solution (*see* **Note 3**). Select Run -> Run sensorgram. A window entitled Detection opens and prompts a choice of flow cells. Choose the relevant cells and hit OK (*see* **Note 4**).

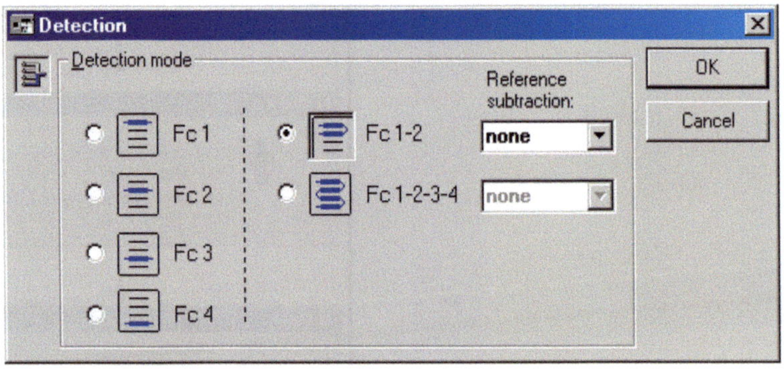

8. A Flow dialog box appears. Input a flow rate of 20 μL/min. -> OK. You will be prompted to set up a .blr file where the results are saved (*see* **Note 5**). As soon as you hit Save the main screen refreshes (as shown) giving a readout for each channel selected in 7 (*see* **Note 6**). The control buttons for manual injection are now active and a Command Queue dialog box opens.

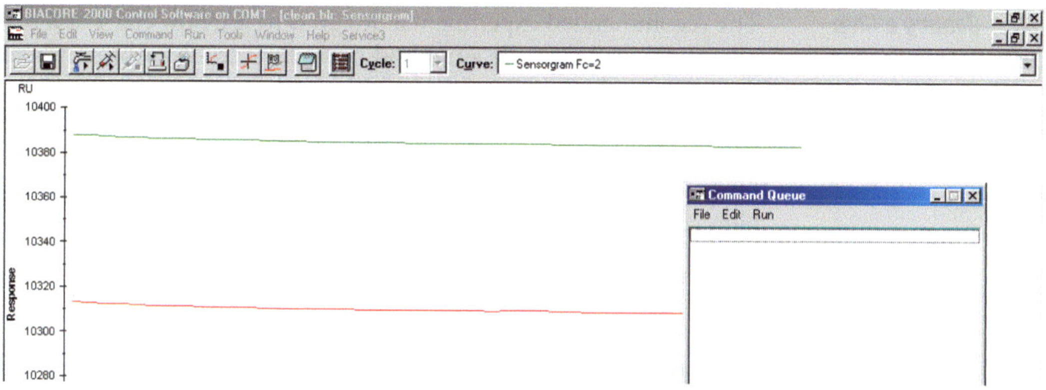

9. Set up three consecutive 1-min injections of Cleaning Solution A (*see* **Note 7**). Select -> Inject; a dialog box opens. Position asks you to nominate the rack position for the cleaning solution vial. Volume -> 20 μL is recommended. The volume entered determines the contact time for a given flow rate. Consumption (the volume of conditioning solution used) is calculated. This includes system dead volumes plus the 20 μL wash volume.

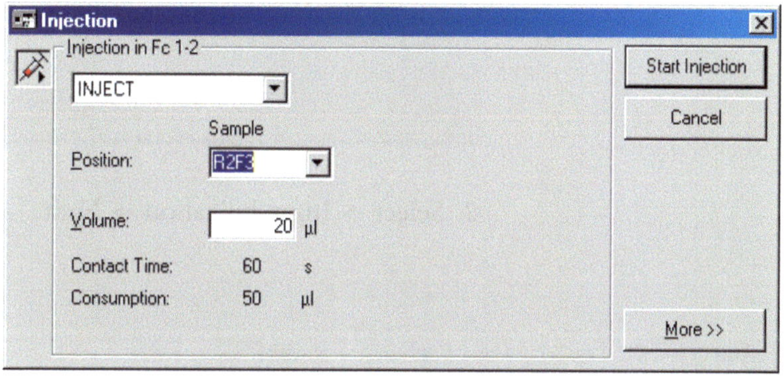

10. Hit -> Start Injection. You will be shown the Command Queue. While the first command is executed (indicated by the cogwheel icon), add two identical injections to the command queue.

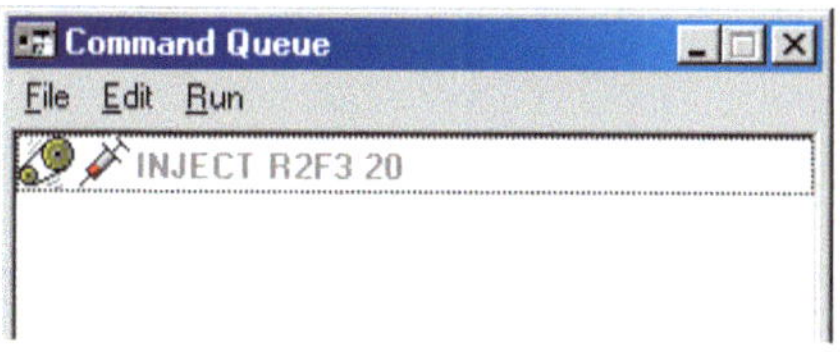

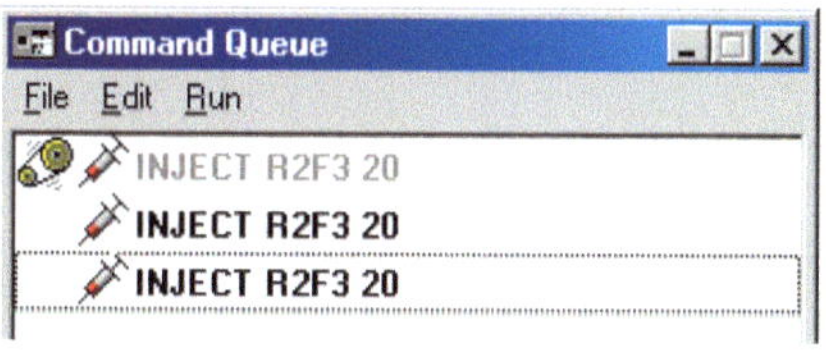

11. After the last injection finishes, hit -> Stop Sensorgram. The queue is cleaned.

Hit -> OK. An example sensorgram is shown in Fig. 3.

3.2 Coating a Chip with Degron Peptide

The recommended experimental design is to use flow cell 1 as a reference. We will describe a method using flow cell 2 for biotinylated degron peptide, keeping flow cells 3 and 4 in reserve (*see* **Note 8**). Coating is best run using the wizard, but may also be done with Manual Injections. You will need to consider surface coating density for kinetic experiments (*see* **Note 9**). Here we give conditions for high coating densities which are suitable for initial screening work.

1. Run -> Run Application Wizard. Select -> Surface Preparation -> Start.

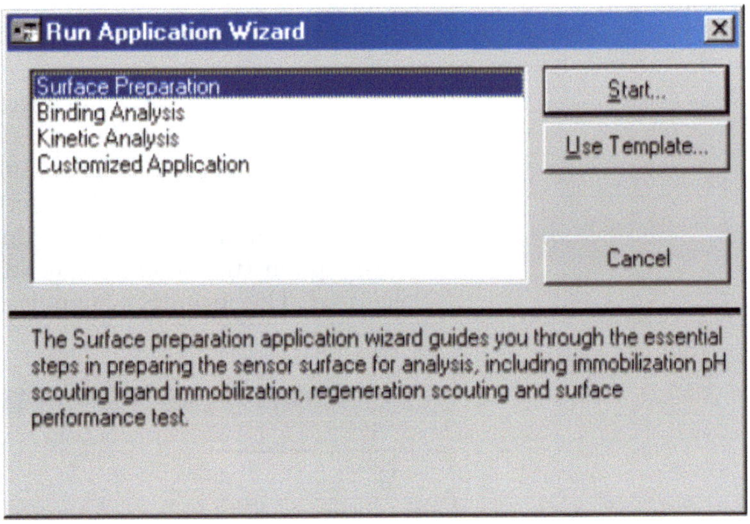

2. Select -> Immobilization -> Next.

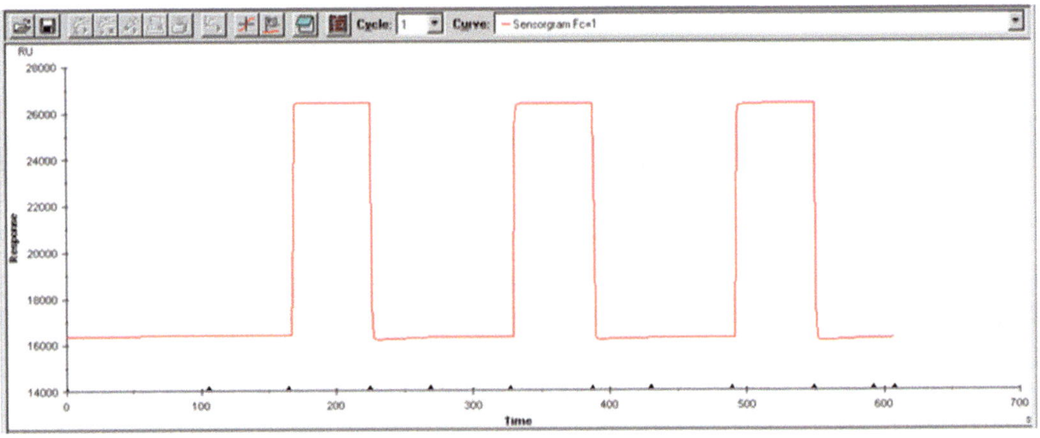

Fig. 3 Manual sensorgram collected during chip surface cleaning. Three 1-min injections are represented by three consecutive peaks. Data are shown for flow cell 1 only. You may also visualize sensorgrams for all channels simultaneously using the menu drop at the top right of the window

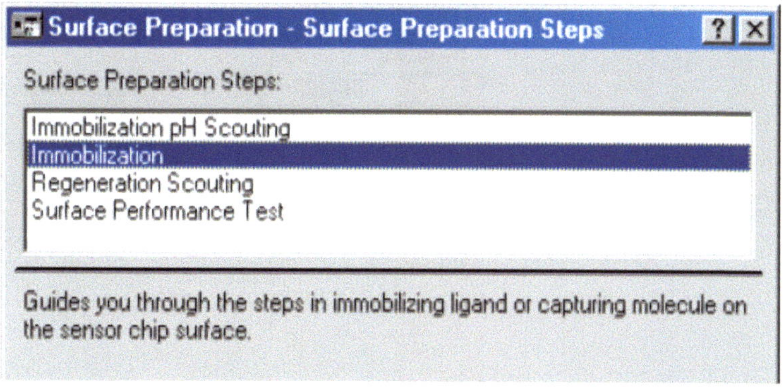

3. Sensor chip, select -> SA -> Next.

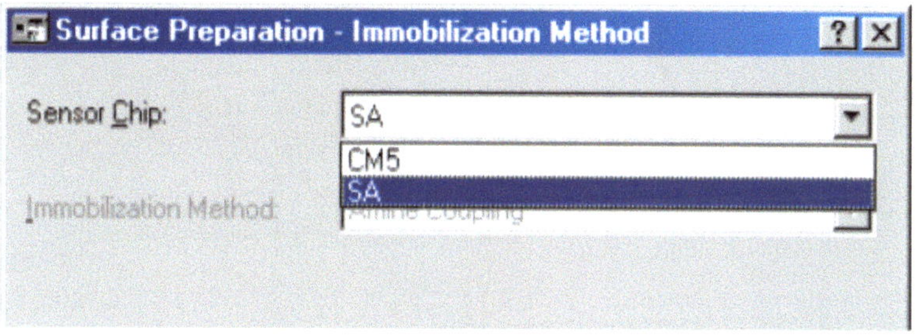

4. Enter the Ligand Name for each cell to be coated (*see* **Note 10**).
5. Enter Injection Time and Flow rate as shown. -> Next.

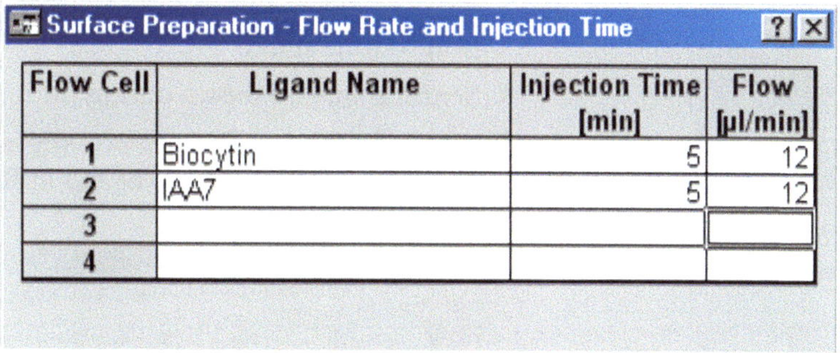

6. The consumption of solutions is calculated and shown in the next dialog box. Prepare the appropriate volumes of biocytin and biotinylated peptide (at 0.05 mg/mL, stock diluted 1 to 20 in water) and place them into the rack positions indicated. -> Next.

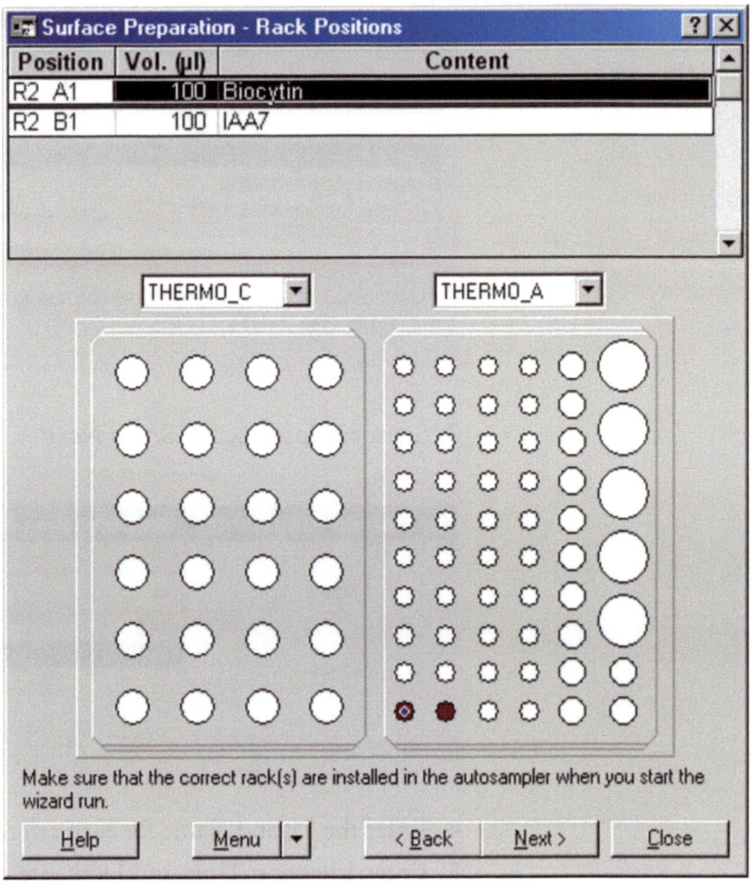

7. Provide a filename when prompted. -> Save. The immobilization routine starts.
8. Sensorgrams are shown during the immobilization run (Fig. 4).
9. When finished, a Results box records the amount of ligand bound for each flow cell/ligand in RUs (*see* **Note 11**).

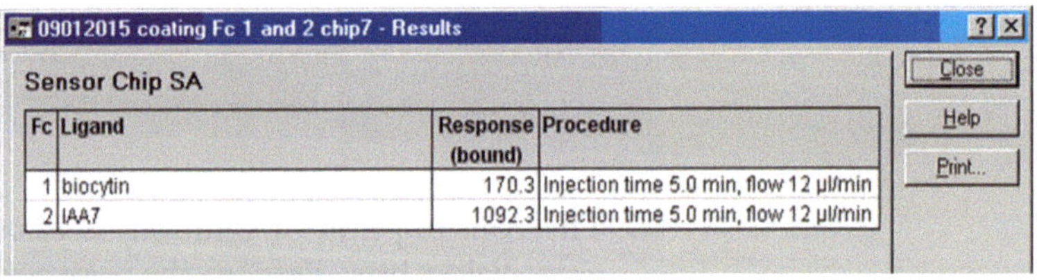
09012015 coating Fc 1 and 2 chip7 - Results

Sensor Chip SA

Fc	Ligand	Response (bound)	Procedure
1	biocytin	170.3	Injection time 5.0 min, flow 12 µl/min
2	IAA7	1092.3	Injection time 5.0 min, flow 12 µl/min

Close
Help
Print...

The bound RU represents the difference between the baselines before and after each immobilization. The report points are marked by an x (Fig. 4). Cells 1 and 2 are now ready for use.

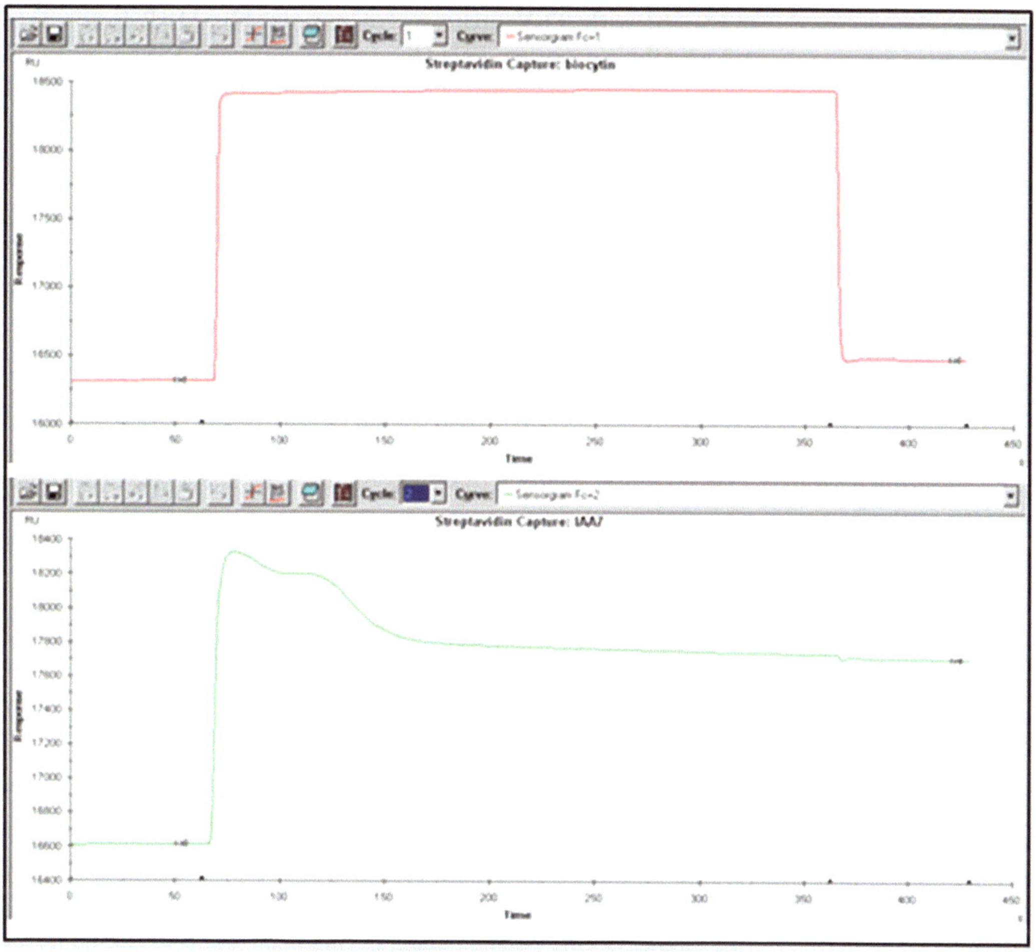

Fig. 4 Example sensorgrams for automated SA chip coating. For both flow cells it is clear that the signal does not return to baseline after the final wash step, but reaches a new steady level to show that the surface was successfully coated. The sensorgram for our degron peptide is typically "wavy"

3.3 TIR1 Activity Assay

The purpose of the assay is to test protein for auxin binding activity. This assay is a simple and effective procedure to confirm successful purification of active TIR1. Some typical analytical experiments are described below.

3.3.1 Sample Preparation

TIR1 protein is co-expressed with ASK1 in insect cells using the baculovirus system [7, 9] and purified using affinity chromatography. In brief, cells 3 days post-infection are harvested by centrifugation. Cells are lysed, the lysate cleared, and proteins purified by nickel affinity chromatography. The eluted protein is incubated overnight with TEV protease (to remove His-MBP fusion tags) before affinity capture of TIR1 (and associated ASK1) by FLAG affinity chromatography. Elution is in SPR buffer plus FLAG peptide.

1. In three labeled vials prepare the following mixes (*see* **Note 12**):

	Volumes (µL)				
Sample	**IAA[a]**	**DMSO**	**SPR buffer**	**TIR1[b]**	**Final volume (µL)**
Buffer blank	0	0	400	0	*400*
TIR1 + IAA	0.5	0	49.5	50	*100*
TIR1 − IAA	0	0.5	49.5	50	*100*

[a]From a 10 mM DMSO working stock this gives 50 µM IAA and DMSO is diluted to 0.5 %.

[b]TIR1 concentration does not need to be measured at this stage because we are looking for yes/no answers.

2. Mix samples thoroughly by gentle vortexing, cap, and keep on ice.

3.3.2 SPR Setup

1. Cease Standby mode -> Stop.
2. If necessary, Dock your SA chip and Prime.
3. Run -> Run Application Wizard. Select -> Binding Analysis -> Start.

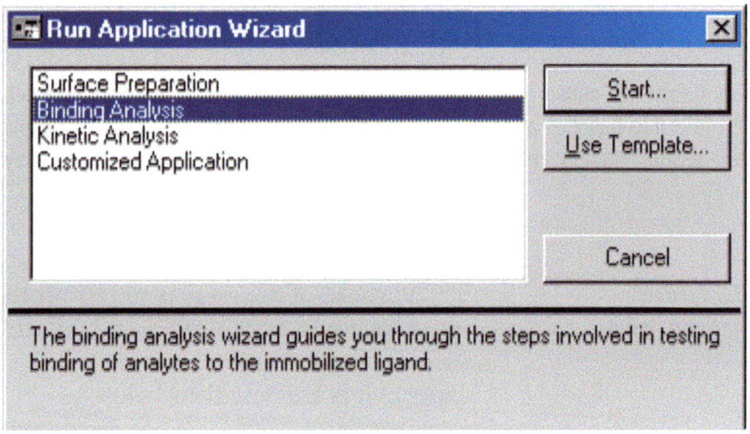

4. Select -> Direct Binding, as we have already prepared the chip. -> Next.

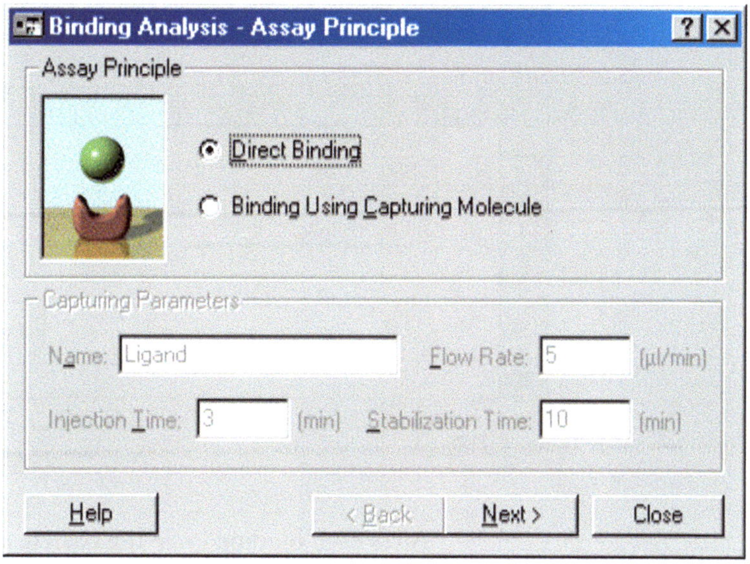

5. Use Flow Cell(s) -> "2 with 1 as reference."
6. Flow Rate -> 20 μL/min. The rest of the parameters are left as defaults. -> Next.

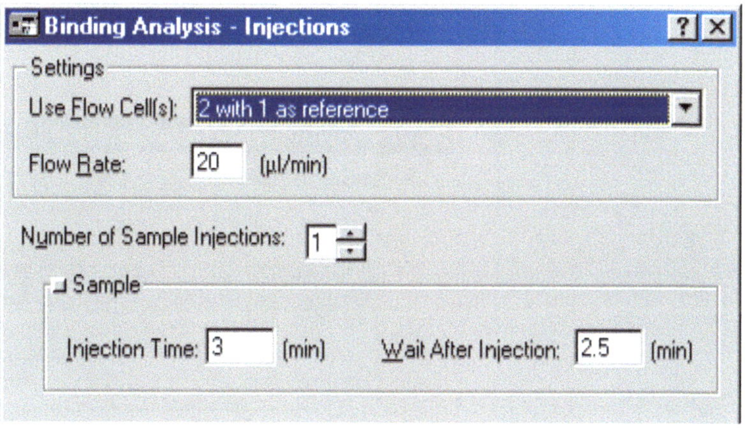

7. Ensure that Run Order is "As Entered" (*see* **Note 13**).
8. You need to specify how many times each sample is assayed. In the column Repl., enter 1 against each sample except the buffer blank (*see* **Note 14**), as shown.
   -> Next.

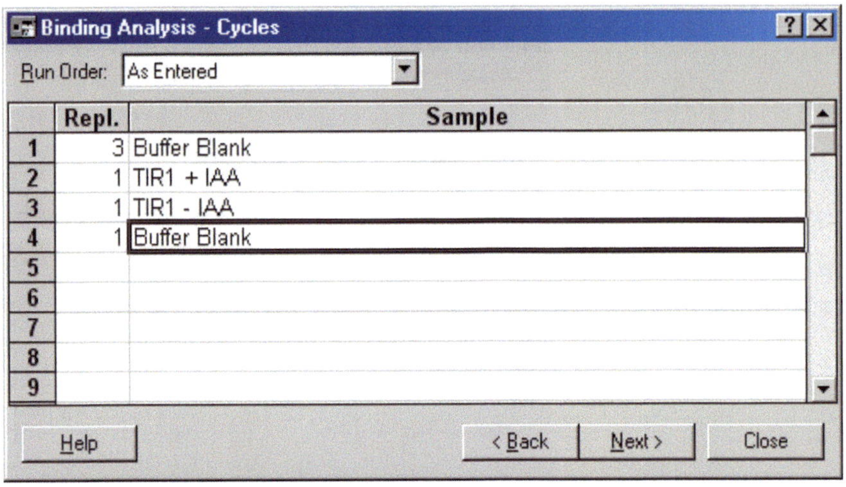

After each binding cycle the coated surfaces need to be cleared of bound analyte, a process referred to as Regeneration. We use 50 mM NaOH. Use default values for the regeneration routine.
Regeneration Method -> Single Injection,
Flow Rate - 30 μL/min,
Solution - 50 mM NaOH,
Injection Time - 30 s.
-> Next.

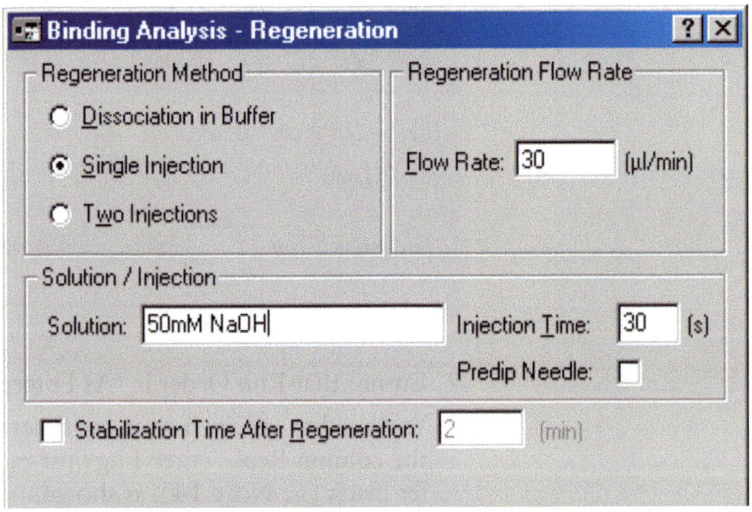

9. Rack Positions for all sample tubes are pre-selected, but you can change them by drag-and-dropping onto the desired positions. Make sure that vial contents correspond to positions. -> Next.

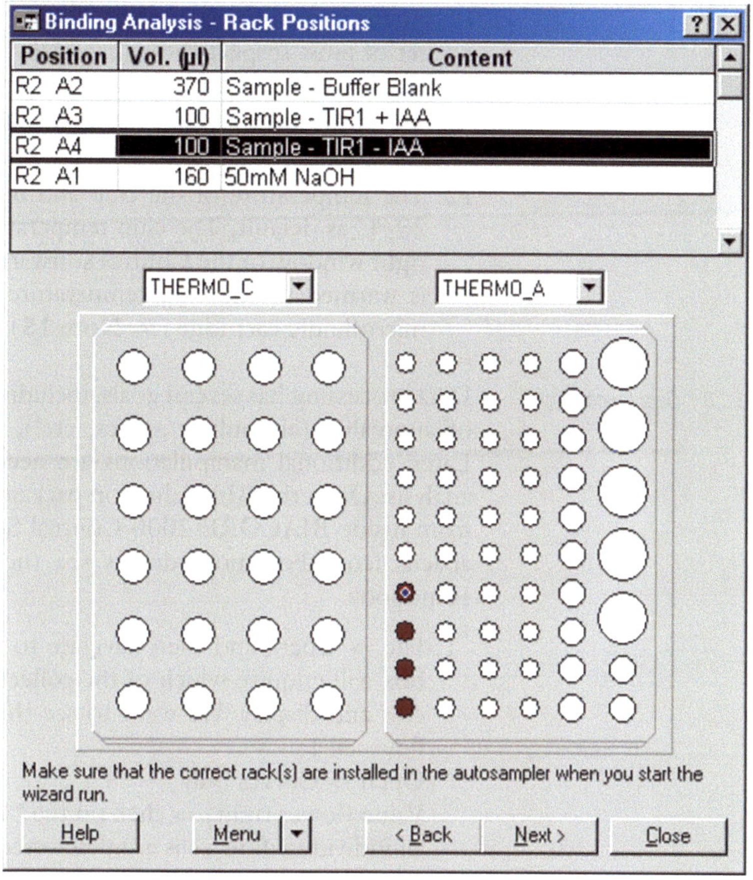

10. This is your last chance to check the vial contents, volumes, and positions.
Ensure the "Standby Flow After Run" box is checked (this is the default).
Click -> Start, and enter a filename when prompted.

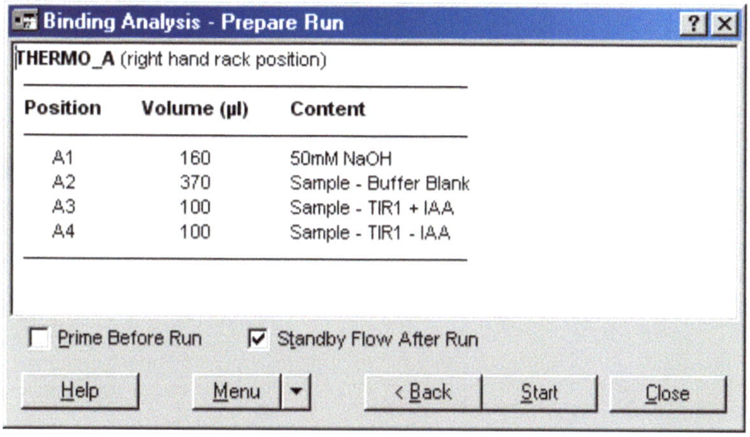

11. At the end of the run you will see a data summary box with a set of tabs. Inspect these (especially the baseline drift report), but you will get a better picture of your data by opening the .blr file in BIAevaluation software (*see* Subheading 3.3.3 below).
12. The temperature of the chip and binding experiment will be 25 °C as default. The chip temperature is shown in the lower right window of the Control software at all times. Each analyte is warmed to the chip temperature as it passes through the microfluidic cartridge (*see* **Note 15**).

3.3.3 Data Processing

Data processing has several goals, including normalization, removal of anomalies (air bubble spikes, etc.), and baseline adjustment. Later, additional manipulations are needed for successful kinetic analysis. Open the BIAevaluation package. This can also be started from inside BIACORE 2000 Control Software by hitting on the abacus icon. For more details see the BIAevaluation Software Handbook.

1. File -> Open, and then navigate to your results file. A dialog box will enquire which of the collected data you want to process and display. We want to see the binding with respect to flow cell 1.
   Open -> Curves only
   Value (lower right) -> click on the "2 - 1." The data files from all individual flow cells are deselected. You see only Fc2 - Fc1.
   Curves -> use the mouse and Ctrl to select all you want to consider, click -> OK.

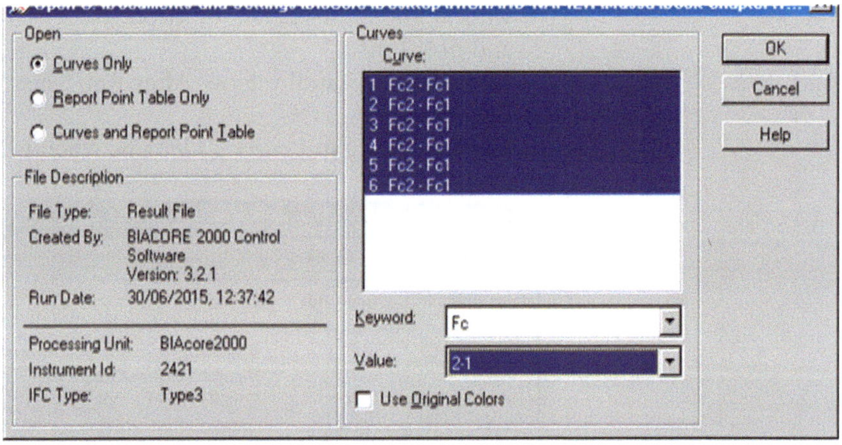

2. You will see in the Project window the names, IDs, and Sources of the data you are going to process.

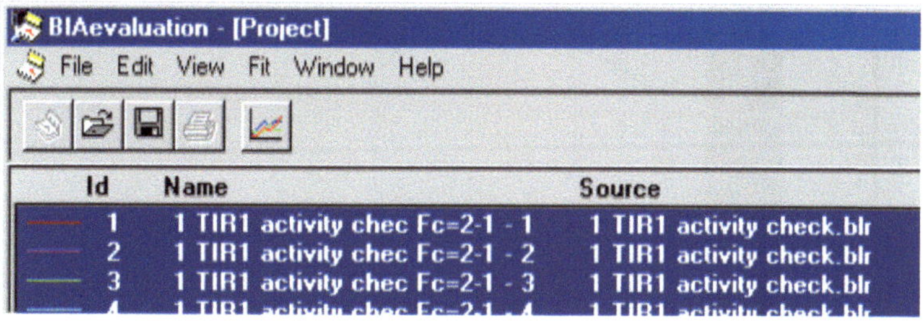

3. Each curve is given an automatic name consisting of the filename of the results file, the cell selection (i.e., Fc=2-1), and the order in which they were run. To edit the name and display or change features of a binding curve, double left click on the respective title and work within the Curve Properties dialog box.

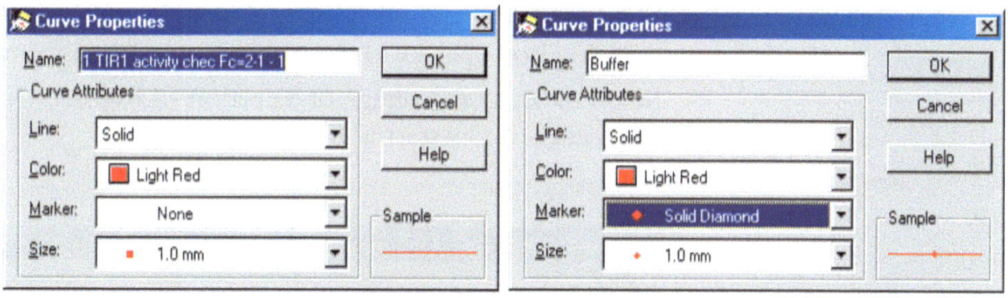

The end result will appear as shown below.

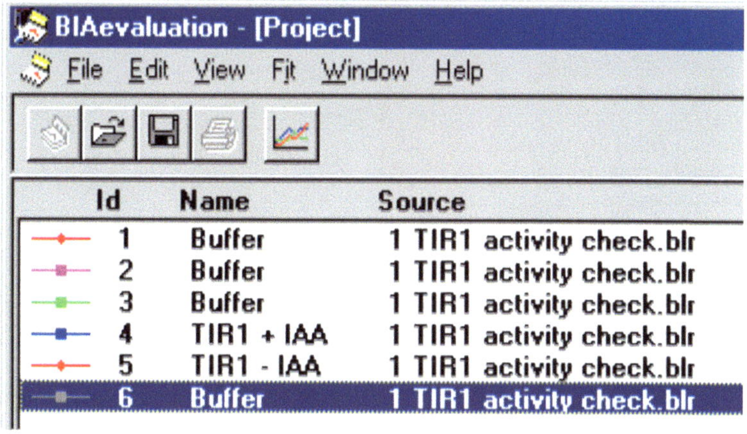

4. Select the curves [TIR1 + IAA], [TIR1 – IAA] and one of the Buffer traces and click on the graph icon button. The graphs that will open would look as follows. The axes auto-scale and spikes from injection start and stop points will dominate. We will cut these from the data after normalizing the graphs.

4.

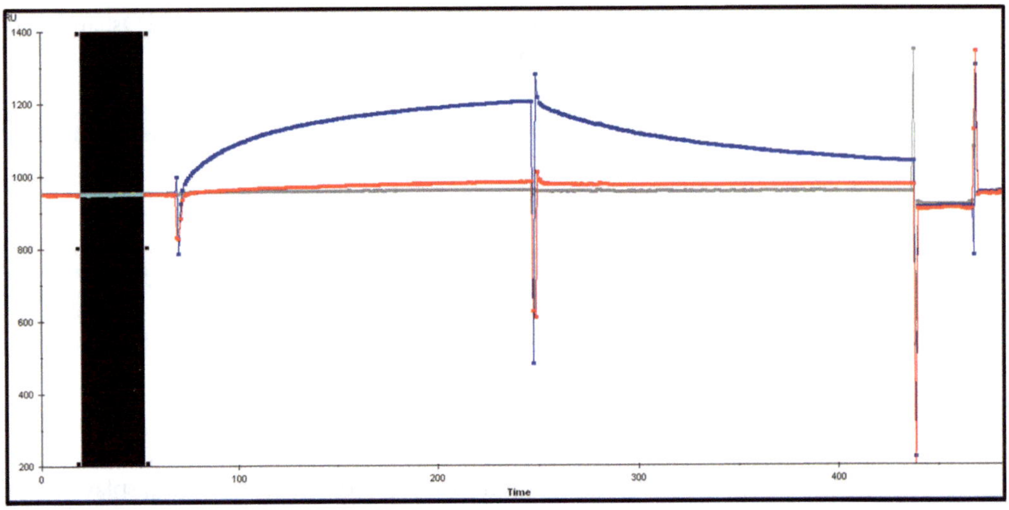

5. To normalize the curves, right click and drag over the appropriate baseline area. Click -> Y-Transform button.
6. A dialog box opens with several choices of transformation.

   Select -> Zero at Average of Selection -> Replace Original. Both curves now start at 0 RU.

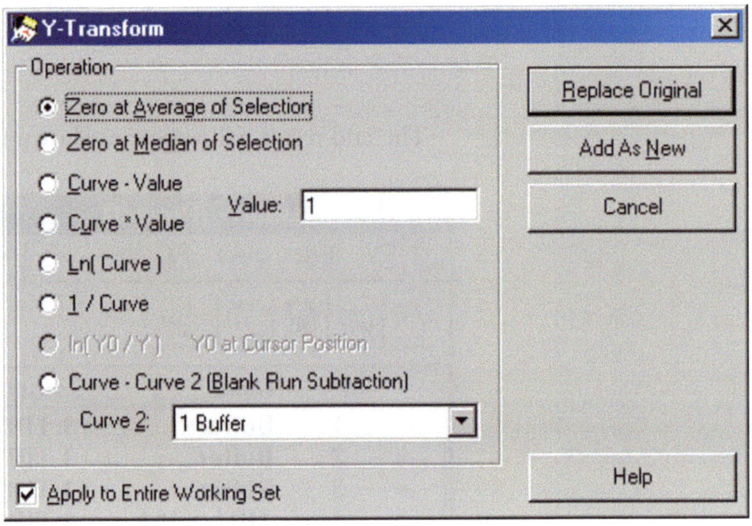

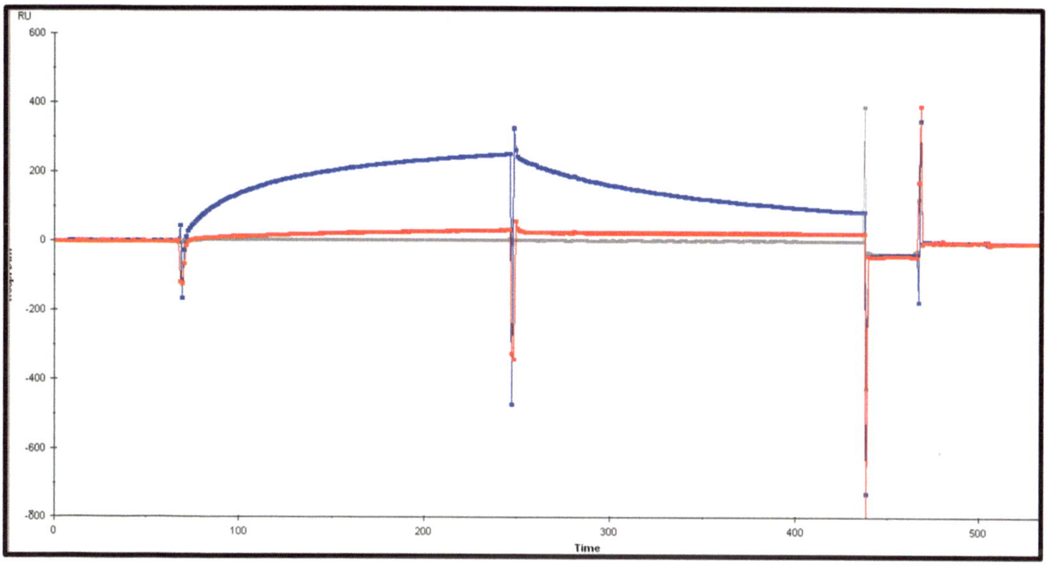

7. Cut out the spikes caused during regeneration and/or at the start and end of injections. Use right mouse button to make a selection, as above.
   Click on Edit -> Cut. Data from all curves will be edited simultaneously.
   Note that this editing requires three separate operations.

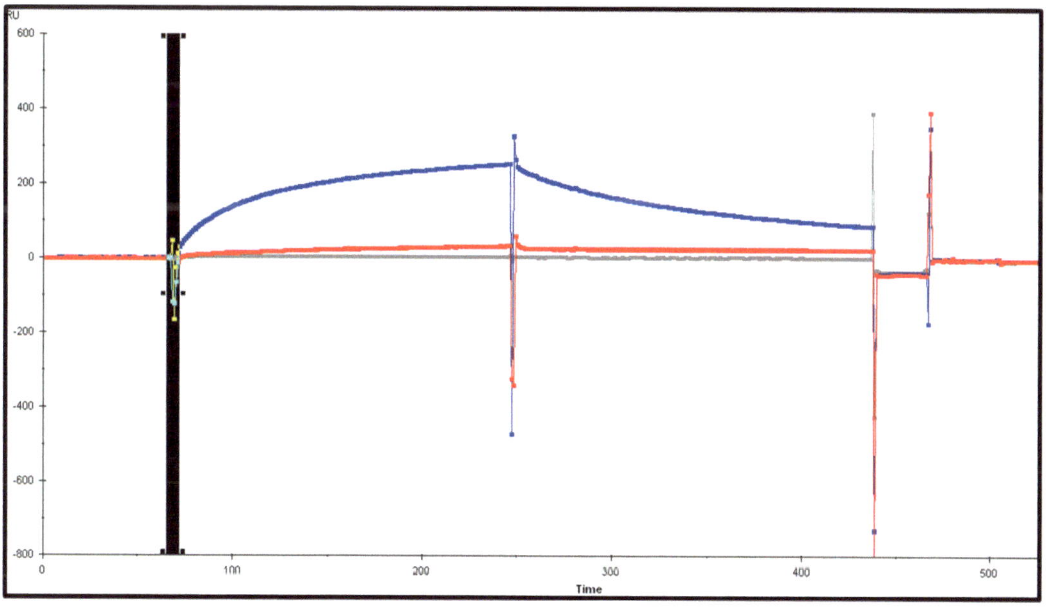

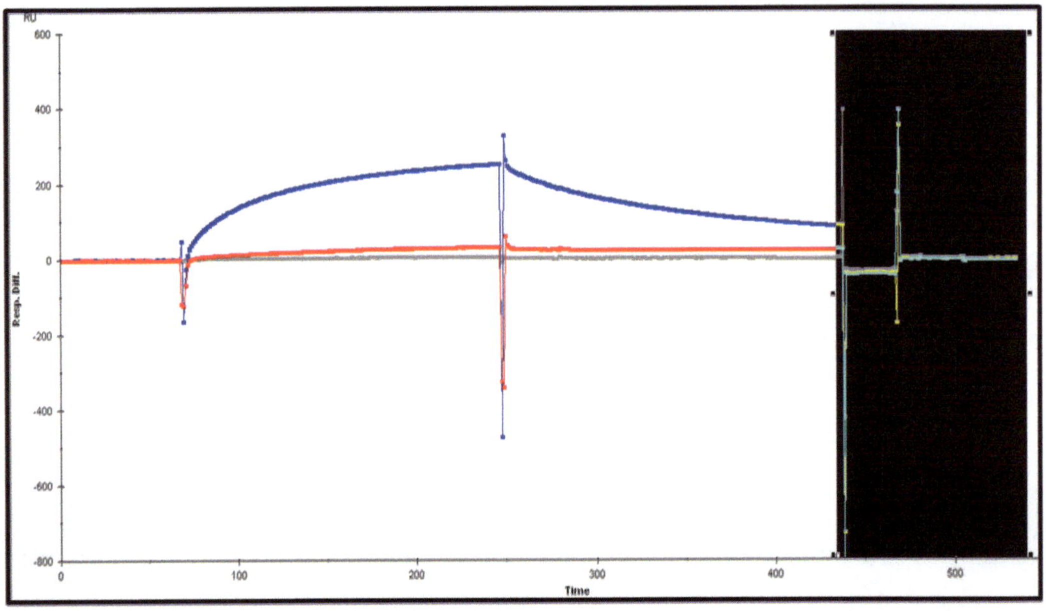

The end result should look as follows.

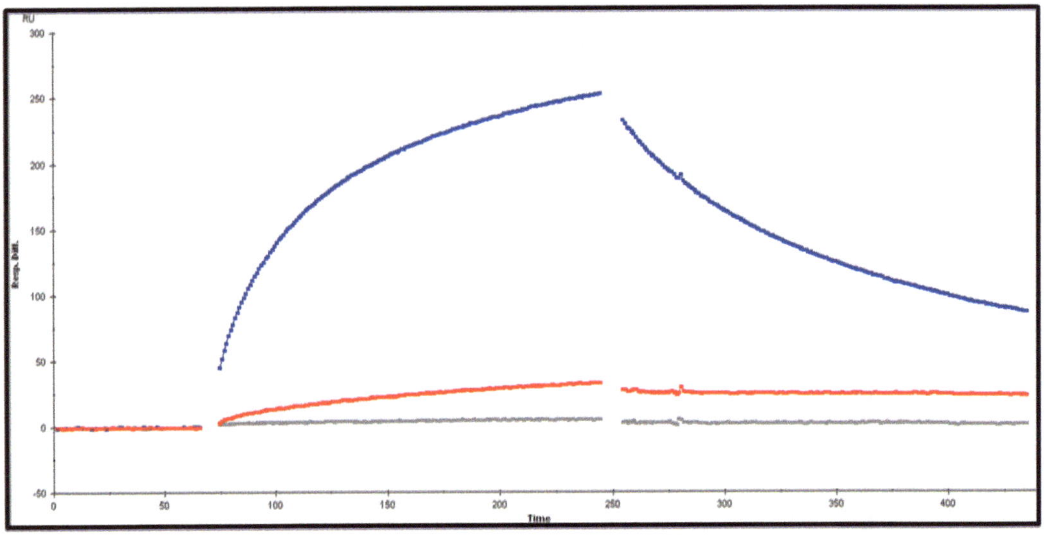

8. If there is some background binding in the absence of IAA, you may wish to subtract this from all the binding curves to leave graphs of auxin-dependent binding.
Click on -> Y-Transform and select "Curve-Curve 2."
Select "TIR1 - IAA" and -> Replace Original.

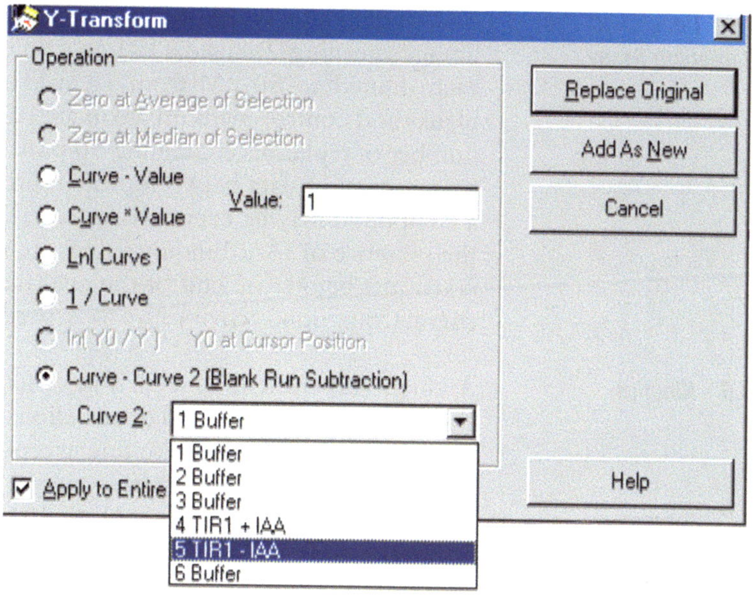

The end result should look as follows.

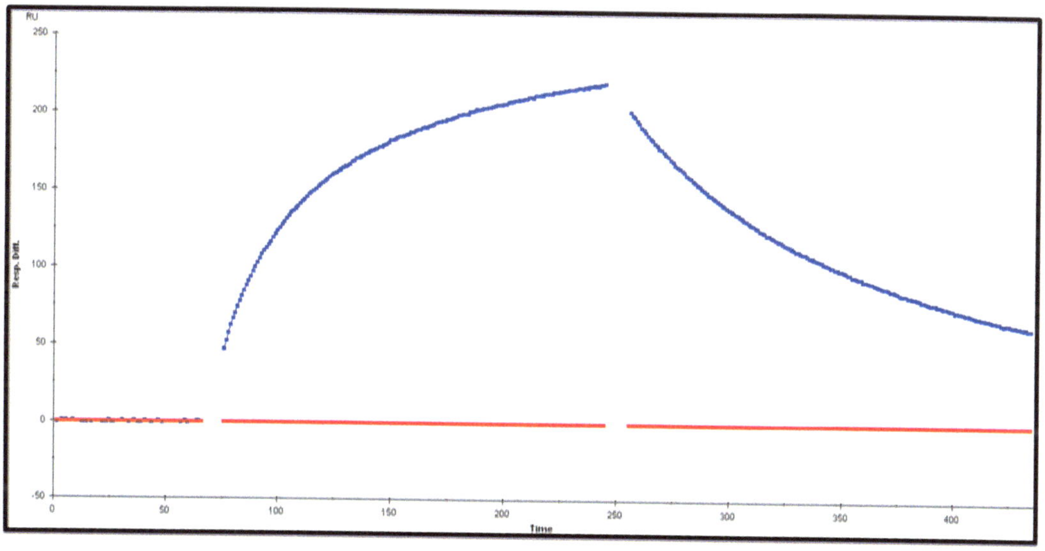

9. The data are now processed and clearly show auxin-dependent TIR1 activity. In this case R_{max} is around 200 RU (*see* **Note 16**). This rapid assay confirms that the protein is suitable for further experimentation, such as compound screening or kinetic analysis (*see* **Notes 17** and **18**).

3.4 Compound Screening Assay

The binding assay described above is readily adapted for compound library screening. Purified TIR1 protein is mixed with compounds, each diluted to 50 μM from 10 mM stocks in DMSO. Buffer blanks and controls with no auxin are run in every set, as well as a number of replicates containing 50 μM IAA as positive controls to help normalize data between protein preparations. The activity of a compound may be expressed as, e.g., a percentage of binding in the presence of IAA. Binding values are taken from a report point taken just before the end of the association phase of the binding curve. Compound activities are loosely classified as shown in Fig. 5.

3.5 Kinetics

A kinetic experiment is needed in order to get values for affinity and rates of association and dissociation. For kinetic work we need to ensure that no factor is limiting except one variable, the concentration of analyte. In particular, we need to ensure that there is no mass transport limitation (*see* **Note 19**), and we start by limiting coating of the chip surface (*see* **Note 9**) such that R_{max} is 300 RU or lower (on Biacore 2000 and 3000 instruments). The kinetic analysis wizard on the Biacore 2000 only allows use of flow cells 1 and 2 (Fc2-1) in a kinetic run (*see* **Note 20**).

1. Run -> Run Application Wizard -> Kinetic Analysis.

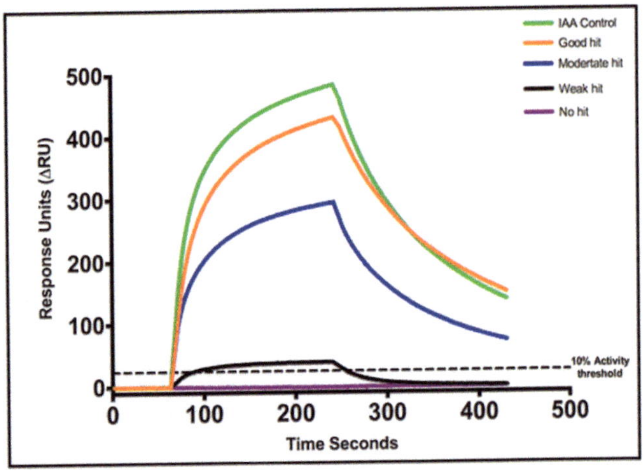

Fig. 5 Classification of binders based on % of IAA's maximum binding. Data has undergone additional processing with Prism version 6.00f for Mac OS X, GraphPad Software, La Jolla, California, USA, www.graphpad.com

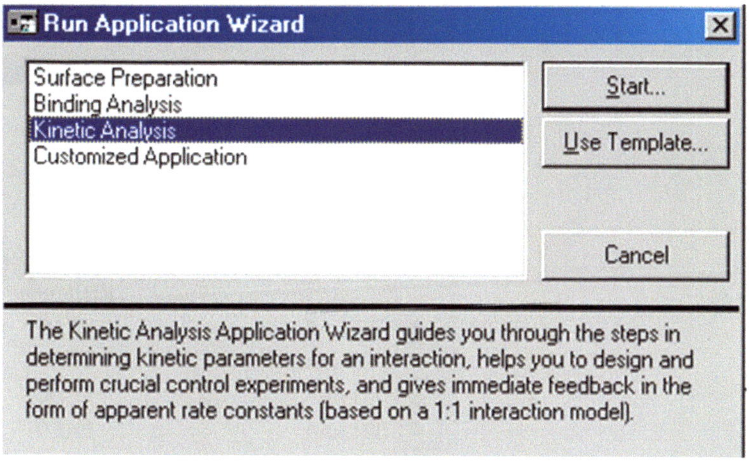

2. -> Concentration Series (*see* **Note 21**).

   You can choose to run two additional control experiments at the same time (*see* **Notes 19** and **22**).

3. -> Direct Binding.

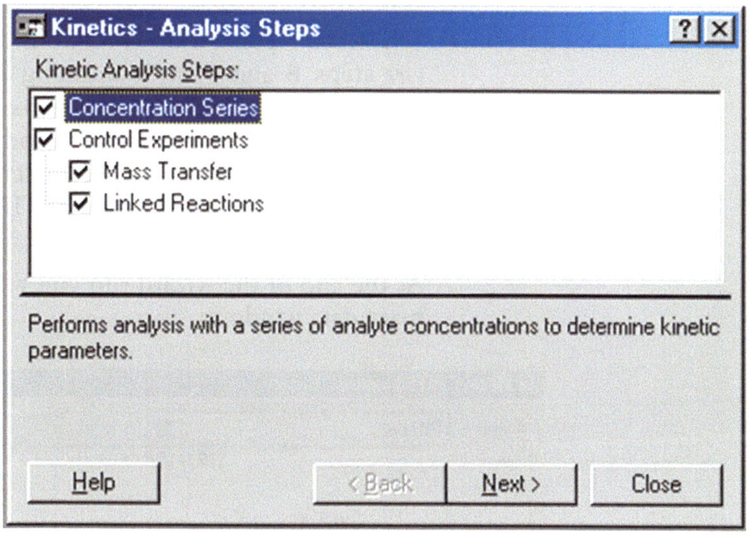

4. You will be prompted to fill in the concentration series for your chosen analyte molecule. Input the molar concentrations and the molecular weight of analyte (*see* **Note 21**).

5. Use Flow Cells -> 2, using 1 as reference (*see* **Note 20**).
   Flow Rate -> 30 μL/min (default) (*see* **Note 19 and 22**).
   Stabilization Time -> 0 min (default).
   Injection Time -> 3 min (default).
   Dissociation Time -> 15 min (default; *see* **Note 23**).

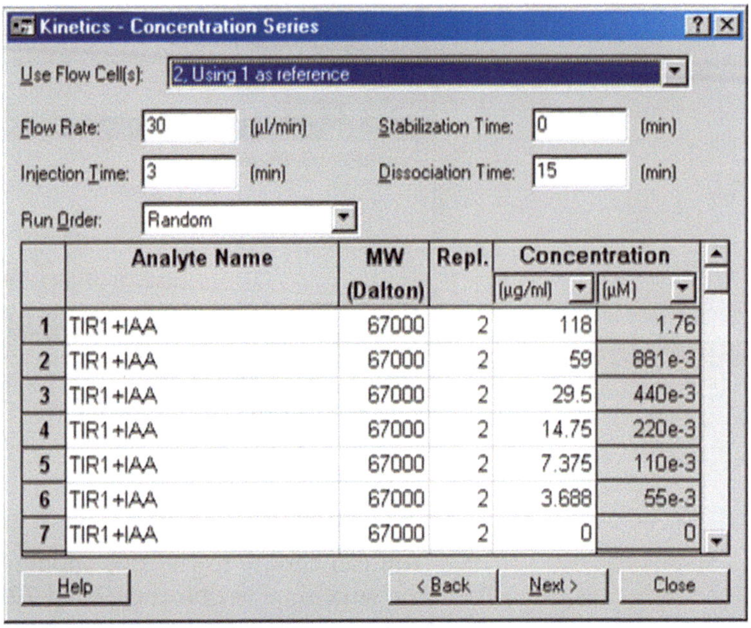

Run Order -> Random (default)

6. Select the regeneration conditions and sample rack positions (*see* **steps 8** and **9** in Subheading 3.3.2). Check the sample summary sheet. -> Start -> Filename.
   The wizard runs the experiment and presents you with a summary table of kinetic parameters, statistics, etc., all based on a 1:1 Langmuir kinetic model. For TIR1 and IAA this should be appropriate.
7. At the end of the wizard run you will be presented with summary data windows.

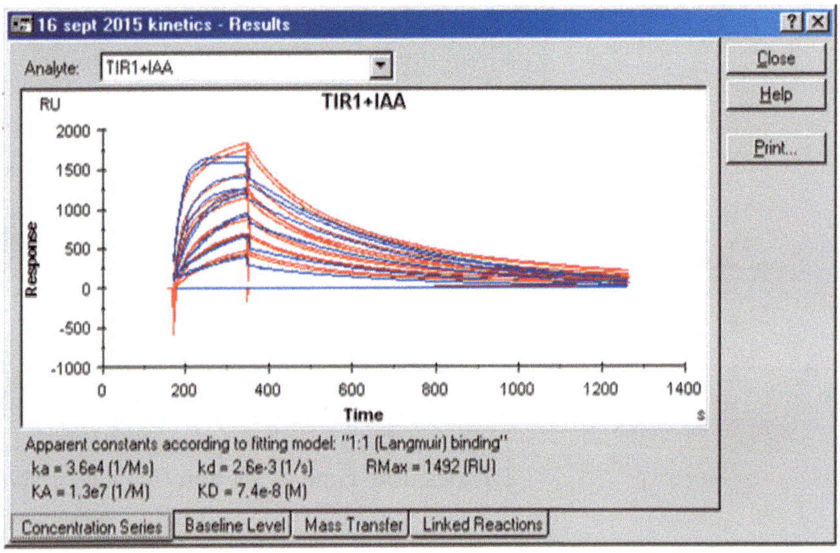

The first summary is the kinetic fit (in blue) to the data (in red). The kinetic parameters of the fit show that the affinity was calculated to be $K_D = 7.4e\text{-}8$ M (74 nM). In this example the TIR1 protein has been mixed with saturating IAA (100 μM) and the protein is titrated in a dilution series against the degron peptide on the chip. You will see that some repetitions were built into the experiment. In this case the raw data and model curves are not perfectly matched. You may use the other report tabs to explore why this might be.

-> tab Baseline Level.

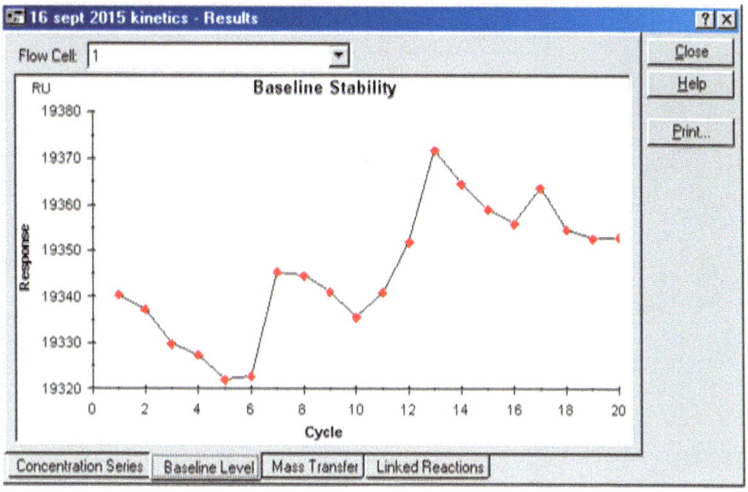

8. You will see that this report shows how much the baseline shifts over all the injections for Fc1. In this case there has been some drift, but this is not systematic, nor large (note RU scale). -> Mass Transfer.

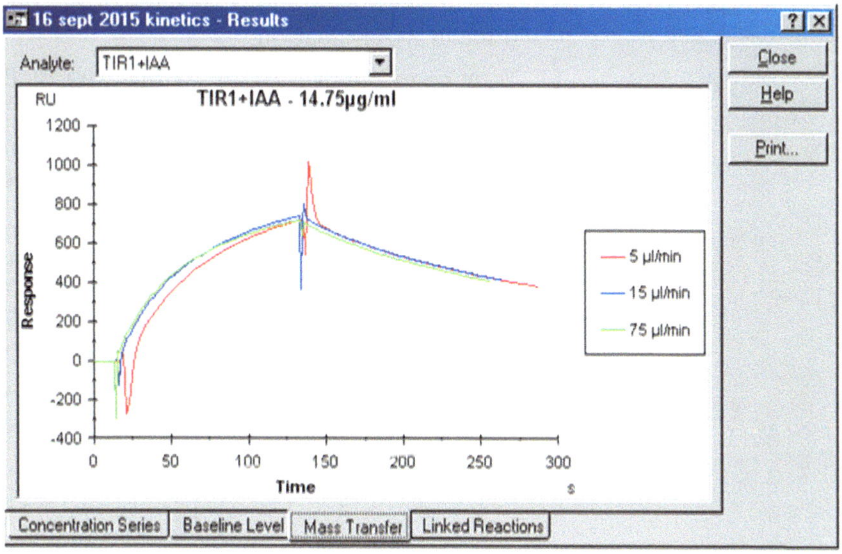

9. This report shows the results from the optional mass transport limitation experiment at three flow rates (*see* **Notes 19** and **22**). In this case the binding curves almost overlap, but the red curve is slightly displaced, indicating some reduced binding early in the injection cycle at this low flow rate. Therefore, there was some limitation to binding at the slowest flow rate, but the curves for 15 and 75 μL/min are essentially identical and so you could choose any flow rate greater than 15 μL/min for kinetic work on this chip. -> Linked Reactions.

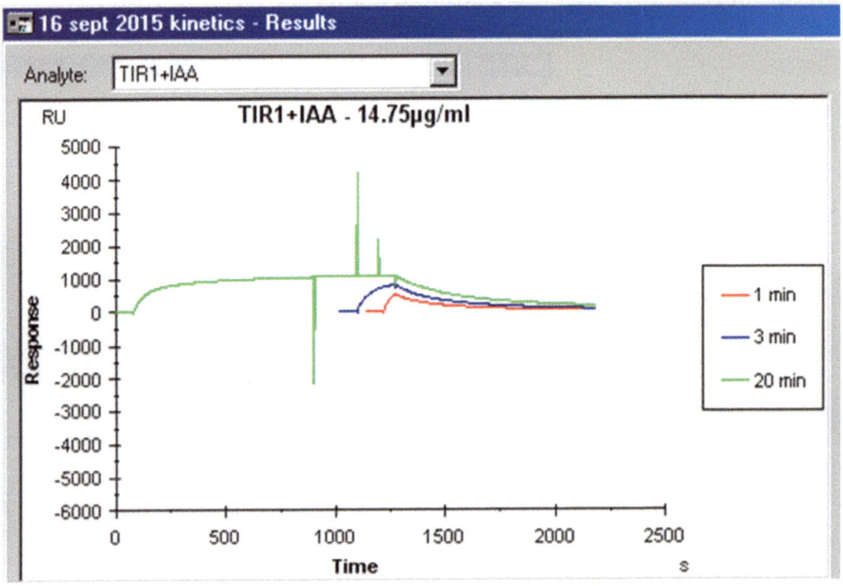

10. Linked reactions are asking whether your binding interaction changes the reagents by comparing off-rates after different reaction times. In this example it is difficult to tell and if you want a more quantitative analysis of the off-rates, this can be done in BIAevaluation.
11. Your own kinetic analysis of the data in BIAevaluation will allow you to clean the data, and select the best time windows for fitting, and the option of selecting alternative kinetic models. BIAevaluation also allows kinetic analysis using flow cells 3 and 4.
12. Data processing for kinetic experiments in BIAevaluation is similar to that shown in Subheading 3.3.3 above. Particularly important is the selection of curves and subtraction of your chosen blank(s). Create a working set as an Overlay plot. There are a number of kinetic models that can be applied to data. TIR1 has a single binding pocket for auxin and so we will use the 1:1 Langmuir model (*see* **Note 22**).

13. Start the kinetic evaluation wizard by choosing Fit -> Kinetics Simultaneous ka/kd, or using the toolbar button with the sensorgram icon.

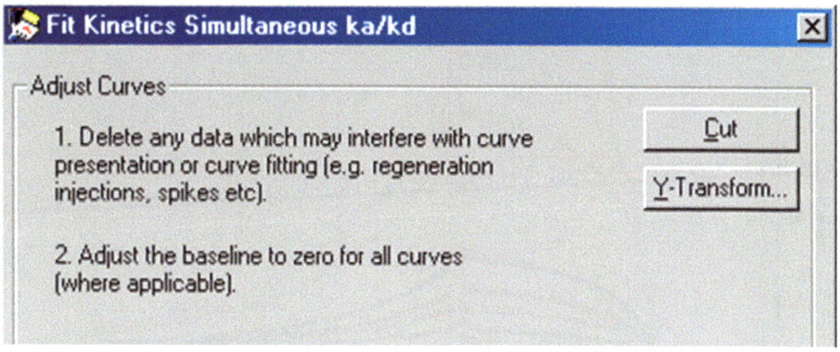

14. Unless you have edited the curves beforehand, use Cut to process the curves, remove regeneration spikes, etc.
15. Subtract the chosen blank -> Y-Transform and select "Curve-Curve 2."

    Select your blank sample without the analyte "IAA 0" and -> Replace Original.
16. You will see a dialog box headed Select Data and a time bar appears at the top of the sensorgram screen.

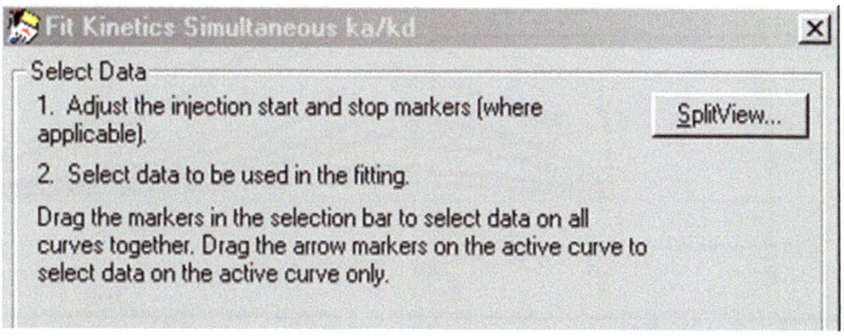

    The black lines should be dragged to the times of the start of the injection and the end of the injection. The grey windows in the bar are then dragged to select the time periods you want to select for kinetic modeling. Use as much of your data as is reasonable. You will see the kinetic model you have selected displayed at the top of the window. You may identify which curve is which by clicking on it and the identifier will be displayed with color code at the top right of the window.
    -> Next will produce a table in which you will see the concentration data for the analyte (as you entered the data in **3.5.5** above). The software is ready to fit.
17. Hit -> Next -> Filename -> Start. You will see the iterative fits on screen.

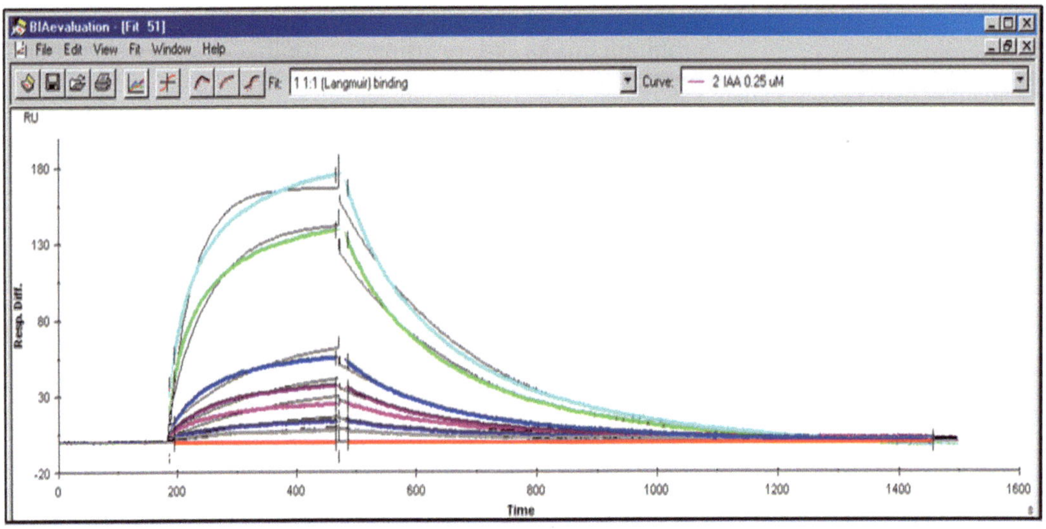

18. At the end of the fit, the first report window shows a table displaying the fitted parameters with statistics. The Selections tab will review the data selection (as above). The fitted graphs are superimposed on the colored data curves. On the Report page you are seeking a low chi^2 value and you may close any fit and change the data selection windows, datasets, etc. until you are satisfied with the result.

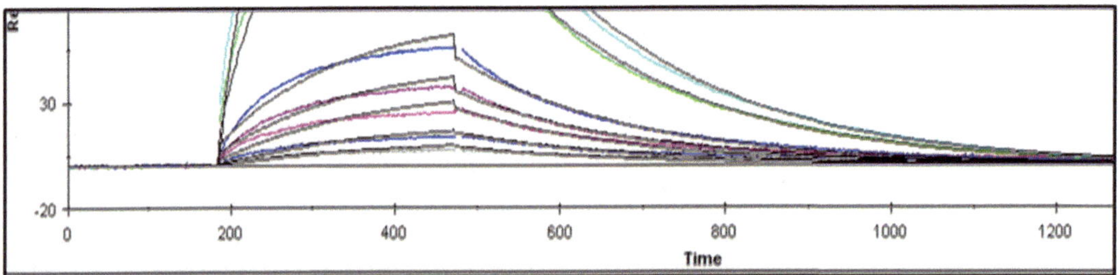

	ka (1/Ms)	kd (1/s)	Rmax (RU)	RI (RU)	Conc of analyte	KA (1/M)	KD (M)	Req (RU)	kobs (1/s)	Chi2
	1.93e3	4.69e-3				4.12e5	2.43e-6			5.37
IAA 0.25 uM			378	3.08	2.5e-7			35.3	5.17e-3	
IAA 5 uM			189	17	5e-6			127	0.0143	
IAA 10 uM			198	7.41	1e-5			159	0.024	
IAA 0.125 uM			428	0.766	1.25e-7			21	4.93e-3	
IAA 0.5 uM			258	6.57	5e-7			44.1	5.65e-3	
IAA 0.06125 uM			495	0.381	6.25e-8			12.4	4.81e-3	
IAA 1 uM			207	10.6	1e-6			60.3	6.62e-3	
IAA 0 uM			0	-8.48e-168	0			0	4.69e-3	

This example shows an experiment in which IAA was titrated against a fixed TIR1 protein concentration. The chi^2 value of 5.37 is reasonable and the model calculates the affinity K_D to be 2.43e − 6 M (2.4 μM), which reflects the affinity of TIR1 for IAA on its own (*see* **Note 24**).

19. -> Residuals shows how well the data fit across the selected windows.

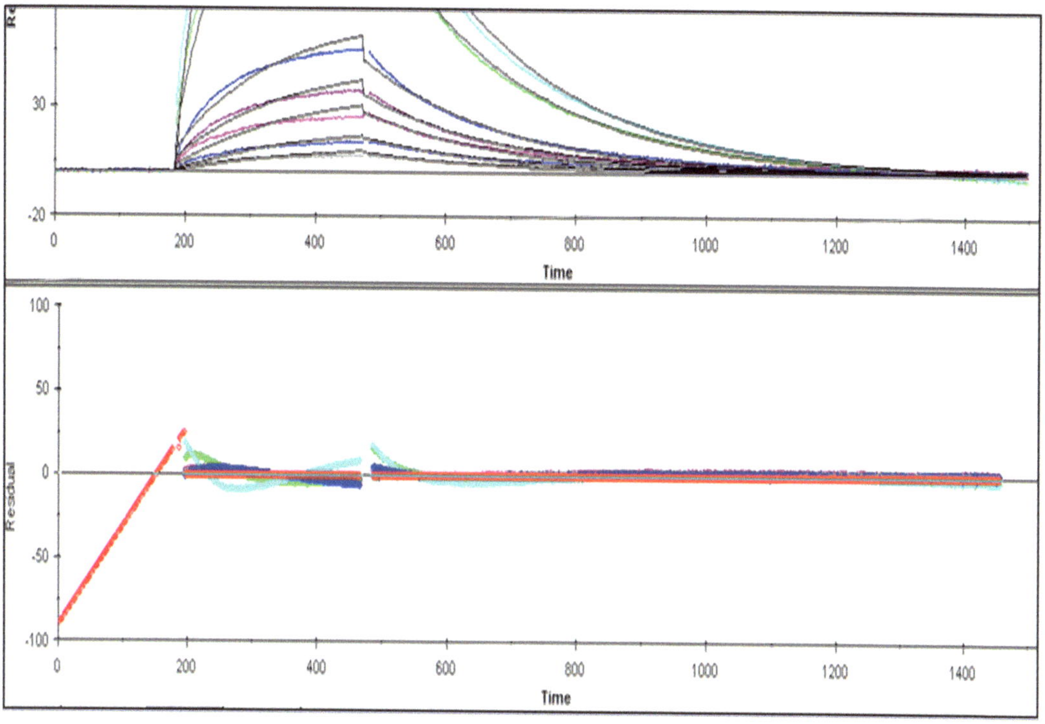

20. The residuals from each binding curve and fit are plotted. Ideally they all fall on the zero line. Here residuals deviate more as the concentration of analyte rises, and more towards the start of each phase (association and dissociation). R_{max} in this example is 378 RU and so all residuals are less than 10 % R_{max}, most below 5 %.

3.6 Thermodynamics with SPR

1. As noted above, thermodynamic values may be derived from SPR experiments although ITC will give values for the reaction in solution. The thermodynamic experiment is a kinetic experiment run at a series of chip temperatures. Biacore instruments can all control microfluidic and chip temperatures to within 0.1 °C over the range of 5–45 °C. Given that the transition from one temperature to the next can take many tens of minutes and that kinetic experiments all require long runs, allow plenty of time for thermodynamic work. Depending on the availability of your protein you may choose as many temperatures for the runs as you wish, but intervals of 10 °C should give reasonable data for Van't Hoff analysis. The wizard will give summary report windows, as for kinetics above.

3.7 System Maintenance and Closure

1. For all the wizard routines, unless you have unchecked the default box for Standby Flow After Run, the SPR buffer solution will be passed over your chip at 5 μL/min for up to 4 days. The degron peptide chips are stable over many rounds of binding and regeneration, but once experiments are concluded the chip is undocked and replaced with a Maintenance Chip.
2. Once undocked, rinse the degron-coated chip with deionized water and then dry the surface under a stream of dry gas. Store in a 50 mL Falcon tube at 4 °C.
3. Replace the SPR buffer with SPR rinse solution and rinse or prime the system. This washes any residual salts from the fluidics. The instrument may be left on Standby with SPR rinse solution, but if Desorb has not been run for 4 weeks run a Desorb routine. Ideally, each user will follow their runs with a Desorb cycle to ensure that the SPR is ready and clean for the next user. Make sure that Desorb is run only with a Maintenance Chip docked.
4. Select -> Desorb from Working Tools (*see* **step 4** in Subheading **3.1**).
5. If longer term storage is anticipated, follow Desorb with Sanitize.

4 Notes

1. The surfactant concentration is 10x higher than the suggested one in the ready-made buffer from GE Healthcare (0.005 % v/v). We found out that while having no negative effects on binding, it does improve the smoothness of the signal and reduces bubble generation especially in instruments where fluidics are not brand new.
2. Always Prime when exchanging the running solution, even between two batches of SPR buffer. This exchanges the fluid in the pumps, flushes air bubbles, and washes the chip.
3. We strongly recommend that you condition the new chip before coating by running this extra cleaning step. It removes from the binding surface debris left from manufacture.
4. The Detection dialog box allows each cell to be selected independently. If you are cleaning flow cells 1 and 2 you select the option Fc 1-2. The reference subtraction-none means that the signal from both cells will be displayed separately. The subtracted 2-1 signal is not helpful when cleaning. Sensorgrams will be recorded only from the flow cells you select.
5. All files with extension .blr contain raw results and can be opened and post-processed with BIAevaluation software as will be shown.
6. Notice that the two flow cells of a brand new chip give different absolute response units (RU) when same fluid is moved

over them. These units are arbitrary. When analyzing the data what matters is the response difference.

7. The manual run can be adapted for other occasions when manual control over the process is preferable. Select combinations of commands via the buttons to create appropriate command queues.
8. Heavy usage with flow cells left uncoated tends to allow debris to accumulate over them and impair subsequent coating. While economical to hold open cells 3 and 4, remember that their performance may deteriorate if held for long periods. If you know what you want to coat 3 and 4 with, we recommend that all flow cells are coated at the same time.
9. You may vary surface coating density by diluting the biotinylated reagent or by reducing coating time. In practice with biotinylated reagents, the affinity between biotin and streptavidin is so high (10^{-15} M) the better strategy is dilution. Our biotinylated peptide stocks at 1 mg/mL are diluted 20-fold for saturating chip surfaces, but diluted up to a further 1000-fold for kinetic experiments for which mass transport limitation (*see* **Note 18**) needs to be avoided. You may also wish to use the Surface Preparation wizard and the facility for Target coating. In this you provide a diluted biotinylated ligand and enter an RU value that you want to achieve. The software will then sip, wash, and check as coating progresses, with short injections used to allow ligand binding up to the target value.
10. The molecule captured on the chip is referred to as the ligand in Biacore terminology. The molecule in solution which binds to the immobilized ligand is referred to as the analyte.
11. Although arbitrary, RUs scale directly with mass and you will see that far more peptide appears immobilized than biocytin. In fact, more moles of biocytin are immobilized than the peptide, probably because the small biocytin can access more available sites on the surface.
12. For TIR1 + IAA and TIR1 - IAA we recommend pipetting the IAA in DMSO first onto the walls of the sample tube followed by the buffer; this will ensure better mixing.
13. In more advanced analyses you may, e.g., randomize sampling.
14. We recommend that assay runs are preceded by at least three buffer injections to ensure that the cells are purged and stabilized. The regeneration of the coated surfaces between assay cycles will remove some loosely associated ligand on the first runs and a few regeneration cycles settle the surface before accurate measurements are taken.
15. We recommend that the sample racks are chilled to 10 °C if you have a series of experiments to run, but for a rapid activity

assessment this may not be necessary. Rack temperature is controlled externally on the Biacore 2000.

16. R_{max} is the value at the top of the binding hyperbola and represents full binding site occupancy. This value is important in kinetic experiments.
17. The data can also be exported for processing in you favorite software. Select File, then -> Export.
18. Files from BIAevaluation are saved as .ble files.
19. Mass transport limitation: If the rate of exchange of analyte from solution with the ligand on the surface is limiting, or binding is so advanced at the front of the flow cell such that the analyte concentration flowing over areas of the flow cell behind the front is depleted, the measured rates of binding will be affected. This happens, for example, when the coating density is too high or flow rates are too slow. Even if you follow best practice with low coating densities we recommend that you check for mass transport limitation by evaluating association rates at different flow rates. The kinetic wizard offers this as an optional extra.
20. The kinetic analysis wizard only allows you to use Fc2-1 in kinetic screens. In order to use Fc4-3 or 2, 3 and 4 with 1 as reference, you can set up the method in the binding analysis wizard, but you will then need to fit kinetic models in the BIAevaluation software.
21. Analyte concentrations should cover a full range of binding curves. The kinetic analysis wizard requires a minimum of five different concentrations of analyte and at least one concentration in duplicate. A minimum of one zero concentration sample (blank) should also be included. There is a calculator built in so that if you enter the MW, entry of either nanomolar or ng/mL concentrations will calculate the other. If you enter the highest concentration first, the wizard will automatically suggest values for the next rows when you press Enter to complete an entry.
22. The Kinetic Analysis dialog box offers you the option of selecting two additional experiments, Mass Transport Limitation and Linked Reactions. If you have already checked that a flow rate of 30 μL/min will not limit the association rate, you do not need to run the mass transport option. However, if you have not checked, we recommend that this option is included. It will ask for more protein sample (positive control mix) and will run the binding at 5, 15, and 75 μL/min and report on these as a cluster at the end of the run. You are looking to see that there is no difference between association curves at the higher flow rates. If there is a difference, increase the flow rate to 75 μL/min. Remember that this will use a lot more of your

analyte sample. The Linked Reactions option will run the binding of your positive control for three reaction times, 1, 3, and 20 min. As above, you will have a report box on this at the end of the run. The curves will be aligned at the point of the start of the dissociation curve. You are looking to see that association time does not affect dissociation rate. If the dissociation curves differ you need to consider the possibility of linked reactions and, hence, more complex binding models than the 1:1 Langmuir. Again, this option requires considerably more analyte sample.

23. In kinetic analyses a long (15 min) dissociation time is offered as default. This may be shortened if you know that dissociation of the bound complex is rapid. Otherwise, the model used to fit the kinetic parameters needs a lot of data if the off-rate is slow (gradient is shallow).

24. We show examples of two different kinetic experiments. First was titrating TIR1 saturated with IAA over the degron. Second was titrating IAA against TIR1 before exposure to the degron. Hence the first experiment reflects the affinity of the TIR1-IAA complex for degron, and the second the affinity of TIR1 for IAA. The former has a higher affinity than the latter, i.e., bound IAA as molecular glue has elevated the affinity for the three-partner interaction over tenfold [7].

Acknowledgement

The authors are grateful for support from BBSRC, UK, Award BB/L009366.

References

1. Napier RM, David KM, Perrot-Rechenmann C (2002) A short history of auxin-binding proteins. Plant Mol Biol 49:339–348
2. Venis MA, Napier RM (1992) Plant hormone receptors: past, present and future. Biochem Soc Trans 20:55–59
3. Dharmasiri N, Dharmasiri S, Estelle M (2005) The F-box protein TIR1 is an auxin receptor. Nature 26:441–445
4. Kepinski S, Leyser O (2005) The Arabidopsis F-box protein TIR1 is an auxin receptor. Nature 26:446–451
5. Woo EJ, Marshall J, Bauly J, Chen JG, Venis M, Napier RM, Pickersgill RW (2002) Crystal structure of auxin-binding protein 1 in complex with auxin. EMBO J 17:2877–2885
6. Tan X, Calderon-Villalobos LI, Sharon M, Zheng C, Robinson CV, Estelle M, Zheng M (2007) Mechanism of auxin perception by the TIR1 ubiquitin ligase. Nature 5:640–645
7. Calderon-Villalobos LI, Lee S, De Oliveira C, Ivetac A, Brandt W, Armitage L, Sheard LB, Tan X, Parry G, Mao H, Zheng N, Napier R, Kepinski S, Estelle M (2012) A combinatorial TIR1/AFB-Aux/IAA co-receptor system for differential sensing of auxin. Nat Chem Biol 1:477–485
8. Woodward AW, Bartel B (2005) Auxin: regulation, action, and interaction. Ann Bot 95:707–735
9. Lee S, Sundaram S, Armitage L, Evans JP, Hawkes T, Kepinski S, Ferro N, Napier RM (2014) Defining binding efficiency and specificity of auxins for SCF(TIR1/AFB)-Aux/IAA co-receptor complex formation. ACS Chem Biol 21:673–682

Chapter 16

Studying Transcription Factor Binding to Specific Genomic Loci by Chromatin Immunoprecipitation (ChIP)

S. Vinod Kumar and Doris Lucyshyn

Abstract

Plant hormone signaling involves complex transcriptional networks, where transcription factors orchestrate the control of specific gene expression. These networks include cross talk between hormone signaling pathways, and the integration of environmental signals and the developmental program. Understanding how particular transcription factors respond and integrate specific signals is crucial in order to understand the basic mechanisms of hormonal signaling and cross talk. Studying transcription factor binding at specific genomic loci by chromatin immunoprecipitation (ChIP) is therefore a valuable technique in order to analyze transcriptional regulation. The method is based on cross-linking proteins to DNA, the isolation of chromatin, and immunoprecipitation of a transcription factor of interest. The attached DNA is then recovered and analyzed by quantitative real-time PCR in order to establish binding sites of the respective transcription factor. Here, we present a relatively simple and short protocol for ChIP on single loci.

Key words Chromatin immunoprecipitation (ChIP), Transcription factor, Gene expression, Chromatin extraction, Binding site

1 Introduction

Plant hormone perception is mostly implemented by complex transcriptional networks, where transcriptional regulators play an important role in orchestrating the downstream control of gene expression. A well-studied example is phytochrome interacting factors (PIFs) [1]. They are involved in gibberellin, auxin, brassinosteroid, and ethylene signaling and respond to environmental triggers, such as light and temperature [2–9]. Chromatin immunoprecipitation (ChIP) has played a pivotal role for determining direct PIF-target genes and hence elucidating the function of these transcription factors in many studies [1–4, 10].

ChIP is commonly used to assess the binding of transcriptional regulators to specific genomic loci. The technique involves cross-linking proteins to DNA and extracting chromatin, which is then fragmented to achieve relatively small DNA fragments with

Jürgen Kleine-Vehn and Michael Sauer (eds.), *Plant Hormones: Methods and Protocols*, Methods in Molecular Biology, vol. 1497, DOI 10.1007/978-1-4939-6469-7_16, © Springer Science+Business Media New York 2017

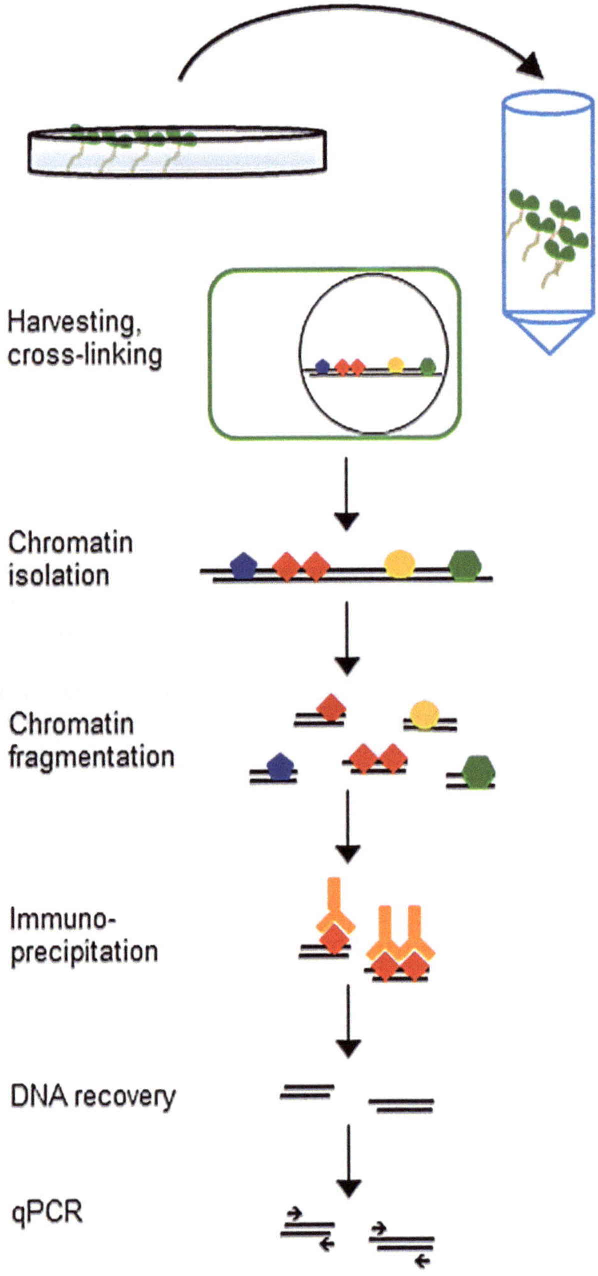

Fig. 1 ChIP workflow

transcriptional regulators attached. The fragmented and cross-linked chromatin is then used for immunoprecipitation (IP) using an antibody directed to an epitope-tagged transcriptional regulator of interest. After this step, any DNA attached to your protein of interest is recovered and analyzed for specific genomic regions by quantitative real-time PCR (*see* Fig. 1). Several protocols for isolating chromatin are available. The method described here is based

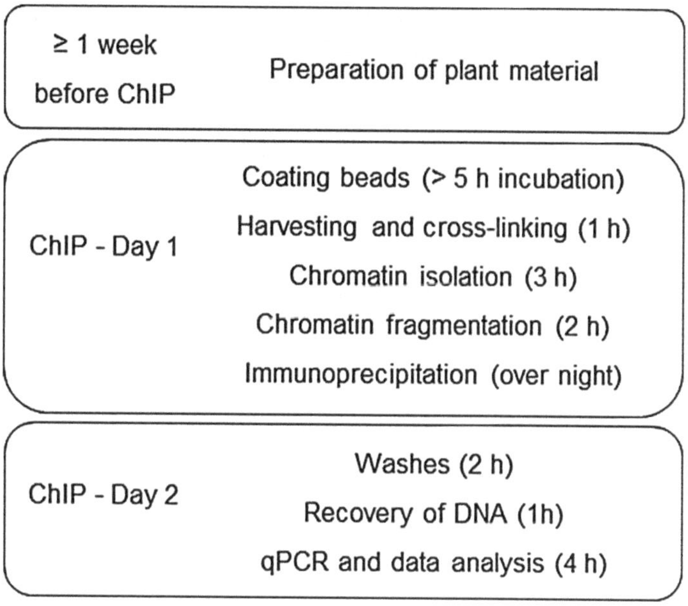

Fig. 2 Principle of ChIP

on enrichment of nuclei by using a sequence of buffers with different sucrose concentrations during extraction [11]. This will also separate chloroplasts from nuclei and give relatively pure chromatin.

Setting up a ChIP experiment requires careful preparation, and its success will also depend on the right choice of starting material. The most important considerations are described below; for more details see for example [11–13]. Here, we describe a simple and short protocol to perform ChIP on specific targets. For the analysis of genome-wide protein-DNA interactions like ChIP-seq, cell-type-specific interactions, or the analysis of locus-specific histone modifications, *see* other protocols such as [14–17].

1.1 Starting Material

Preparing or choosing a suitable plant line is essential to achieve clear results. Using a small tag for fusion to your protein of interest, such as HA or FLAG tag, is a good choice, as they most likely will not interfere with the activity of your protein of interest, and reliable good-quality antibodies are available from a range of suppliers. Using a protein-specific antibody requires extensive optimization and careful design of negative controls, and might give poor or false-positive results. Another important factor is the choice of promoter for your constructs. Placing the epitope-tagged gene of interest under the endogenous promotor would be ideal and is certainly worth trying. However, the efficiency of ChIP is often very low, and using endogenous promoters might not result in protein levels that are sufficient to get clear results. ChIP for transcriptional regulators is therefore often performed using over-expression constructs. If possible, transform your epitope-tagged construct to a mutant (loss-of-function) background and test for complementation. It is also advisable to check protein levels by Western blot on plant material grown in the same conditions that you are planning for ChIP. This will give confidence about the plant line, tissue, growth conditions, and antibody before you start with the ChIP protocol.

1.2 Primer Design

Design primers suitable for qPCR-based analysis of recovered DNA after ChIP. PCR fragments should be 100–200 bp in length; fragments longer than 500 bp are not suitable. For primer design, online tools such as primer3 (http://bioinfo.ut.ee/primer3-0.4.0/) or primer-BLAST (http://www.ncbi.nlm.nih.gov/tools/primer-blast/) are very helpful. Use these tools also to look for potential off-targets and adjust primers accordingly. The PCR fragments should cover predicted binding sites, and it is recommended to design several sets of primers covering a region of interest. Negative control primers are necessary to assess background—choose regions in the coding sequence or 3′-UTR of your gene of interest (keep at least 1–2 kb distance to the predicted binding sites), as well as promoters of unrelated genes for negative control fragments. Keep in mind that transcriptional regulators might bind in coding regions, and predictions about negative control fragments within the coding region of your gene of interest are not always accurate.

2 Materials

2.1 Equipment

1. Desiccator.
2. Miracloth.
3. Liquid nitrogen.
4. Mortar and pestle.
5. Sonication device.
6. Photometer or NanoDrop.
7. DNA-electrophoresis equipment.
8. Water bath.
9. Magnetic stand.
10. Rotating wheel (in a cold room).
11. Real-time PCR machine.

2.2 Material and Buffers

1. Plant growth media.
2. Cross-linking buffer: 0.4 M Sucrose, 10 mM Tris–HCl pH 8, 10 mM $MgCl_2$.
3. Formaldehyde.

4. 2 M Glycine.
5. EB1, extraction buffer 1 (prepare freshly from stock solutions): 0.4 M Sucrose, 10 mM Tris–HCl pH 8, 10 mM $MgCl_2$, 5 mM 2-mercaptoethanol, 0.1 mM PMSF, 1× protease inhibitor mix.
6. EB2, extraction buffer 2 (prepare freshly from stock solutions): 0.25 M Sucrose, 10 mM Tris–HCl pH 8, 10 mM $MgCl_2$, 5 mM 2-mercaptoethanol, 1 % Triton X-100, 0.1 mM PMSF, 1× protease inhibitor mix.
7. EB3, extraction buffer 3 (prepare freshly from stock solutions): 1.7 M Sucrose, 10 mM Tris–HCl pH 8, 10 mM $MgCl_2$, 5 mM 2-mercaptoethanol, 0.1 mM PMSF, 1× protease inhibitor mix.
8. Nuclei lysis buffer: 50 mM Tris–HCl pH 8, 10 mM EDTA, 1× protease inhibitor mix.
9. Triton X-100.
10. 20 % Chelex® 100 Resin.
11. Elution buffer: 0.1 M $NaHCO_3$, 1 % SDS (prepared freshly, incubate at 65 °C immediately before use).
12. Proteinase K (10 mg/ml).
13. RNaseA (10 mg/ml).
14. Phenol-chloroform-isoamyl alcohol (PCI, 25:24:1) (optional, *see* **step 9** of Subheading 3.4).
15. Glycogen 1 μg/μl (optional, *see* **step 9** of Subheading 3.4).
16. 3 M NaOAc (optional, *see* **step 9** of Subheading 3.4).
17. Antibody against epitope-tagged protein.
18. Magnetic beads (Protein A or Protein G beads, according to the data sheets of your antibody and the magnetic beads).
19. BSA (5 mg/ml).
20. ChIP dilution buffer: 5 mM EDTA, 10 mM Tris–HCl, pH 8, 160 mM NaCl, 1× protease inhibitor mix.
21. Low-salt buffer: 150 mM NaCl, 0.1 % SDS, 1 % Triton X-100, 2 mM EDTA, 20 mM Tris–HCl pH 8.
22. High-salt buffer: 500 mM NaCl, 0.1 % SDS, 1 % Triton X-100, 2 mM EDTA, 20 mM Tris–HCl pH 8.
23. LiCl wash buffer: 0.25 M LiCl, 1 % IGEPAL, 1 % Na-deoxycholate, 1 mM EDTA, 10 mM Tris–HCl pH 8.
24. TE buffer: 10 mM Tris–HCl pH8, 1 mM EDTA.
25. DNA purification or PCR cleanup kit.
26. SYBR Green real-time PCR master mix.

3 Methods

3.1 Plant Growth

1. Sterilize an amount of seeds that will give at least 2.5 g of material (*see* **Note 1**).
2. Prepare plant growth media.
3. Sow the seeds and stratify for 2–4 days in the dark at 4 °C.
4. Grow plants according to the requirements of your experiment.

3.2 Harvesting and Fixing

1. Prepare 60 ml cross-linking buffer for each sample (for cross-linking and rinsing).
2. Punch holes in the lids of 50 ml tubes, aliquot 30 ml of cross-linking buffer for each sample, and add 0.85 ml of 35 % formaldehyde (1 % final concentration).

   Caution: Formaldehyde is toxic—work under the fume hood.
3. Harvest 2.5–5 g of plant material directly in the prepared tubes (*see* **Note 2**).
4. Apply vacuum in a desiccator for max. 20 min (*see* **Note 3**).
5. To stop cross-linking, add 2.5 ml of 2 M glycine in each tube, and mix well.
6. Apply vacuum for 5 min.
7. Carefully remove solution from the tubes and rinse well with cross-linking buffer (without formaldehyde).
8. Drain tissue on lab tissue paper, keeping the tubes inverted for 2–5 min. Do not let the plant material dry.
9. Snap freeze plant material in liquid nitrogen.

   Stopping point: The material can be stored at −80 °C at this point (see Note 4*).*

3.3 Preparation of Chromatin

Caution: 2-Mercaptoethanol is toxic, work in the fume hood.

1. Grind tissue in liquid nitrogen to a fine powder using mortar and pestle.
2. Weigh ground material and use roughly same amounts for chromatin preparation for each sample (*see* **Note 5**).
3. Resuspend samples in 25 ml of EB1 (*see* **Note 6**).
4. Filter through two layers of miracloth.
5. Centrifuge filtrate at 4000 rpm (1500 ×*g*), 4 °C, for 15 min.
6. Resuspend pellet in 1.5 ml EB2, and incubate on ice for 3–5 min with intermittent swirling.
7. Centrifuge at 14,000 rpm (15800 ×*g*), 4 °C, for 10 min (*see* **Note 7**).

8. Prepare a fresh 2 ml tube containing 500 μl EB3 for each sample.
9. Resuspend the pellet in 500 μl of EB3 and slowly load on top of fresh EB3 prepared before.
10. Centrifuge at 14,000 rpm (15800×*g*), and 4 °C for 20–30 min (*see* **Note 8**).
11. Completely remove supernatant.

    Stopping point: The pellet can be frozen and stored at −80 °C for months.

3.4 Chromatin Fragmentation

1. Resuspend the crude nuclei pellet in 500 μl nuclei lysis buffer, avoid bubble formation.
2. Sonicate the samples at 4 °C to shear chromatin to fragments of 200–1000 bp (the optimal fragment size is 500 bp). This step will have to be optimized depending on the sonication device (*see* **Note 9**).
3. Prepare an aliquot of sheared chromatin for the control of fragment size on a gel and DNA quantification: transfer 25 μl of the chromatin to a fresh tube, add an equal volume of 20% Chelex, vortex briefly, and boil for 10 min with intermittent mixing (**Note 10**).
4. Bring tubes to room temperature, add 0.1 μl proteinase K (10 mg/ml), and incubate at 50 °C for 30 min with intermittent swirling.
5. Boil again for 10 min.
6. Spin the samples at maximum speed for 1 min, and recover the supernatant carefully without taking Chelex.
7. Add 0.1 μl RNaseA (10 mg/ml) and incubate at 37 °C for 20 min.
8. Load samples on a 1.5% agarose gel (*see* **Note 11**).
9. *Alternative de-cross-linking protocol (see* Note 12*): After RNase treatment at* step 7*, add 50 μl elution buffer to the chromatin, mix well, add 4 μl 5 M NaCl, and incubate at 65 °C overnight. Extract DNA by adding an equal volume of PCI, vortex, and centrifuge at full speed for 5 min at room temperature. Recover supernatant, add 5 μl glycogen (1 μg/μl) as carrier and 1/10 of sample volume 3 M NaOAc, and precipitate with 2.5 vol EtOH at −20 °C. Wash the pellet once with 70% ethanol, dissolve the pellet in TE, and load on an agarose gel as before.*
10. De-cross-linked DNA can be used for quantification using a spectrophotometer or NanoDrop.
11. If sonication is complete add Triton X-100 to a final concentration of 1% and centrifuge at 13,000 rpm (18400×*g*), 4 °C, for 10 min.

12. Recover the supernatant and discard the pellet to remove unsoluble debris.

    Stopping point: Chromatin can be stored at –80 °C for months. Snap freeze in liquid nitrogen and avoid multiple freeze-thaws. Freeze the input-aliquot separately (see Subheading 3.5.1).

3.5 Immuno-precipitation (IP)

3.5.1 Prepare Beads and Chromatin

1. Wash 45 μl magnetic beads for the following three samples: 15 μl beads each for two IPs (technical repeats) and one no-antibody control (mock) (*see* **Note 13**).
2. Block the beads: Add 500 μl of 5 mg/ml BSA in PBS, and incubate on a rotating wheel for 2 min at 4 °C. Bring the tubes to the magnetic rack, let beads attach to sides of the tube, discard the buffer, and repeat washing two more times.
3. Wash the beads twice in 1 ml ChIP dilution buffer, remove the buffer completely, and resuspend the beads in 45 μl ChIP dilution buffer.
4. Take out 30 μl of washed beads, resuspend in 1 ml ChIP dilution buffer, and add antibody (2–5 μg per 25 μg of chromatin, *see* **Note 14**).
5. Incubate at least for 5 h or overnight at 4 °C on a rotating wheel to couple the antibody to the beads. Alternatively this can be done for 1 h at room temperature.
6. Store remaining 15 μl of washed beads in ChIP dilution buffer at 4 °C until needed for the mock sample.
7. Wash freshly antibody-coated beads twice with ChIP dilution buffer.

3.5.2 Immuno-precipitation

1. Aliquot chromatin: Use equal amounts (ca. 25 μg) for each IP, the no-antibody control, and the input-sample. Store input at –80 °C until de-cross-linking with the other samples after immunoprecipitation (*see* **Note 15**).
2. Dilute chromatin with ChIP dilution buffer to an SDS concentration of 0.1 %.
3. Mix diluted chromatin with the beads: two samples with 25 μg chromatin + 15 μl antibody-coated beads, one sample with 25 μg chromatin + 15 μl beads (mock), and bring samples to a final volume of ca. 1.5 ml with ChIP dilution buffer (*see* **Note 16**).
4. Incubate on a rotating wheel at 4 °C overnight.

3.5.3 Washes

For each washing step add 1 ml of respective buffer, invert gently, incubate on ice for 5 min, and let the beads settle in the magnetic stand. Remove supernatant carefully, and do not let the beads dry out.

1. Wash twice with low-salt wash buffer.

2. Wash twice with high-salt wash buffer.
3. Wash once with LiCl wash buffer.
4. Wash twice with TE (*see* **Note 17**).
5. Resuspend beads in 100 μl TE.

3.5.4 De-cross-linking and DNA Recovery

At this step, include input sample for de-cross-linking.

1. To recover DNA, add 100 μl of 20% Chelex and boil for 10 min.
2. Bring tubes to room temperature, add 0.1 μl proteinase K (10 mg/ml), incubate at 50 °C for 30 min, and boil again for 10 min. Mix regularly during incubation steps.
3. Spin the tubes at maximum speed for 1 min, and recover the supernatant carefully.
4. Optional: In case Chelex particles remain in the samples, purify recovered DNA using a PCR product cleanup kit. Follow the instructions of the kit and elute DNA with 100 μl TE. The eluate from this step can be directly used for PCR (*see* **Note 18**).

   Stopping point: The recovered DNA can be stored at −20 °C.

3.6 qPCR and Data Analysis

1. Dilute eluted DNA 1:10 with water and use 5 μl as template in 15 μl reactions (*see* **Note 19**).
2. For analyzing data using the percentage of input method (recommended), use the following method:
   $\%\text{ of input} = 100 \times 2^{\Delta(\text{ct(input)} - \text{ct(sample)})}$ (*see* **Note 20**).
3. For analyzing data using the fold enrichment method:

   Fold enrichment = (fraction of input$^{\text{sample)}}$/(fraction of input$^{\text{control)}}$ (*see* **Note 20**).

4 Notes

1. Any healthy plant tissue is suitable, such as seedlings, roots, rosette leaves, or flowers, depending on the expression patterns of your transcription factor and your experimental setup.
2. For harvesting, work quickly. Be sure to always harvest at the same time of day for biological repeats and take potential circadian regulation of your transcription factor into account.
3. Apply vacuum until the tissue sinks and gets translucent. Keep the exposure to formaldehyde as short as possible and work quickly during cross-linking.
4. If the material is stored, wrap it in aluminum foil.
5. Make sure that the ground material does not thaw at any stage, and use spatula chilled in liquid nitrogen while transfer.

6. Adjust the amount of buffer to the amount of ground material. 25 ml of EB1 works well for 2–3 g of ground material.
7. If the pellet looks yellow and clean, proceed with chromatin fragmentation (Subheading 3.4). If there is a green rim, continue with the next step.
8. At this step, nuclei will settle in the pellet, while the chloroplasts keep floating in the top layer of EB3. This is a crude separation enriching for nuclei.
9. This is a crucial step for the success of the ChIP protocol, and should be optimized in order to get reproducible results. Check fragmentation on an agarose gel. If necessary, reduce or increase the number of sonication cycles to optimize fragment size. During sonication it is essential that the samples do not heat up. Try to avoid foaming, as it might cause uneven sonication.
10. Always mix Chelex well before pipetting, as it settles very quickly. Intermittent swirling is crucial to keep Chelex in suspension during incubation steps.
11. DNA fragments should run between 200 and 1000 bp of size, with the majority of signal at around 500 bp. If the samples do not migrate into the gel, either sonication or de-cross-linking was not efficient. Try more sonication cycles and de-cross-linking by using the alternative protocol described at point **step 9** of Subheading 3.4 of this protocol to determine the cause of the problem.
12. Chelex offers the advantage of speeding up and simplifying the de-cross-linking protocol, as well as being nonhazardous compared to classical DNA precipitation methods [18]. If you encounter problems with de-cross-linking due to Chelex, use this alternative protocol. Chelex is also not compatible with sequencing of recovered DNA, as most of it is denatured after the treatment.
13. See antibody and magnetic bead data sheets for choosing protein A or G magnetic beads.

    Never let the beads dry out; always resuspend before pipetting.
14. The amount of antibody might have to be optimized—start with the recommendations of the antibody supplier.
15. In case the amount of chromatin is limiting, keep less for the input sample (i.e., 1/10 of the total chromatin). Be sure to take a corresponding dilution factor into account when analyzing the qPCR data.
16. Use safe-lock 2 ml tubes with round bottom.

17. Transferring the beads to fresh tubes between TE washes helps to reduce background.
18. The alternative de-cross-linking protocol described in **step 9** of Subheading 3.4 can be used at this point.
19. Use this dilution as a starting point. If the qPCR does not give good results, try using more template. In some cases too much template might also be inhibitory, depending on the purity of the DNA.
20. For more details and other possibilities of data analysis and the differences between these methods *see* Ref. [13].

References

1. Pfeiffer A et al (2014) Combinatorial complexity in a transcriptionally centered signaling hub in Arabidopsis. Mol Plant 7(11):1598–1618
2. de Lucas M et al (2008) A molecular framework for light and gibberellin control of cell elongation. Nature 451(7177):480–484
3. Franklin KA et al (2011) Phytochrome-interacting factor 4 (PIF4) regulates auxin biosynthesis at high temperature. Proc Natl Acad Sci U S A 108(50):20231–20235
4. Kumar SV et al (2012) Transcription factor PIF4 controls the thermosensory activation of flowering. Nature 484(7393):242–245
5. Bai MY et al (2012) Brassinosteroid, gibberellin and phytochrome impinge on a common transcription module in Arabidopsis. Nat Cell Biol 14(8):810–817
6. Sakuraba Y et al (2014) Phytochrome-interacting transcription factors PIF4 and PIF5 induce leaf senescence in Arabidopsis. Nat Commun 5:4636
7. Feng S et al (2008) Coordinated regulation of Arabidopsis thaliana development by light and gibberellins. Nature 451(7177):475–479
8. Huq E, Quail PH (2002) PIF4, a phytochrome-interacting bHLH factor, functions as a negative regulator of phytochrome B signaling in Arabidopsis. EMBO J 21(10):2441–2450
9. Koini MA et al (2009) High temperature-mediated adaptations in plant architecture require the bHLH transcription factor PIF4. Curr Biol 19(5):408–413
10. Hornitschek P et al (2012) Phytochrome interacting factors 4 and 5 control seedling growth in changing light conditions by directly controlling auxin signaling. Plant J 71(5):699–711
11. Bowler C et al (2004) Chromatin techniques for plant cells. Plant J 39(5):776–789
12. Yamaguchi N et al (2014) PROTOCOLS: chromatin Immunoprecipitation from Arabidopsis Tissues. Arabidopsis Book 12:e0170
13. Haring M et al (2007) Chromatin immunoprecipitation: optimization, quantitative analysis and data normalization. Plant Methods 3:11
14. Deal RB, Henikoff S (2011) The INTACT method for cell type-specific gene expression and chromatin profiling in Arabidopsis thaliana. Nat Protoc 6(1):56–68
15. Kaufmann K et al (2010) Chromatin immunoprecipitation (ChIP) of plant transcription factors followed by sequencing (ChIP-SEQ) or hybridization to whole genome arrays (ChIP-CHIP). Nat Protoc 5(3):457–472
16. Song J, Rutjens B, Dean C (2014) Detecting histone modifications in plants. Methods Mol Biol 1112:165–175
17. Weinhofer I, Kohler C (2014) Endosperm-specific chromatin profiling by fluorescence-activated nuclei sorting and ChIP-on-chip. Methods Mol Biol 1112:105–115
18. Nelson JD et al (2006) Fast chromatin immunoprecipitation assay. Nucleic Acids Res 34(1), e2

Chapter 17

Hormone Receptor Glycosylation

Ulrike Vavra, Christiane Veit, and Richard Strasser

Abstract

Glycosylation is essential for all trees of life. N-glycosylation is one of the most common covalent protein modifications and influences a large variety of cellular processes including protein folding, quality control and protein-receptor interactions. Despite recent progress in understanding of N-glycan biosynthesis, our knowledge of N-glycan function on individual plant proteins is still very limited. In this respect, plant hormone receptors are an interesting group of proteins as several of these proteins are present at distinct sites in the secretory pathway or at the plasma membrane and have numerous potential N-glycosylation sites. Identifying and characterization of N-glycan structures on these proteins is essential to investigate the functional role of this abundant protein modification. Here, a straightforward immunoblot-based approach is presented that enables the analysis of N-glycosylation on endogenous hormone receptors like the brassinosteroid receptor BRI1.

Key words N-glycosylation, Glycoprotein, Brassinosteroid receptor, Secretory pathway, Endoplasmic reticulum, Golgi apparatus

1 Introduction

Glycosylation is a ubiquitous protein modification in eukaryotic cells. Dependent on the linkage of the sugar or oligosaccharide to the amino acid side chain there are two main types of glycosylation: N- and O-glycosylation. N-glycosylation and early N-glycan processing steps are highly conserved between plants and animals. N-glycans are crucial for protein folding in the endoplasmic reticulum and for quality control processes including the glycan-mediated degradation of terminally misfolded glycoproteins [1]. A wide variety of other functions ranging from the modulation of protein–receptor interactions and regulation of subcellular targeting to protein stability have been described in mammals. In plants, comparatively little is known about the contribution of distinct N-glycans to protein function [2]. However, studies with the heavily N-glycosylated pathogen recognition receptor EFR, for example, revealed an important role of N-glycosylation and N-glycan-dependent quality control during plant immunity [3–7]. In contrast

Jürgen Kleine-Vehn and Michael Sauer (eds.), *Plant Hormones: Methods and Protocols*, Methods in Molecular Biology, vol. 1497, DOI 10.1007/978-1-4939-6469-7_17, © Springer Science+Business Media New York 2017

to animals, O-glycosylation of proteins is fundamentally different. Plants lack a functional mucin-type O-glycosylation pathway which involves the transfer of a single N-acetylgalactosamine (GalNAc) residue to serine or threonine [8]. Instead, plants can attach galactose to serine and modify hydroxyproline residues in proline-rich proteins either with small arabinose chains or with larger arabinogalactans [9].

Hormone receptors in the nucleus or cytoplasm and cytoplasmic domains of membrane-anchored receptors are typically neither N-glycosylated nor contain any extensive O-glycan-like modifications because these proteins and their cytoplasmic regions are never in contact with the ER/Golgi-located glycosylation machinery. However, there is the possibility for the attachment of a single N-acetylglucosamine (GlcNAc) to serine or threonine residues of cytosolic or nuclear proteins. This posttranslational modification is well described in mammals, where it plays a major role in diverse signaling reactions [10]. Even though O-linked GlcNAc modifications have been described for plant proteins including histones [11], most of the substrate proteins are currently unknown. Interestingly, one of the two Arabidopsis O-GlcNAc-transferases (SPINDLY) is also involved in gibberellin and cytokinin dependent processes [12, 13]. Moreover, the recent identification of the specific nucleotide sugar transporter ROCK1 suggests that an unknown GlcNAc or GalNAc modification in the ER is important for regulation of cytokinin levels [14].

Here, we focus on N-glycosylation and N-glycan processing of hormone receptors in Arabidopsis, because the machinery for this frequent glycosylation reaction is already very well understood [2] and N-glycosylation has been clearly demonstrated for some of them. N-glycosylation is initiated in the lumen of the ER by the oligosaccharyltransferase (OST) complex. OST recognizes the Asn-X-Ser/Thr (X can be any amino acid except proline) consensus sequence on nascent polypeptide chains and transfers a preassembled oligosaccharide *en bloc* to the asparagine of the acceptor substrate in the lumen of the ER. In addition to this cotranslational protein modification, which plays a major role for protein folding, some skipped N-glycosylation sites are glycosylated posttranslationally by a distinct OST complex with a different subunit composition [15]. As soon as the glycan precursor is transferred, ER-resident glucosidases and mannosidases remove terminal sugar residues and the glycoprotein with partially processed N-glycans can undergo glycan-dependent quality control to ensure that only fully folded proteins are released to other compartments of the secretory pathway [1].

Which hormone receptors are possible targets for N-glycosylation? Primary candidates for this protein modification are all receptors that reside in the ER or along the secretory pathway including the plasma membrane and extracellular space.

Auxin transport, binding or metabolic enzymes have been detected in the ER membrane and the ER lumen [16, 17]. Ethylene perception is supposed to take place in the ER [18] and studies with fluorescent cytokinin receptor fusions suggest localization in this organelle [19, 20]. Brassinosteroids are perceived at the plasma membrane by the receptor BRASSINOSTEROID INSENSITIVE 1 (BRI1) [21]. Likewise, abscisic acid and cytokinin binding may be mediated by plasma membrane anchored receptors [22, 23]. For many of the aforementioned hormones the N-glycosylation status of their putative receptors is unknown or not very well investigated. N-glycosylation of the cytokinin receptor AHK3 has been reported by Endoglycosidase H (Endo H) digestion of transiently expressed GFP-AHK3 [19]. AUXIN BINDING PROTEIN1 (ABP1) contains one conserved N-glycosylation site [24] and an insect cell produced ABP1 was found to be N-glycosylated [25]. In ER-resident auxin-transport related proteins like PILS5 an N-glycosylation motif is present in one of the ER-lumen facing loops. However, this motif is not conserved in all PILS proteins [26] and only found in PILS5 and PILS7. In contrast to most of these ER-to-plasma membrane located hormone binding proteins, N-glycosylation of BRI1 is very well established [27–29]. Furthermore, it has been demonstrated that the N-glycans of misfolded BRI1 variants play a crucial role during ER-mediated quality control processes and glycan-dependent ER-associated degradation (ERAD) [29–32].

In this chapter we describe simple immunoblot-based protocols for the N-glycosylation analysis of hormone receptors. These procedures are designed to discriminate between glycosylated and non-glycosylated endogenous proteins without using tagged or overexpressed proteins. Based on the presence of distinct N-glycan structures additional information can be gathered about the subcellular localisation and trafficking route of a given plant hormone receptor. To identify the precise glycosylation profile of these proteins more sophisticated mass spectrometry (MS)-based methods have to be applied which in most cases require at least partial purification of the protein of interest (*see* **Note 1**). While this is still challenging for many low expressed or tightly regulated endogenous proteins, advances in MS-based technologies will provide a more complete picture of plant hormone receptor glycosylation in the future.

2 Materials

1. *Arabidopsis thaliana stt3a-2* seeds [33] can be purchased from the European Arabidopsis Stock Center (NASC ID: N800052). The *fut11 fut12* seeds [34] can be obtained upon request from the Strasser group.

2. Mixer mill (e.g., from Retsch) with steel beads.
3. Container with liquid nitrogen.
4. Refrigerated benchtop centrifuge with a relative centrifugal force of 9400 or higher and a rotor for 1.5/2 ml microcentrifuge tubes (such as Eppendorf Centrifuge 5414R).
5. Thermo block.
6. Orbital shaker.
7. Vertical electrophoresis system (such as Mini-Protean, Bio-Rad).
8. Tank transfer system (such as Mini-Trans Blot, Bio-Rad).
9. Power supply with at least 100 V and 400 mA (such as Power Pac Universal Power Supply, Bio-Rad).
10. SDS-PAGE (10%) gels.
11. Tris–glycine–SDS PAGE running buffer (25 mM Tris, 192 mM glycine, 0.1% (w/v) SDS).
12. 3× Laemmli sample buffer (187.5 mM Tris pH 6.8, 30% (w/v) glycerol, 6% (w/v) SDS 15% (v/v) β-mercaptoethanol, 0.15% bromophenol blue).
13. Protein standard (we use Prestained protein marker, Fermentas).
14. Tris–glycine–methanol transfer-buffer (25 mM Tris, 192 mM glycine, 20% (v/v) methanol).
15. Blotting paper (we use Whatman 3 MM).
16. Blotting membrane: we use Amersham Protran Premium nitrocellulose membrane (GE Healthcare).
17. 1× PBS: phosphate buffered saline (137 mM NaCl, 2.7 mM KCl, 10 mM Na_2HPO_4, 2 mM KH_2PO_4, pH 7.4).
18. PBST: 1× PBS + 0.1% (v/v) Tween 20.
19. 1× TBS: Tris buffered saline (25 mM Tris–HCl, pH 7.5, 150 mM NaCl, 2 mM KCl).
20. TBST: 1× TBS + 0.1% (v/v) Tween 20.
21. Blocking solution for BRI1 detection on immunoblots: TBST + 5% (w/v) skimmed milk powder.
22. Blocking solution for AHK2 detection on immunoblots: PBST + 3% (w/v) BSA.
23. Western blotting detection reagent: we use SuperSignal West PICO Chemiluminescent Substrate (Pierce) or Amersham ECL Prime Western Blotting Detection Reagent (GE Healthcare).
24. Sensitive films for detection: we use Amersham Hyperfilm ECL (GE Healthcare).
25. 1× PBS + 1% (v/v) Triton X-100.
26. Film developer and fixer solutions.

27. Polyclonal anti-BRI1 antibody (Agrisera: AS12 1859): 1:5000 diluted in TBST + 5 % (w/v) skimmed milk powder.
28. Polyclonal anti-AHK2 antibody (Agrisera: AS12 2113): 1:2500 diluted in PBST + 3 % (w/v) BSA.
29. Anti-rabbit IgG-peroxidase (such as Sigma, A0545): 1:50,000 diluted in PBST for detection of AHK2 or 1:5000 diluted in TBST + 5 % (w/v) skimmed milk powder for BRI1 detection.
30. Endoglycosidase H (Endo H, we use 500,000 U/mL, New England Biolabs) + Glyco Buffer 3 (New England Biolabs).
31. Peptide-N-Glycosidase F (PNGase F, we use 500,000 U/mL, New England Biolabs) + 1× Glycoprotein Denaturing Buffer (New England Biolabs).
32. NP-40 (10% solution, such as provided by New England Biolabs).
33. Kifunensine (such as Sigma: K1140): class I α-mannosidase inhibitor, dissolved in ultrapure water.

3 Methods

3.1 N-glycosylation Prediction

The prerequisite for monitoring of hormone receptor N-glycosylation is the presence of the consensus sequence Asn-X-Ser/Thr (where X can be any amino acid except proline) within the protein of interest. N-glycosylation site prediction tools like NetNGlyc (http://www.cbs.dtu.dk/services/NetNGlyc/) [35] (*see* **Note 2**) provide a list of all potential sites present in a given amino acid sequence (*see* **Note 3**). Examples are shown for Arabidopsis AHK2 and BRI1 in Figs. 1 and 2, respectively. Using artificial neural networks and a set of human proteins the NetNGlyc predictor is trained to discriminate glycosylated from non-glycosylated sites. This is relevant, since not all N-glycosylation sites are glycosylated (*see* **Note 4**).

Next, it is important to predict the subcellular localization and topology of the protein. The presence of a signal peptide—e.g., predicted using SignalP (http://www.cbs.dtu.dk/services/SignalP/)—indicates targeting to the secretory pathway. The detection of transmembrane regions (e.g., using TMHMM: http://www.cbs.dtu.dk/services/TMHMM/ or HMMTOP: http://www.enzim.hu/hmmtop/) and their position provides information on the accessibility of certain protein domains for the glycosylation machinery in the lumen of the ER. While these predictions are informative for some proteins like BRI1 (Fig. 2) or the proposed abscisic acid receptors GTG1 and GTG2 [22] which do not have any potential N-glycosylation site at all (not shown), for others (e.g., AHK2, Fig. 1) the number of accessible potential N-glycosylation sites is ambiguous. Regardless of the quality of these predictions, experimental confirmation of the

MSITCELLNLTSKKAKKSSSSDKKWLKKPLFFLILCGSLVIVLVMFLRLGRSQKEETDSCNGEEKVLYRHQNVTRSEIHD
80
LVSLFSDSDQVTSFECHKESSPGMWTNYGITCSLSVRSDKQETRGLPWNLGLGHSISSTSCMCGNLEPILQQPENLEEEN
160
HEEGLEQGLSSYLRNAWWCLILGVLVCHKIYVSHSKARGERKEKVHLQEALAPKKQQQRAQTSSRGAGRWRKNILLLGIL
240
GGVSFSVWWFWDTNEEIIMKRRETLANMCDERARVLQDQFNVSLNHVHALSILVSTFHHGKIPSAIDQRTFEEYTERTNF
320
ERPLTSGVAYALKVPHSEREKFEKEHGWAIKKMETEDQTVVQDCVPENFDPAPIQDEYAPVIFAQETVSHIVSVDMMSGE
400
EDRENILRARASGKGVLTSPFKLLKSNHLGVVLTFAVYDTSLPPDATEEQRVEATIGYLGASYDMPSLVEKLLHQLASKQ
480
TIAVDVYDTTNTSGLIKMYGSEIGDISEQHISSLDFGDPSRNHEMHCRFKHKLPIPWTAITPSILVLVITFLVGYILYEA
560
INRIATVEEDCQKMRELKARAEAADIAKSQFLATVSHEIRTPMNGVLGMLKMLMDTDLDAKQMDYAQTAHGSGKDLTSLI
640
NEVLDQAKIESGRLELENVPFDMRFILDNVSSLLSGKANEKGIELAVYVSSQVPDVVVGDPSRFRQIITNLVGNSIKFTQ
720
ERGHIFISVHLADEVKEPLTIEDAVLKQRLALGCSESGETVSGFPAVNAWGSWKNFKTCYSTESQNSDQIKLLVTVEDTG
800
VGIPVDAQGRIFTPFMQADSSTSRTYGGTGIGLSISKRLVELMQGEMGFVSEPGIGSTFSFTGVFGKAETNTSITKLERF
880
DLAIQEFTGLRALVIDNRNIRAEVTRYELRRLGISADIVSSLRMACTCCISKLENLAMILIDKDAWNKEEFSVLDELFTR
960
SKVTFTRVPKIFLLATSATLTERSEMKSTGLIDEVVIKPLRMSVLICCLQETLVNGKKRQPNRQRRNLGHLLREKQILVV
1040
DDNLVNRRVAEGALKKYGAIVTCVESGKAALAMLKPPHNFDACFMDLQMPEMDGFEATRRVRELEREINKKIASGEVSAE
1120
MFCKFSSWHVPILAMTADVIQATHEECMKCGMDGYVSKPFEEEVLYTAVARFFEPC

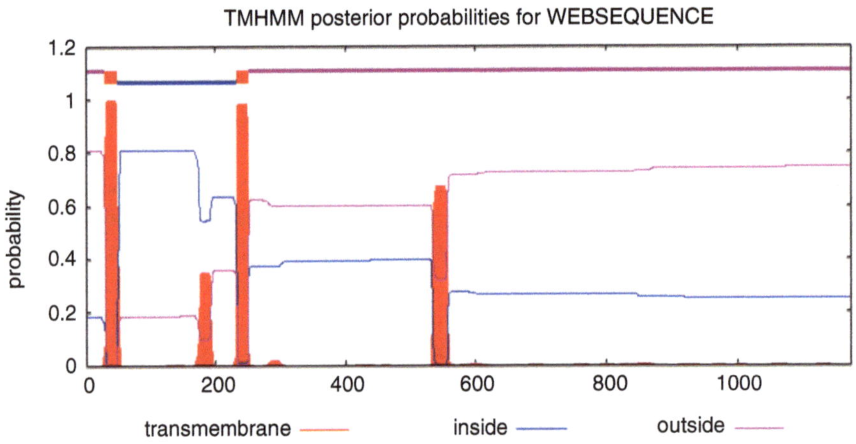

Fig. 1 *N*-glycosylation site and transmembrane domain predictions for AHK2 (At5g35750, Q9C5U2 Uniprot database, 1176 amino acids). Putative *N*-glycosylation sites are depicted in blue. Two to four transmembrane domains are predicted by TMHMM. Based on this prediction the topology and the number of potential N-glycosylation in luminal (extracellular—marked as "*outside*") domains are unclear

N-glycosylation status of an individual protein is always required to identify novel glycoproteins.

3.2 Monitoring of Protein Underglycosylation

N-glycosylation shows that the hormone receptor is present in the secretory pathway or extracellular environment (*see* **Note 5**). A simple way to monitor N-glycosylation is the analysis of differences in mobility upon SDS-PAGE and immunoblotting in underglycosylation mutants which display a reduced N-glycosylation efficiency. The most suitable

```
MKTFSSFFLSVTTLFFFSFFSLSFQASPSQSLYREIHQLISFKDVLPDKNLLPDWSSNKNPCTFDGVTCRDDKVTSIDLS
80
SKPLNVGFSAVSSSLLSLTGLESLFLSNSHINGSVSGFKCSASLTSLDLSRNSLSGPVTTLTSLGSCSGLKFLNVSSNTL
160
DFPGKVSGGLKLNSLEVLDLSANSISGANVVGWVLSDGCGELKHLAISGNKISGDVDVSRCVNLEFLDVSSNNFSTGIPF
240
LGDCSALQHLDISGNKLSGDFSRAISTCTELKLLNISSNQFVGPIPPLPLKSLQYLSLAENKFTGEIPDFLSGACDTLTG
320
LDLSGNHFYGAVPPFFGSCSLLESLALSSNNFSGELPMDTLLKMRGLKVLDLSFNEFSGELPESLTNLSASLLTLDLSSN
400
NFSGPILPNLCQNPKNTLQELYLQNNGFTGKIPPTLSNCSELVSLHLSFNYLSGTIPSSLGSLSKLRDLKLWLNMLEGEI
480
PQELMYVKTLETLILDFNDLTGEIPSGLSNCTNLNWISLSNNRLTGEIPKWIGRLENLAILKLSNNSFSGNIPAELGDCR
560
SLIWLDLNTNLFNGTIPAAMFKQSGKIAANFIAGKRYVYIKNDGMKKECHGAGNLLEFQGIRSEQLNRLSTRNPCNITSR
640
VYGGHTSPTFDNNGSMMFLDMSYNMLSGYIPKEIGSMPYLFILNLGHNDISGSIPDEVGDLRGLNILDLSSNKLDGRIPQ
720
AMSALTMLTEIDLSNNNLSGPIPEMGQFETFPPAKFLNNPGLCGYPLPRCDPSNADGYAHHQRSHGRRPASLAGSVAMGL
800
LFSFVCIFGLILVGREMRKRRRKKEAELEMYAEGHGNSGDRTANNTNWKLTGVKEALSINLAAFEKPLRKLTFADLLQAT
880
NGFHNDSLIGSGGFGDVYKAILKDGSAVAIKKLIHVSGQGDREFMAEMETIGKIKHRNLVPLLGYCKVGDERLLVYEFMK
960
YGSLEDVLHDPKKAGVKLNWSTRRKIAIGSARGLAFLHHNCSPHIIHRDMKSSNVLLDENLEARVSDFGMARLMSAMDTH
1040
LSVSTLAGTPGYVPPEYYQSFRCSTKGDVYSYGVVLLELLTGKRPTDSPDFGDNNLVGWVKQHAKLRISDVFDPELMKED
1120
PALEIELLQHLKVAVACLDDRAWRRPTMVQVMAMFKEIQAGSGIDSQSTIRSIEDGGFSTIEMVDMSIKEVPEGKL
```

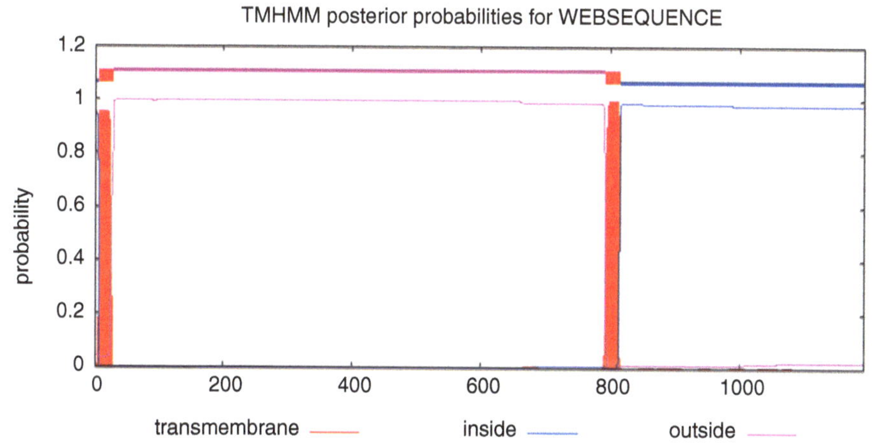

Fig. 2 *N*-glycosylation site and transmembrane domain predictions for Arabidopsis BRI1 (At4g39400, 022476 Uniprot database, 1196 amino acids). The 14 Asn-residues from *N*-glycosylation sites (*N*-X-S/T, shown in *blue*) that face the lumen of the ER and are located in the BRI1 extracellular domain are highlighted in *red*. Asn-residues in *N*-glycosylation sites that face the cytoplasmic side are shown in *blue*. The first transmembrane domain prediction overlaps with the signal peptide sequence

Arabidopsis thaliana mutant for this approach is *stt3a* [33] (*see* **Note 6**). This line has a mutation in one of the two catalytic subunits of the OST complex. As a consequence, glycoproteins can have either no or a reduced number of N-glycans compared to wild-type plants. A faster migrating band indicates the presence of underglycosylation suggesting that the protein is N-glycosylated in wild-type plants (Fig. 3a).

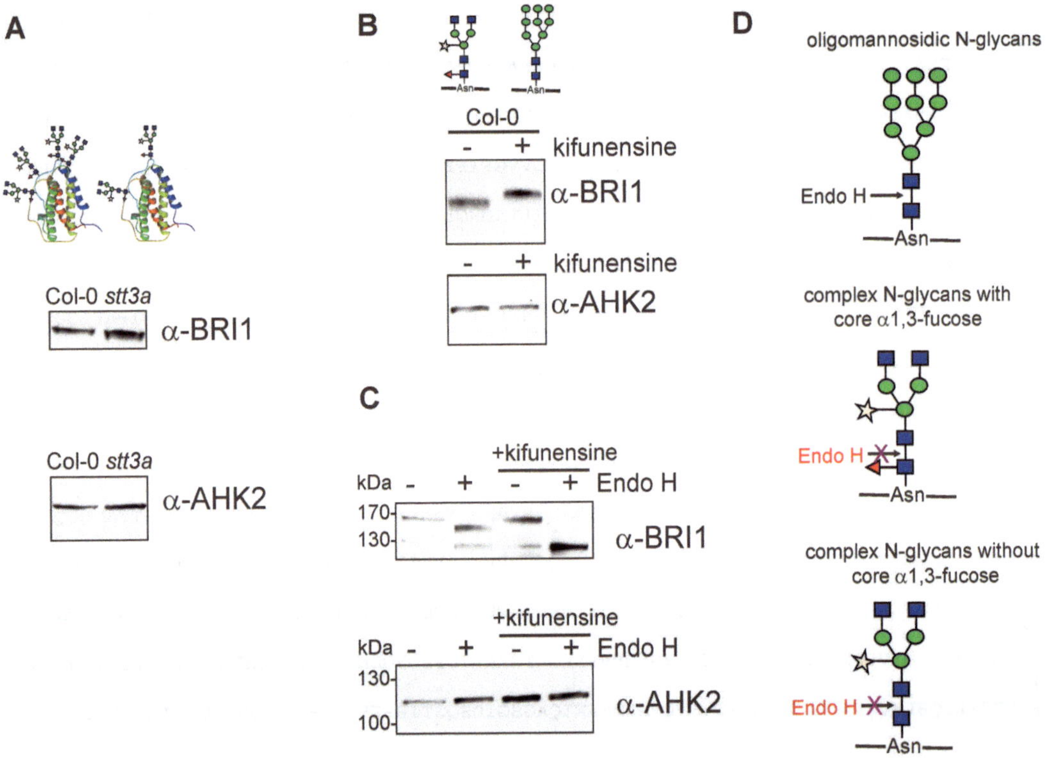

Fig. 3 (**a**) Immunoblot analysis of BRI1 and AHK2 in the *stt3a* underglycosylation mutant which has a defect in the catalytic subunit of the oligosaccharyltransferase complex [33]. BRI1 displays a faster migrating band indicating a reduced number of N-glycans. (**b**) Col-0 seedlings were incubated with the α-mannosidase inhibitor kifunensine and protein extracts were subjected to immunoblotting with BRI1- or AHK2-specific antibodies. Due to the blocked N-glycan processing BRI1 displays a shift in mobility. The major *N*-glycan structures are indicated. (**c**) Immunoblot of Endo H digested protein extracts from Col-0 seedlings that were incubated in the presence or absence of kifunensine. The faster migrating band (<130 kDa) in the Endo H-treated sample represents BRI1 lacking any *N*-glycans. No clear shift is observed for AHK2 suggesting that it is not *N*-glycosylated. (**d**) Schematic representation showing the cleavage specificity of Endo H. Note: a single GlcNAc is still present after Endo H digestion. For description of sugar symbols *see* Fig. 4

1. Harvest 100 mg *stt3a* and Col-0 wild-type seedlings (or leaf material) and transfer to 2 mL Eppendorf tubes containing 2 steel beads (5 mm diameter) per tube.
2. Submerge 2 mL tubes with plant material in container with liquid nitrogen.
3. Mount 2 mL tubes in mixer mill and run 2 min at 50–60 amplitude.
4. Add 4 μL extraction buffer (e.g., PBST or RIPA) per mg of leaf material, vortex shortly, transfer liquid into a 1.5 mL tube and incubate on ice for 15 min, invert tube every 3 min.
5. Centrifuge two times 15 min, 9600 × *g* at 4 °C, transfer supernatant each time to a new tube.

6. Mix samples with 3× Laemmli sample buffer and heat to 95 °C for 5 min.
7. Load SDS-PAGE gel with 10–20 μl and run protein separation for 1.5 h at 100 V.
8. Soak nitrocellulose membrane, sponges and blotting paper in transfer buffer.
9. Perform the gel-membrane assembly according to the user manual from Bio-Rad.
10. Blot for 1 h at 100 V.
11. Disassemble gel-membrane sandwich and carefully rinse the membrane with ultrapure water.
12. Incubate in blocking solution for 1 h at room temperature.
13. Rinse briefly with PBST (AHK2) or TBST (BRI1).
14. Incubate membrane on a shaker for 1.5 h at room temperature in the antibody solution.
15. After incubation, wash 4 times 5 min with PBST or TBST.
16. Add second antibody solution to membrane and incubate for 1.5 h at room temperature on shaker.
17. Wash four times 5 min in PBST or TBST.
18. Perform detection using the chemiluminescent substrate.
19. Develop the film.

3.3 Kifunensine Treatment Followed by Immunoblotting

Due to the current lack of understanding how the OST complex works in plants, it is possible that N-glycosylation is not detected using the *stt3a*-dependent approach, e.g., when N-glycosylation is solely dependent on STT3B activity (*see* **Note 7**). Therefore another practicable approach is the use of the N-glycan processing inhibitor kifunensine. Kifunensine is a class I α-mannosidase inhibitor that blocks the removal of α-linked mannose residues from oligomannosidic N-glycans [36] (Fig. 4). Since this trimming is absolutely required for further processing and complex N-glycan formation [37] all glycoproteins will essentially have the same N-glycan structures upon kifunensine treatment. The change in N-glycan processing can be seen by a mobility shift towards a slower (due to the presence of larger oligomannosidic N-glycans) migrating band (Fig. 3b). Even more important for determination of the glycosylation status, all N-glycans from kifunensine-treated plants are fully sensitive to deglycosylation by Endo H (Fig. 3c, d) (*see* **Note 8**).

1. Seedlings were grown in 0.5× MS medium (Duchefa) supplemented with 0.8% (w/v) agar and 1% (w/v) sucrose or on soil at 22 °C under long day conditions (16 h light/8 h dark).
2. Transfer 10–12-day-old seedlings to liquid 0.5× MS medium supplemented with 1% sucrose and 20 μM kifunensine (*see* **Note 9**).

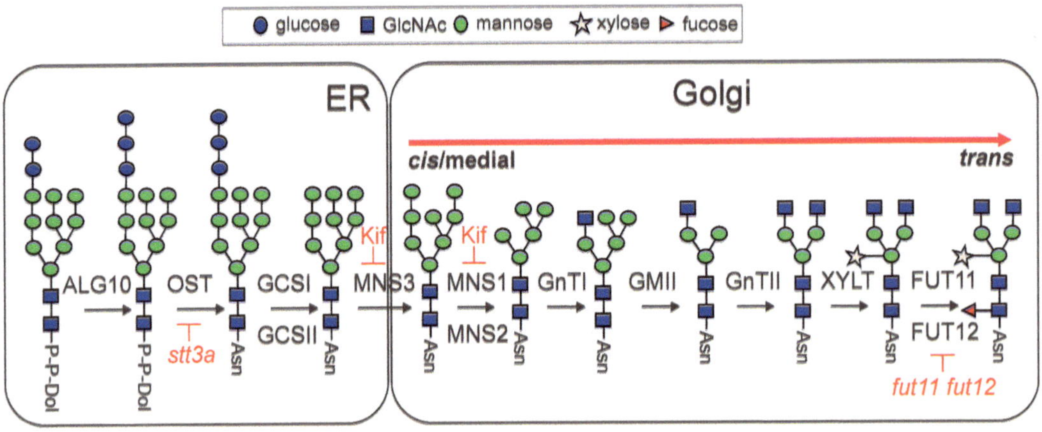

Fig. 4 Major N-glycan processing pathway in plants. STT3A is one of the two catalytic subunits of the oligosaccharyltransferase (OST) complex. Kifunensine (Kif) blocks the activity of class I α-mannosidases (MNS1 to MNS3) resulting in the formation of oligomannosidic *N*-glycans with $Man_9GlcNAc_2$ structures [37]. The symbols for representation of the glycan structures follow the style of the Consortium for Functional Glycomics (www.functionalglycomics.org). *ALG10* α1,2-glucosyltransferase, *GCSI* α-glucosidase I, *GCSII* α-glucosidase II, *MNS3* ER-α-mannosidase I, *MNS1/MNS2* Golgi-α-mannosidase I, *GnTI* *N*-acetylglucosaminyltransferase I, *GMII* Golgi-α-mannosidase II, *GnTII* *N*-acetylglucosaminyltransferase II, *XYLT* β1,2-xylosyltransferase, *FUT11/FUT12* core α1,3-fucosyltransferase

3. Incubate at 22 °C with gentle shaking under long day conditions for 24 h (*see* **Note 10**).
4. Harvest seedlings, remove excess liquid and extract protein as described under Subheading 3.2.
5. Perform immunoblotting as described before.

3.4 Endo H Digestion of Kifunensine Treated Plant Material

1. Incubate 22.5 μL of the protein extract (1–20 μg of protein) with 2.5 μL 10× glycoprotein denaturing buffer for 10 min at 95 °C, transfer to ice and cool for 5 min (*see* **Note 11**).
2. Mix 22.5 μL of the denatured glycoproteins with 3 μL 10× Glyco 3 buffer (NEB), 3 μL ultrapure water, and 1.5 μL Endo H (*see* **Note 12**). For the control reaction replace Endo H with ultrapure water.
3. Incubate for 120 min at 37 °C and stop the reaction by heating to 95 °C for 5 min.
4. Load samples on the SDS-PAGE gel and perform immunoblotting, for example, with anti-BRI1 antibody as described before in Subheading 3.2.

3.5 PNGase F Digestion in Core Fucose Deficient Plants

This protocol shows that the protein is N-glycosylated and can help to answer the question whether the hormone receptor is trafficking through the Golgi. Golgi processed glycoproteins commonly carry N-glycans with core α1,3-fucose which are resistant to PNGase F digestion. A difference in electrophoretic mobility

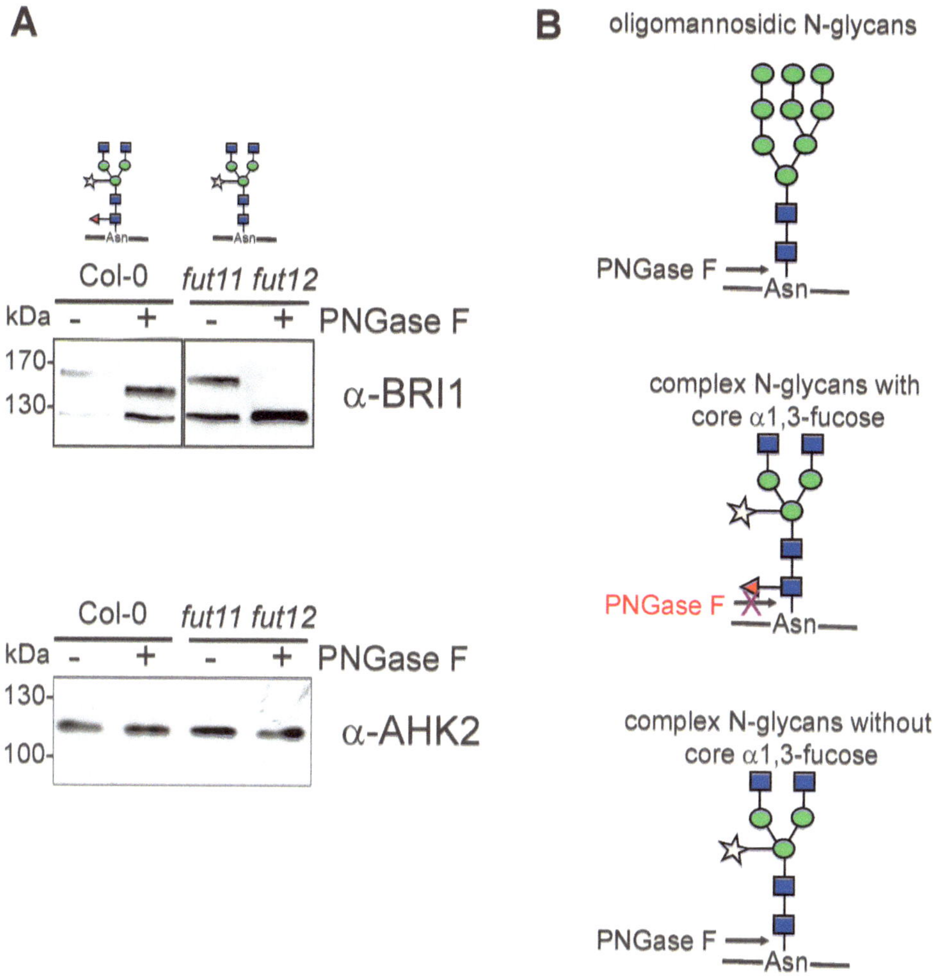

Fig. 5 (**a**) PNGase F digestion of BRI1 and AHK2 in Col-0 and in the *fut11 fut12* double mutant which is devoid of core α1,3-fucose {Strasser, 2004}. BRI1 in Col-0 is only partially deglycosylated (only oligomannosidic and complex *N*-glycans devoid of core fucose are removed) by PNGase F as visible by the minor shift in mobility (band >130 kDa). PNGase F treatment of BRI1 in *fut11 fut12* results in complete deglycosylation (band <130 kDa). The major complex *N*-glycan structure is shown. Again no shift is detectable for AHK2. (**b**) Schematic illustration showing the cleavage specificity of PNGase F. Note: PNGase F deaminates the asparagine to aspartic acid. For description of sugar symbols *see* Fig. 4

between PNGase F-digested wild-type and *fut11 fut12* protein extracts indicates processing in the Golgi (Fig. 5a, b).

1. Perform solubilization and denaturation as described for the Endo H digestion (*see* **Note 13**).
2. Mix 22.5 μL of the denatured glycoproteins with 3 μL 10× Glyco 2 buffer, 3 μL NP-40, and 1.5 μL PNGase F. For the control reaction replace PNGase F with ultrapure water.

3. Incubate for 120 min at 37 °C and stop the reaction by heating to 95 °C for 5 min.
4. Load samples on the SDS-PAGE gel and perform immunoblotting for example with anti-BRI1 antibody as described before in Subheading 3.2.

4 Notes

1. Liquid-chromatography-tandem mass spectrometry-based analysis of glycoprotein glycosylation is the method of choice to obtain information on the glycoforms present at a specific N-glycosylation site [38]. This method which allows also the simultaneous identification of other posttranslational modifications has been described in detail recently [39].
2. There are also other N-glycosylation site prediction tools such as GlycoMine (http://www.structbioinfor.org/Lab/GlycoMine/) which provide the probability for N-glycosylation.
3. Apart from Asn-X-Ser/Thr there is the rare possibility for the use of noncanonical glycosylation sites. The best described is Asn-X-Cys. In addition, on a small number of glycoproteins from diverse model organisms N-glycosylation of Asn-X-Val and Asn-Gly has been described in a recent proteomics approach [40, 41]. While the presence of N-glycosylation at Asn-X-Cys has been reported for a recombinant protein expressed in different plant cells [42], the existence of N-glycosylation at such noncanonical sites remains to be shown for endogenous plant proteins [41].
4. The sequon is necessary but not sufficient for N-glycosylation. N-glycosylation efficiency is dependent on many factors involving the amino acid sequence close to the N-glycosylation site, the secondary structure, the positioning of the site within the protein [15] as well as organism-specific factors. The latter are dependent on the composition and function of the OST complex, which is known to differ between species [43]. Absence or altered function of a specific OST subunit may cause skipping of individual N-glycosylation sites resulting in underglycosylation. Since the OST complex is not very well studied in plants [44] cell- or tissue-specific variations of N-glycosylation on a given glycoprotein cannot be ruled out. Organ-specific changes in N-glycan processing have been reported for the Lewis a epitope formation on complex N-glycans in Arabidopsis [45].
5. N-glycosylation is spatially restricted to the lumen of the ER. Per definition only proteins that enter the secretory pathway can be N-glycosylated. Some studies have reported the presence of N-glycosylated proteins in chloroplasts [46]. These proteins very likely are transported from the ER/Golgi to

chloroplasts via a largely unknown trafficking route. The presence of N-glycosylated byproducts of the ERAD pathway in the cytosol is also possible, but this proteasomal degradation route which requires retrotranslocation from the ER has not been clearly shown for endogenous plant glycoproteins.

6. The *stt3a* mutant grows like wild-type under standard growth conditions and does not show any severe developmental phenotype.
7. Arabidopsis STT3A and STT3B are the two different catalytic subunits of the OST complex. Genetic and biochemical data suggest that they have overlapping as well as distinct functions in N-glycosylation. The mechanism of glycosylation site selection is currently unknown [33]. It is therefore necessary to experimentally test whether a protein is subjected to STT3A-dependent glycosylation. Underglycosylation in the *stt3b* single mutant can be tested in the same way.
8. Crucial for this kind of analysis is the removal of all N-glycans from the glycoprotein. Unfortunately, there is no endoglycosidase available that allows this cleavage. The problem arises from the presence of core α1,3-linked fucose on complex N-glycans which blocks the activity of the commonly used deglycosylation enzyme PNGase F [47]. For subsequent analysis it is therefore essential to prevent the attachment of the core α1,3-fucose. Two strategies are shown here: (i) generation of oligomannosidic N-glycans by kifunensine treatment and (ii) analysis of N-glycosylation in the *fut11 fut12* double knockout [34] which is devoid of active core α1,3-fucosyltransferases.
9. Tunicamycin treatment of plant cells or other plant material is another possibility to examine the N-glycosylation status of a protein [7]. However, in our opinion tunicamycin treatment has several disadvantages: (a) it completely blocks N-glycosylation and therefore can cause extensive misfolding of glycoproteins with subsequent triggering of the unfolded protein response in the ER which eventually leads to cell death. (b) As a consequence of the misfolding and absence of N-glycans the protein of interest may be more instable or aggregate and the subsequent detection by immunoblots may be impossible. By contrast, kifunensine leads only to a modest activation of the unfolded protein response and glycan-dependent folding is less affected.
10. Kifunensine treatment can be done in different ways. Germination and growth of seedlings on MS agar supplemented with 10 μM kifunensine is possible [37]. Alternatively, leaves can be detached from soil grown plants, cut into small pieces and incubated for 24 h in 20 μM kifunensine.
11. Samples in Laemmli buffer can also be used to perform Endo H and PNGase F digestions.

12. For endoglycosidase treatment it is important to find a balance between protein denaturation (to enable access of the deglycosylation enzyme) and protein stability. Prolonged incubation at 37 °C should be avoided to prevent the loss of the glycoprotein.
13. Exoglycosidase digestions are less frequently used to investigate N-glycosylation of plant glycoproteins. Typically, these enzymes remove only individual sugar residues. Unless using MS-based approaches their effect on glycoprotein glycosylation is difficult to monitor. One useful exception is the digestion of glycoproteins with jack bean α-mannosidase. Jack bean α-mannosidase removes up to eight α-linked mannose residues from oligomannosidic N-glycans and allows to distinguish between complex and oligomannosidic N-glycans or between oligomannosidic and glucosylated oligomannosidic N-glycans [30].

Acknowledgements

This work was supported by a grant from the Austrian Science Fund (FWF): P23906.

References

1. Aebi M, Bernasconi R, Clerc S et al (2010) N-glycan structures: recognition and processing in the ER. Trends Biochem Sci 35:74–82
2. Strasser R (2014) Biological significance of complex N-glycans in plants and their impact on plant physiology. Front Plant Sci 5:363
3. Lu X, Tintor N, Mentzel T et al (2009) Uncoupling of sustained MAMP receptor signaling from early outputs in an Arabidopsis endoplasmic reticulum glucosidase II allele. Proc Natl Acad Sci U S A 106:22522–22527
4. Nekrasov V, Li J, Batoux M et al (2009) Control of the pattern-recognition receptor EFR by an ER protein complex in plant immunity. EMBO J 28(21):3428–3438
5. Li J, Zhao-Hui C, Batoux M et al (2009) Specific ER quality control components required for biogenesis of the plant innate immune receptor EFR. Proc Natl Acad Sci U S A 106:15973–15978
6. Saijo Y, Tintor N, Lu X et al (2009) Receptor quality control in the endoplasmic reticulum for plant innate immunity. EMBO J 28(21): 3439–3449
7. Häweker H, Rips S, Koiwa H et al (2010) Pattern recognition receptors require N-glycosylation to mediate plant immunity. J Biol Chem 285:4629–4636
8. Strasser R (2012) Challenges in O-glycan engineering of plants. Front Plant Sci 3:218
9. Hijazi M, Velasquez SM, Jamet E et al (2014) An update on post-translational modifications of hydroxyproline-rich glycoproteins: toward a model highlighting their contribution to plant cell wall architecture. Front Plant Sci 5:395
10. Bond MR, Hanover JA (2015) A little sugar goes a long way: the cell biology of O-GlcNAc. J Cell Biol 208:869–880
11. Delporte A, De Zaeytijd J, De Storme N et al (2014) Cell cycle-dependent O-GlcNAc modification of tobacco histones and their interaction with the tobacco lectin. Plant Physiol Biochem 83:151–158
12. Olszewski NE, West CM, Sassi SO et al (2010) O-GlcNAc protein modification in plants: evolution and function. Biochim Biophys Acta 1800:49–56
13. Steiner E, Efroni I, Gopalraj M et al (2012) The Arabidopsis O-linked N-acetylglucosamine transferase SPINDLY interacts with class I TCPs to facilitate cytokinin responses in leaves and flowers. Plant Cell 24:96–108
14. Niemann MC, Bartrina I, Ashikov A et al (2015) Arabidopsis ROCK1 transports UDP-GlcNAc/UDP-GalNAc and regulates ER pro-

tein quality control and cytokinin activity. Proc Natl Acad Sci U S A 112:291–296
15. Shrimal S, Cherepanova NA, Gilmore R (2015) Cotranslational and posttranslocational N-glycosylation of proteins in the endoplasmic reticulum. Semin Cell Dev Biol 41:71–78
16. Friml J, Jones AR (2010) Endoplasmic reticulum: the rising compartment in auxin biology. Plant Physiol 154:458–462
17. Barbez E, Kleine-Vehn J (2013) Divide Et Impera—cellular auxin compartmentalization. Curr Opin Plant Biol 16:78–84
18. Qiao H, Shen Z, Huang SS et al (2012) Processing and subcellular trafficking of ER-tethered EIN2 control response to ethylene gas. Science 338:390–393
19. Caesar K, Thamm AM, Witthöft J et al (2011) Evidence for the localization of the Arabidopsis cytokinin receptors AHK3 and AHK4 in the endoplasmic reticulum. J Exp Bot 62:5571–5580
20. Wulfetange K, Lomin SN, Romanov GA et al (2011) The cytokinin receptors of Arabidopsis are located mainly to the endoplasmic reticulum. Plant Physiol 156:1808–1818
21. Li J, Chory J (1997) A putative leucine-rich repeat receptor kinase involved in brassinosteroid signal transduction. Cell 90:929–938
22. Pandey S, Nelson DC, Assmann SM (2009) Two novel GPCR-type G proteins are abscisic acid receptors in Arabidopsis. Cell 136:136–148
23. Kim HJ, Ryu H, Hong SH et al (2006) Cytokinin-mediated control of leaf longevity by AHK3 through phosphorylation of ARR2 in Arabidopsis. Proc Natl Acad Sci U S A 103:814–819
24. Napier RM, David KM, Perrot-Rechenmann C (2002) A short history of auxin-binding proteins. Plant Mol Biol 49:339–348
25. Massotte D, Fleig U, Palme K (1995) Purification and characterization of an auxin-binding protein from *Arabidopsis thaliana* expressed in baculovirus-infected insect cells. Protein Expr Purif 6:220–227
26. Feraru E, Vosolsobě S, Feraru MI et al (2012) Evolution and structural diversification of PILS putative auxin carriers in plants. Front Plant Sci 3:227
27. She J, Han Z, Kim TW et al (2011) Structural insight into brassinosteroid perception by BRI1. Nature 474:472–476
28. Hothorn M, Belkhadir Y, Dreux M et al (2011) Structural basis of steroid hormone perception by the receptor kinase BRI1. Nature 474:467–471
29. Hüttner S, Veit C, Vavra U et al (2014) Arabidopsis class I α-mannosidases MNS4 and MNS5 are involved in endoplasmic reticulum-associated degradation of misfolded glycoproteins. Plant Cell 26:1712–1728
30. Jin H, Yan Z, Nam K et al (2007) Allele-specific suppression of a defective brassinosteroid receptor reveals a physiological role of UGGT in ER quality control. Mol Cell 26:821–830
31. Hong Z, Jin H, Tzfira T et al (2008) Multiple mechanism-mediated retention of a defective brassinosteroid receptor in the endoplasmic reticulum of Arabidopsis. Plant Cell 20:3418–3429
32. Hong Z, Kajiura H, Su W et al (2012) Evolutionarily conserved glycan signal to degrade aberrant brassinosteroid receptors in Arabidopsis. Proc Natl Acad Sci U S A 109:11437–11442
33. Koiwa H, Li F, McCully M, Mendoza I et al (2003) The STT3a subunit isoform of the Arabidopsis oligosaccharyltransferase controls adaptive responses to salt/osmotic stress. Plant Cell 15:2273–2284
34. Strasser R, Altmann F, Mach L et al (2004) Generation of *Arabidopsis thaliana* plants with complex N-glycans lacking beta1,2-linked xylose and core alpha1,3-linked fucose. FEBS Lett 561:132–136
35. Blom N, Sicheritz-Pontén T, Gupta R et al (2004) Prediction of post-translational glycosylation and phosphorylation of proteins from the amino acid sequence. Proteomics 4:1633–1649
36. Elbein AD, Tropea JE, Mitchell M et al (1990) Kifunensine, a potent inhibitor of the glycoprotein processing mannosidase I. J Biol Chem 265:15599–15605
37. Liebminger E, Hüttner S, Vavra U et al (2009) Class I alpha-mannosidases are required for N-glycan processing and root development in *Arabidopsis thaliana*. Plant Cell 21:3850–3867
38. Liebminger E, Grass J, Jez J et al (2012) Myrosinases TGG1 and TGG2 from *Arabidopsis thaliana* contain exclusively oligomannosidic N-glycans. Phytochemistry 84:24–30
39. Gruber C, Altmann F (2015) Site-specific glycosylation profiling using Liquid Chromatography-Tandem Mass Spectrometry (LC-MS). Methods Mol Biol 1321:407–415
40. Zielinska DF, Gnad F, Wiśniewski JR et al (2010) Precision mapping of an *in vivo* N-glycoproteome reveals rigid topological and sequence constraints. Cell 141:897–907

41. Zielinska DF, Gnad F, Schropp K et al (2012) Mapping N-glycosylation sites across seven evolutionarily distant species reveals a divergent substrate proteome despite a common core machinery. Mol Cell 46:542–548
42. Matsui T, Takita E, Sato T et al (2011) N-glycosylation at noncanonical Asn-X-Cys sequences in plant cells. Glycobiology 21:994–999
43. Kelleher D, Gilmore R (2006) An evolving view of the eukaryotic oligosaccharyltransferase. Glycobiology 16:47R–62R
44. Farid A, Malinovsky FG, Veit C et al (2013) Specialized roles of the conserved subunit OST3/6 of the oligosaccharyltransferase complex in innate immunity and tolerance to abiotic stresses. Plant Physiol 162:24–38
45. Strasser R, Bondili J, Vavra U et al (2007) A unique beta1,3-galactosyltransferase is indispensable for the biosynthesis of N-glycans containing Lewis a structures in *Arabidopsis thaliana*. Plant Cell 19: 2278–2292
46. Villarejo A, Burén S, Larsson S et al (2005) Evidence for a protein transported through the secretory pathway en route to the higher plant chloroplast. Nat Cell Biol 7: 1224–1231
47. Tretter V, Altmann F, März L (1991) Peptide-N4-(N-acetyl-beta-glucosaminyl)asparagine amidase F cannot release glycans with fucose attached alpha 1-3 to the asparagine-linked N-acetylglucosamine residue. Eur J Biochem 199:647–652

Chapter 18

Highly Sensitive Salicylic Acid Quantification in Milligram Amounts of Plant Tissue

Víctor Carrasco Loba and Stephan Pollmann

Abstract

Technical advances in mass spectrometry constantly raise the bar for analyzing trace amounts of plant hormones in only very small amounts of tissue. Here, a highly sensitive and accurate method is described for the quantitative analysis of the plant hormone salicylic acid not only in the model plant *Arabidopsis thaliana* but also in other plant species. The presented method is optimized for the working up of as little as 20 to 50 mg of plant tissue. The discussed protocol and the utilized laboratory equipment facilitate the implementation of the method into other laboratories that possess access to adequate state-of-the-art gas chromatography-mass spectrometry (GC-MS) equipment.

Key words Salicylic acid, Electron-impact tandem mass spectrometry (EI-GC-MS/MS), Solid-phase extraction, Stable isotopes, Derivatization, Plant hormone analysis

1 Introduction

Plant hormones, for instance, auxins, cytokinins, gibberellins, abscisic acid, salicylic acid, jasmonates, and ethylene, possess pivotal functions in regulating plant growth and development [1]. In this context, salicylic acid plays a particularly important role in the induction of plant defense responses against a variety of biotic and abiotic stresses by orchestrating a multitude of morphological, physiological, and biochemical mechanisms. Among the numerous functions of salicylic acid its central role in triggering plant immune responses to pathogens has to be highlighted [2].

By using large-scale molecular genetic approaches, producing a multitude of publicly available mutant collections and a wealth of different transgenic plants, very valuable insight into the function and signaling of plant hormones has been obtained over the last three decades. However, bearing the essential function of plant hormones in general and salicylic acid in particular in mind, the quantitative analysis of those signaling molecules still remains a challenging task, due to their low abundance in plant tissues.

Jürgen Kleine-Vehn and Michael Sauer (eds.), *Plant Hormones: Methods and Protocols*, Methods in Molecular Biology, vol. 1497, DOI 10.1007/978-1-4939-6469-7_18, © Springer Science+Business Media New York 2017

In order to overcome this problem, modern methods for the quantitative analysis of such trace compounds nowadays rely on highly sensitive mass spectrometric techniques. In combination with the employment of stable-isotope labeled internal standards, mass spectrometry offers not only the unambiguous identification of plant hormones but also their absolute quantitation within a given sample. It has to be remarked that the extraction and pre-purification protocol presented here is suitable to enrich all kind of acidic plant hormones at once, which in fact makes multiplex analysis of several acidic phytohormones, such as auxins, jasmonates, salicylic acid, abscisic acid, and with modifications of gibberellins, from one sample possible if necessary GC-MRM-MS/MS equipment is available [3–5]. Apart from very few entry-level GC-MS machines, modern GC-MS setups generally possess tandem MS (MS/MS or MS^n) capacities. In brief, working in MS/MS mode refers to the selection of suitable parent or precursor ions in the first MS step, which are then fragmented through the collision with a noble gas, most commonly argon. The resulting so called fragment, daughter or product ions are finally recorded in a second MS step. The upside of this methodology is the substantial improvement of the signal to noise ratio that offers the possibility to obtain clean spectra for given target compounds even from very complex samples. On the downside, utilizing a very sensitive technique such as GC-MS renders the establishment of an efficient pre-purification/sample preparation protocol mandatory, because primary extracts are generally not considered suitable for direct assessment by mass spectrometry, even though exactly this would be desirable. In this respect, it has to be remarked that there is no one method that facilitates the simultaneous pre-purification of all plant hormones at once, because of their very different chemical properties. However, mainly because of major advances in sensitivity, liquid chromatography-mass spectrometry (LC-MS) moves more and more into the limelight, also in the field of plant hormone quantification. This is due to the fact that the LC-MS setup is less prone to malfunction and long-lasting contamination of the hardware when the sample quality is slightly on the low side. However, it has to be underlined that also in LC-MS comprehensive sample preparation protocols are sometimes required to facilitate analysis of the desired target substances [6–10].

The method presented in this chapter facilitates the quantitative analysis of salicylic acid in small tissue samples. As stated above, the described purification procedure is also suitable for several other acidic phytohormones and related substances. The method has been validated and successfully used for a number of plant species, for instance Arabidopsis, barley, tobacco, potato, tomato, corn, and rice. It is, however, not appropriate to purify basic plant hormones, such as cytokinins and related derivatives. In this case, the reader is referred to advanced literature for further reading [11–16].

2 Material

2.1 Salicylic Acid Extraction, Pre-purification, and Derivatization

Prepare all solvents using reagents in pro analysis or at least high-performance liquid chromatography (HPLC) grade. If not directly using the pure compounds, it is recommended to prepare all solvents freshly and to use them at room temperature (if not explicitly stated otherwise). The reagents should be stored at room temperature.

1. Acetic acid.
2. Acetone.
3. Methanol.
4. Diethyl ether, dry, free of peroxides.
5. Chloroform.
6. 2-Propanol.
7. (Trimethylsilyl)diazomethane solution, 2.0 M in diethyl ether,
8. *N*,*O*-Bis(trimethylsilyl)trifluoroacetamide (BSTFA) + Trimethylsilyl chloride (TMCS), 99:1 [v/v],
9. Washing solvent for solid-phase extraction: chloroform–2-propanol [2:1, v/v],
10. Elution solvent for solid-phase extraction: diethyl ether + 2% acetic acid [v/v],
11. Thermomixer.
12. Aminopropyl solid-phase extraction columns (Chromabond NH_2 shorty 10 mg, Macherey-Nagel GmbH, Düren, Germany).
13. Centrifuge and corresponding rotor for 1.5 ml microcentrifuge tubes.
14. SpeedVac concentrator.
15. Extraction vacuum manifold.
16. Membrane vacuum pump stand with 1.5 mbar minimal vacuum.
17. Ball mill with Eppendorf cup adaptor (such as Retsch MM300).
18. Stainless steel balls, 3 mm diameter.
19. Ultrasonic bath (these vary widely in their specifications. We use a Bansonic 5510E-DTH, Branson Ultrasonics Corporation, Danbury, CT, USA).
20. 600 μl crimp top conical microvial (Chromacol Uni-VL Supelco #27312), plus appropriate 8 mm crimp seal PTFE/rubber for closing.

2.2 Stable-Isotope Labeled Internal Standard

Stable-isotope labeled salicylic acid, [$^2H_4$]-salicylic acid, to be used as internal standard in the mass spectrometric analysis is commercially available and can be purchased, e.g., from CDN isotopes (Pointe-Claire, Canada) or OlChemIm Ltd. (Olomouc, Czech

Republic). There is the possibility to use standards from other vendors (*see* **Note 1**) given that the isotopic enrichment of the substance is higher than 96%, and the compound does neither decay nor detectably exchange the heavy isotopes during the sample preparation and analysis process. Please note that it is highly recommended that the mass difference between internal standard and analyte is not less than two AMU.

3 Methods

3.1 Extraction of Plant Tissue

If not explicitly specified otherwise, all steps can be carried out at room temperature.

1. Weigh between 20 to 50 mg of plant tissue (fresh weight) without previous homogenization into a 1.5 ml disposable reaction tube and immediately add 1 ml of methanol, 30 pmol of [$^2H_4$]-salicylic acid internal standard (*see* **Note 2**), and three steel balls.
2. Use a thermomixer to heat the sample to 60 °C and incubate it under gentle agitation for 3 min, then homogenize by heavy vortexing or shaking for several minutes. Optimally, a vibrating-ball micro-mill (Retsch MM300) can be utilized, subjecting the sample to two rounds of 3 min, each at 30 Hz (*see* **Note 3**). Further extraction will take place at room temperature for at 1 h. The sample should be vortexed or inverted once in a while.
3. Remove the steel balls by using a magnetic tool and centrifuge the sample (14,000 × *g*, 1 min) to sediment cell wall debris and other floating particles. Then, carefully transfer the supernatant to a new reaction tube, not picking up parts of the pellet. Take the sample to complete dryness using a SpeedVac set to 60 °C (*see* **Note 4**) and maximum vacuum. Please note that after drying the sample can be stored at −80 °C until further processing.

3.2 Pre-purification and Enrichment of Acidic Compounds by Solid-Phase Extraction

1. Thoroughly dissolve the crude residue in 50 μl methanol, then add 200 μl of diethyl ether. Sonify the sample for 5 min in an ultrasonic bath, before centrifuging it for 5 min at 14,000 × *g*.
2. During the preparation of the sample, already start with the equilibration of the aminopropyl solid-phase extraction column (*see* **Note 5**). To do so, wash the column with 2 × 200 μl diethyl ether.
3. Transfer the entire sample onto the conditioned column and let the sample pass the matrix by gravity flow without applying vacuum to the SPE manifold. This is likely to improve the association of acidic compounds to the NH_2-matrix.
4. After passage of the sample, wash the column twice with 200 μl freshly prepared washing solvent (chloroform:2-propanol). Clean the tip of the plastic gasket holding the column with some white laboratory paper.

5. Elute the sample from the matrix into a fresh reaction vessel by adding 2 × 200 μl of the elution solvent (acidified diethyl ether, containing 2 % acetic acid [v/v]). Finally, dry the combined eluates using a SpeedVac concentrator (10 mbar, 60 °C). Remove residual acetic acid in a gentle stream of nitrogen.

3.3 Derivatization of the Sample Prior to GC-MS Analysis

The contained salicylic acid is preferentially analyzed by gas chromatography-mass spectrometry (GC-MS) in form of its methyl ester, which is characterized by a substantially lower melting point and, thus, better vaporization properties during the subsequent GC-MS analysis.

1. Prepare 10 ml of acetone–methanol (9:1, [v/v]) solution. When kept in darkness and in an airtight vessel, the mixture can be stored at room temperature over a longer period of time.
2. To prepare the derivatization solvent for the methylation of organic acids, mix 220 μl of the acetone–methanol solution with 27 μl diethyl ether and 3 μl of the (trimethylsilyl)diazomethane solution (*see* **Note 6**). The resulting 250 μl are sufficient for the chemical modification of at least ten samples (*see* **Note 7**).
3. Dissolve sample in 25 μl methanol and transfer the solution into a 600 μl crimp top conical microvial (Chromacol Uni-VL Supelco #27312). Dry completely in a gentle stream of nitrogen.
4. Add 20 μl of the freshly prepared derivatization solvent to the sample. Immediately close the vial with a suitable 8 mm crimp seal with PTFE/rubber septa. Let the sample rest for 30 min at room temperature before proceeding with the GC-MS analysis.

3.4 Optional Derivatization Protocol

In case that handling diazomethane containing substances in the laboratory is not desired or permitted for whatever reason, salicylic acid can alternatively and very effectively be chemically modified by trimethyl silylation.

1. Dissolve sample in 25 μl methanol and transfer the solution into a 600 μl crimp top conical microvial (Chromacol Uni-VL Supelco #27312). Dry completely in a gentle stream of nitrogen.
2. Add 20 μl *N*,*O*-Bis(trimethylsilyl)trifluoroacetamide (BSTFA) + Trimethylsilyl chloride (TMCS), 99:1 [v/v] to the sample. Immediately close the vial with a suitable 8 mm crimp seal with PTFE/rubber septa.
3. Incubate the sample over 30 min at 70 °C before proceeding with the GC-MS analysis.

3.5 Gas Chromatographic-Mass Spectrometric Assessment

From the derivatized sample, 1 μl aliquots are injected into the GC-MS system for gas chromatographic separation and subsequent mass spectrometric analysis.

3.5.1 GC Setup

Injection of the sample was carried out in splitless mode. A pressure pulse of 25 psi over 1 min was used to force the transfer of compounds from the injector into the column. After that time, the split was fully opened for 1.5 min before it was set to a split ratio of 20% for the remaining run time. The injector temperature was 250 °C and the column temperature was held at 50 °C for 1.2 min. Then, it was increased by 30 °C/min to 120 °C, followed by a further increase to 325 °C by 10 °C/min. Finally, a temperature of 325 °C was held for another 4 min. Separation was achieved by using a 30 m×0.25 mm i.d. fused silica capillary column with a chemical bond 0.25-μm ZB35 stationary phase (Phenomenex, Torrance, USA) (*see* **Note 8**). Helium at a flow rate of 1 mL/min served as the mobile phase.

3.5.2 Mass Spectrometer Setup

The method described here is optimized for triple-quadrupole mass spectrometers run with positive polarity. The mass spectrometer was operated in EI-MRM mode. The transfer line temperature was set to 250 °C and the ion source temperature to 200 °C. Ions were generated with −70 eV at a filament emission current of 80 μA. The dwell time was 175 ms. Argon set at 2.0 mTorr was used as the collision gas. For both target compounds, derivatized salicylic acid and derivatized [$^2H_4$]-salicylic acid, at least two transitions were recorded. One transition was used as the quantifier ion, whereas the others served as qualifier ions providing additional information about the analyte as well as indicating the presence of possible impurities. The selected precursor ions and corresponding diagnostic product ions are listed in Table 1.

The amount of the endogenous compound was calculated from the signal ratio of the unlabeled over the stable isotope-containing mass fragment observed in the parallel measurements (Table 1).

4 Notes

1. Stable-isotope labeled salicylic acid is also available from Cambridge Isotopes (www.isotope.com), Isotec (Sigma-Aldrich; www.isotec.com, www.sigma-aldrich.com), and Medical Isotopes (www.medicalisotopes.com).
2. Note that it is mandatory to determine the appropriate amount of internal standard to be used in preliminary experiments. The necessary standard content can change for different tissues and developmental stages, respectively. It is recommended that the amount of standard added to the sample should neither exceed the endogenous content of the analyte by five times, nor should it be less than 1/5 of the respective analyte to be analyzed.

Table 1
Characteristic precursor and product ions used for the detection of derivatized salicylic acid and its corresponding internal standard

Compound	Retention time(min)	Retention time window (min)	Precursor ion (*m/z*)	Product ion (*m/z*)	Collision energy	Scan time (%)	Quantifier ion
MeSA	6.31	1	152	120	10	33.33	X
	6.31	1	120	91	10	33.33	
	6.31	1	120	59	15	33.33	
[$^2H_4$]-MeSA	6.31	1	156	124	10	33.33	X
	6.31	1	124	95	10	33.33	
	6.31	1	124	63	25	33.33	
SA-diOTMS	10.39	10.39	267	209	10	33.33	X
	10.39	10.39	267	149	10	33.33	
	10.39	10.39	267	73	10	33.33	
[$^2H_4$]-SA-diOTMS	10.39	10.39	271	213	10	33.33	X
	10.39	10.39	271	153	10	33.33	
	10.39	10.39	271	73	10	33.33	

It has to be noted that the retention time for the target compound may moderately shift due to system dependent parameters. It is recommended to use reference compounds to determine retention times for the actual setup prior to the analysis (*see* **Note 8**)

3. More rigid plant organs, e.g., more lignified tissues such as stems or seeds, may require longer vibration times. If no vibrating-ball micro-mill is available, the tissue can alternatively be ground in liquid nitrogen using micro-pistils and standard reaction tubes. The resulting powder must not thaw before the addition of the methanol.
4. It is important that the sample is completely dried with no traces of water left. Unnecessary processing in the SpeedVac concentrator, however, should be avoided as overdrying may result in analyte loss. Given the fact that salicylic acid has its melting point at approximately 159–161 °C the risk of analyte loss by evaporation is low, though. To our experience, processing times of up to 2 h do not negatively affect analyte contents.
5. It is possible to use commercially available microtips from other vendors, e.g., from Glygen Corp., Columbia, USA or Chromacol Ltd., Welwyn Garden City, UK, for solid-phase extraction. Please note, however, that the presented proto-

col has not been optimized for such type of pipet microtips and, thus, needs to be adapted in case that usage of microtips is desired.

6. Alternatively, ethereal diazomethane can be used for the methylation. Ethereal diazomethane can be prepared from *N*-nitrosomethylurea recrystallized from methanol. *N*-nitrosomethylurea can be obtained from Sigma-Aldrich. However, in any case, when preparing and/or handling diazomethane containing solutions safety precautions should strictly be obeyed. All steps have to be performed in a well-ventilated laboratory fume hood.
7. It is recommended to prepare only the amount of derivatization solution required for immediate use. The solvent should only be used freshly prepared.
8. GC-capillary columns with an intermediate polarity similar to that of the here used ZB-35 column (35 % phenyl- 65 % dimethylpolysiloxane, low bleeding) are standard for acidic phytohormone profiling. However, other stationary phases (e.g., containing 5 % or 50 % phenyl) that provide either lower or higher polarity can also be used. It has to be noted that such columns may need a slightly different temperature program than the one reported here to achieve satisfying separation of the analytes.

References

1. Davies PJ (2010) Plant hormones. Biosynthesis, signal transduction, action! 3rd edn. Kluwer Academic, London, Revised
2. Yan S, Dong X (2014) Perception of the plant immune signal salicylic acid. Curr Opin Plant Biol 20:64–68
3. Müller A et al (2002) A multiplex GC-MS/MS technique for the sensitive and quantitative single-run analysis of acidic phytohormones and related compounds, and its application to *Arabidopsis thaliana*. Planta 216:44–56
4. Schmelz EA et al (2003) Simultaneous analysis of phytohormones, phytotoxins, and volatile organic compounds in plants. Proc Natl Acad Sci U S A 100:10552–10557
5. Birkemeyer C et al (2003) Comprehensive chemical derivatization for gas chromatography-mass spectrometry-based multi-targeted profiling of the major phytohormones. J Chromatogr A 993:89–102
6. Chiwocha SD et al (2003) A method for profiling classes of plant hormones and their metabolites using liquid chromatography-electrospray ionization tandem mass spectrometry: an analysis of hormone regulation of thermodormancy of lettuce (*Lactuca sativa* L.) seeds. Plant J 35:405–417
7. Ziegler J et al (2014) Simultaneous analysis of apolar phytohormones and 1-aminocyclo propan-1-carboxylic acid by high performance liquid chromatography/electrospray negative ion tandem mass spectrometry via 9-fluorenylmethoxycarbonyl chloride derivatization. J Chromatogr A 1362:102–109
8. Strehmel N et al (2014) Profiling of secondary metabolites in root exudates of *Arabidopsis thaliana*. Phytochemistry 108:35–46
9. Wang X et al (2014) Quantitative profiling method for phytohormones and betaines in algae by liquid chromatography electrospray ionization tandem mass spectrometry. Biomed Chromatogr 28:275–280
10. Novák O et al (2012) Tissue-specific profiling of the *Arabidopsis thaliana* auxin metabolome. Plant J 72:523–536
11. Astot C et al (1998) Precolumn derivatization and capillary liquid chromatographic/frit-fast atom bombardment mass spectrometric analysis of cytokinins in *Arabidopsis thaliana*. J Mass Spectrom 33:892–902

12. Jameson PE et al (2000) Cytokinins. Extraction, separation, and analysis. Methods Mol Biol 141:101–121
13. Farrow S, Emery RN (2012) Concurrent profiling of indole-3-acetic acid, abscisic acid, and cytokinins and structurally related purines by high-performance-liquid-chromatography tandem electrospray mass spectrometry. Plant Methods 8:42
14. Kojima M, Sakakibara H (2012) Highly sensitive high-throughput profiling of six phytohormones using MS-probe modification and liquid chromatography–tandem mass spectrometry. In: Normanly J (ed) High-throughput phenotyping in plants. Humana, New York, pp 151–164
15. Takei K et al (2003) A method for separation and determination of cytokinin nucleotides from plant tissues. J Plant Res 116:265–269
16. Novák O et al (2008) Cytokinin profiling in plant tissues using ultra-performance liquid chromatography–electrospray tandem mass spectrometry. Phytochemistry 69:2214–2224

Chapter 19

High-Resolution Cell-Type Specific Analysis of Cytokinins in Sorted Root Cell Populations of *Arabidopsis thaliana*

Ondřej Novák, Ioanna Antoniadi, and Karin Ljung

Abstract

We describe a method combining fluorescence-activated cell sorting (FACS) with one-step miniaturized isolation and accurate quantification of cytokinins (CKs) using ultra-high performance liquid chromatography–tandem mass spectrometry (UHPLC-MS/MS) to measure these phytohormones in specific cell types of *Arabidopsis thaliana* roots. The methodology provides information of unprecedented resolution about spatial distributions of CKs, and thus should facilitate attempts to elucidate regulatory networks involved in root developmental processes.

Key words Cytokinins, Fluorescence-activated cell sorting (FACS), Solid-phase extraction (SPE), Liquid chromatography–mass spectrometry (LC-MS), Roots, Arabidopsis, Protoplasts

1 Introduction

Hormonal interactions affect most (if not all) aspects of plant growth and development. Dissection of the interactive regulatory networks involved frequently requires accurate quantitative information on plants' hormonal contents at a fine tissue level. Thus, methods for measuring cell-type specific auxin levels have been developed and applied [1], but developing corresponding methods for cytokinin (CK) quantification has proved to be much more challenging [2]. This is partly because CK levels are at least tenfold lower than auxin levels, and thus more difficult to measure in small amounts of tissue. Furthermore, CK metabolite profiles are much more complex, requiring simultaneous measurements of multiple compounds. The methodology presented here increases the resolution of CK quantification by permitting targeted isolation of specific cell populations, efficient pre-concentration of cytokinin metabolites and their detailed profiling from sets as small as just 20,000 specific cells.

The method involves use of flow cytometry-based fluorescence-activated cell sorting (FACS), which enables the recognition of

Jürgen Kleine-Vehn and Michael Sauer (eds.), *Plant Hormones: Methods and Protocols*, Methods in Molecular Biology, vol. 1497, DOI 10.1007/978-1-4939-6469-7_19, © Springer Science+Business Media New York 2017

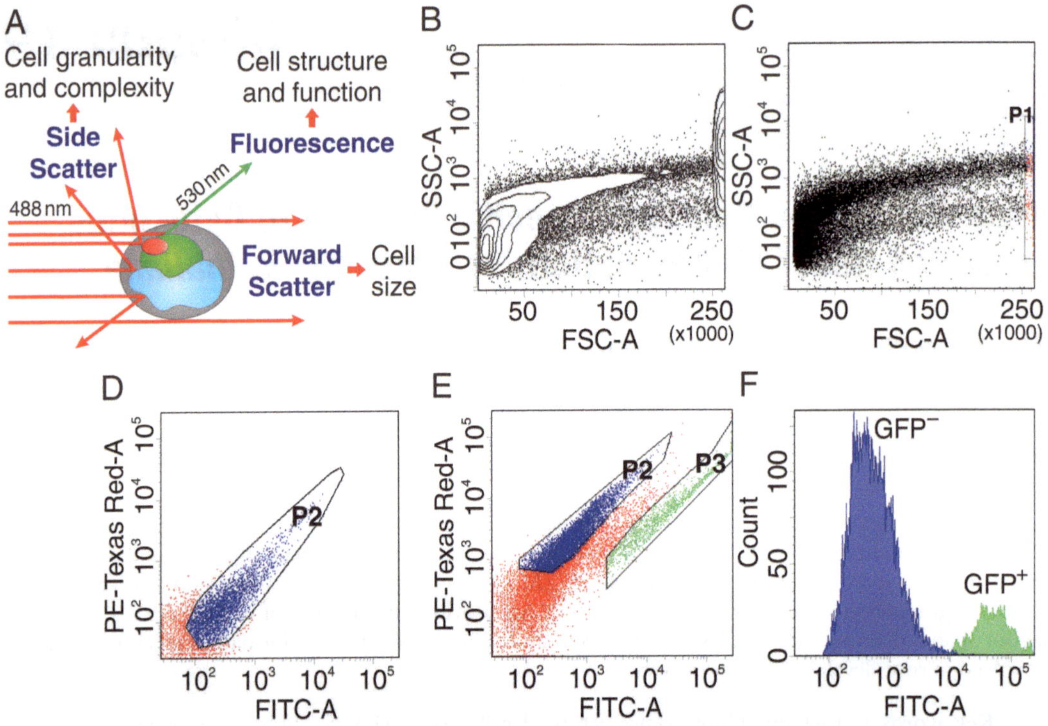

Fig. 1 Isolation of GFP-expressing protoplasts using flow cytometry. (**a**) The scatter of the laser beam by the cells is used to count the cells and measure cell size and intracellular granularity/homogeneity. (**b**) Cell distribution based on FSC-A (Forward-scattered light) vs SSC-A (Side-scattered light). (**c**) Dot plot with a gate encompassing the protoplast population (P1). (**d**, **e**) Plots of green fluorescence intensity (FITC-A; excitation 488 nm/emission 530/30 nm) and red fluorescence intensity (PA-Texas Red-A; excitation 488 nm/emission 576/26 nm). (**d**) Scatter plot of protoplasts from wild-type *Arabidopsis* plants not expressing GFP with most cells having a typical autofluorescence pattern (P2 population). (**e**) Sorting of protoplasts from an Arabidopsis GFP-expressing line with a gate (P3) set to collect the GFP tail where cells display more green than red fluorescence. (**f**) Single-parameter histogram displaying relative fluorescence intensity (FITC-A) on the *x*-axis and the cell count on the *y*-axis of the gated protoplast population (*P2* blue-GFP^{-}; *P3* green-GFP^{+})

isolated protoplasts of targeted size, granularity, and/or complexity, followed by their sorting into single-cell units or homogenous groups according to the presence or absence of internal fluorochromes. A large collection of transgenic plant lines expressing fluorescent reporter genes, especially green fluorescent protein (GFP), under specific promoters is now available, enabling the labeling of spatially well-defined cell types (NASC, the European Arabidopsis Stock Centre; http://arabidopsis.info/). As shown in Figs. 1 and 2, GFP-based isolation of protoplasts from an organized plant tissue of such a transgenic line enables separation of single cells while maintaining their cellular identities [3, 4]. Cell populations isolated using flow cytometry into GFP-expressing (GFP^{+}) and non-GFP-expressing (GFP^{-}) protoplasts (Fig. 1) can be collected via FACS (Fig. 2) and used to analyze cell type-specific metabolite profiles ([5–7], *see* **Note 1**).

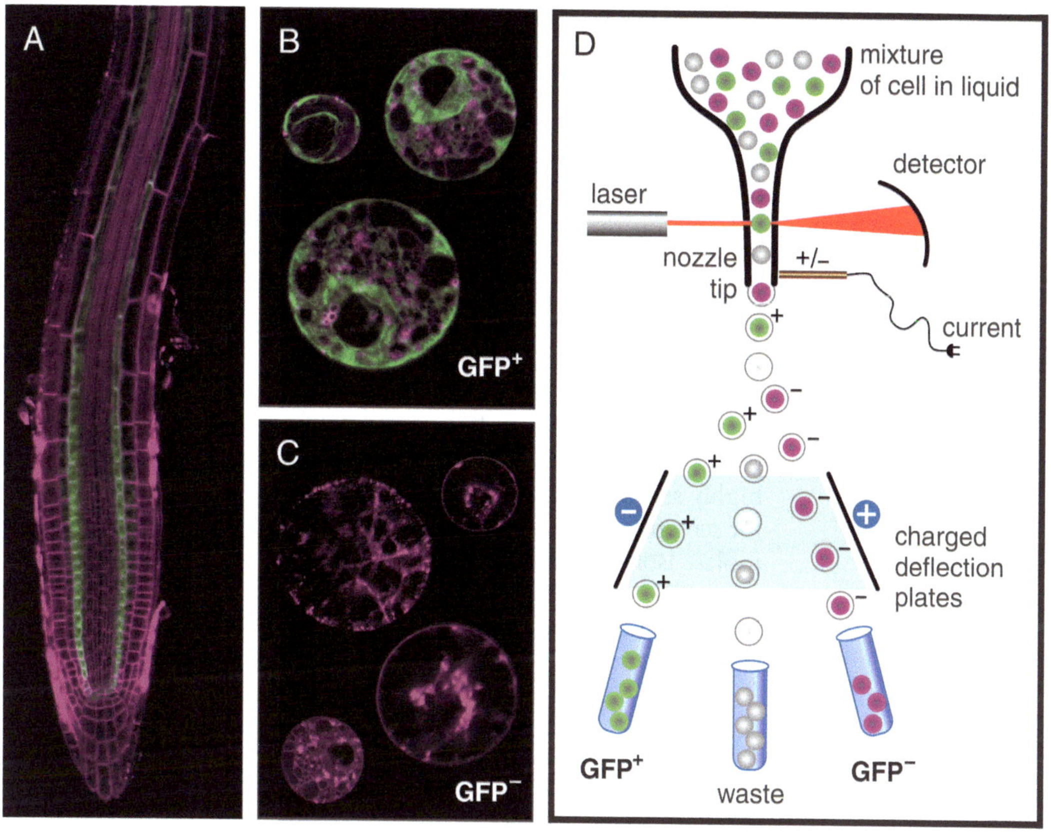

Fig. 2 Fluorescence-activated cell sorting (FACS) of plant protoplasts expressing green fluorescent protein (GFP). (**a**) Fluorescent image of the root tip of the *Arabidopsis thaliana pSCR:GFP* line [3]. (**b**, **c**) The resulting GFP-expressing (GFP$^+$; **b**) and non-GFP-expressing (GFP$^-$; **c**) protoplasts after digestion of the root with cell wall-degrading enzymes (FM4-64 staining in *magenta*; GFP fluorescence in *green*). (**d**) Schematic illustration of fluorescence-activated cell sorting. The cells are individually hit by a laser and consequently light is scattered and fluorescence is emitted. These light signals are perceived by detectors and enable characterization of each cell according to its size, granularity, complexity, and fluorescence. The desired cell populations are then selected by gates designed to indicate the GFP$^-$ and GFP$^+$ cells to be sorted and collected. Stream vibration causes the formation of single droplets (each containing a characterized cell), which are separated from the main stream. An electrical charge is applied to the stream and the newly formed drops become positively or negatively charged according to the gating strategy while the cells/particles excluded from the gates remain uncharged. The charged droplets are then deflected to the left or right by charged electrodes into the appropriate GFP$^+$ and GFP$^-$ cell collection tubes, while the uncharged droplets go straight to the waste aspirator

The methodology also exploits recent advances in mass spectrometry that have sharply increased the sensitivity and selectivity of final analyses of complex plant samples, particularly ultra-high performance liquid chromatography–electrospray tandem quadrupole mass spectrometry (UHPLC-MS/MS), which has been widely used in plant hormone profiling [8]. As shown here, in combination with recently developed micro-purification techniques for cytokinin isolation, such as multilayer solid-phase microextraction [9], batch immunoaffinity chromatography [10], magnetic solid-phase extraction [11], and monolithic molecularly imprinted solid-phase extraction [12], UHPLC-MS/MS has the potential to provide cell-specific quantification of CK metabolites.

The methodology presented here also includes a powerful one-step high-throughput micro-purification (microSPE) approach for extracting cytokinins from only a few thousand sorted cells. It is based on the so-called StageTip (Stop and Go Extraction Tip) technology recently employed for isolating CKs from minute amounts of fresh plant tissues (1–5 mg) using a combination of reverse-phase and cation-exchange polymer-based sorbents [8]. Exploiting the high selectivity, affinity and capacity of the selected phases, we subsequently optimized the microSPE protocol for efficient isolation and enrichment of CKs from a minimal quantity of protoplasts [2]. Finally, the whole procedure was coupled with a highly selective and extremely sensitive UHPLC-MS/MS method for screening 26 isoprenoid CK metabolites. In summary, the protoplast isolation and cell sorting procedure combined with rapid microSPE and ultrasensitive UHPLC-MS/MS can be used for high-resolution measurements of CKs in specific *Arabidopsis* root cell populations.

2 Materials

All solutions are prepared at room temperature using ultrapure water (e.g., Milli-Q water with ≤18 M W /cm resistance at 25 °C). All chemicals and solvents should be at least of analytical reagent grade.

2.1 Protoplast Isolation for One FACS Experiment

1. 1 × MS growth media (3 L): 0.44 % (w/v) Murashige and Skoog salt mixture, 1 % (w/v) sucrose, 0.5 % (w/v) MES (pH 5.7, adjusted with KOH), 1 % (w/v) plant agar.
2. Thirty square petri dishes (120 mm × 120 mm).
3. Thirty sterile square-cut nylon meshes (Sefar Nitex 03-110/47).
4. Sterilized seeds of a transgenic line expressing GFP in a specific cell population (~9000 seeds or ~500 μL volume measured in a microcentrifuge tube).
5. 100 mL PB solution: 600 mM mannitol, 2 mM $CaCl_2$, 10 mM KCl, 2 mM MES, 2 mM $MgCl_2$, 0.1 % (w/v) BSA, pH 5.7 (*see* **Note 2**).
6. 25 mL PBI solution: 45.45 mg pectolyase (0.3 U/mL) and 450 mg cellulysin (45 U/mL) in 25 mL PB solution (*see* **Note 3**).
7. Surgical blade and forceps.
8. Aluminum foil.
9. Orbital shaker.

10. 40 μm cell strainer (BD Falcon).
11. Refrigerated swing-out centrifuge.

2.2 Fluorescent Activated Cell Sorting (FACS)

1. 5 L of 0.7% (w/v) sodium chloride (Sheath Fluid) in sterile container.
2. Sterile 500 mL Bottle Top Filter with 75 mm Supor® machV PES Membrane, 0.2 μm pore diameter (Nalgene® 595-4520 Rapid-Flow™, 45 μm, 33 mm Blue Neck, Thermo Scientific).
3. FACS (BD FACS Aria I Flow Cytometer, BD Biosciences).
4. Software: BD FACSDiva version 6.1.2.
5. FACS Rinse and FACS Clean solutions (Cat. Nos. 340346 and 340345, respectively, BD Biosciences).
6. CS&T and Accudrop beads (BD Biosciences).
7. Cooling system connected to FACS.
8. 100 μm nozzle (P/N 342909, BD Biosciences).
9. Rubber O-Ring (70SL, Apple Rubber Products).
10. 50 μm cup filter (Cat. No. 340631, BD Biosciences).
11. Round-bottom tubes 12×75 mm with lid for sample collection (Cat. No. 352054, BD Falcon) and without lid for sample loading and cleaning processes (Cat. No. 14-961-26, Fisher Scientific).

2.3 Cytokinin Purification

The volumes in parenthesis are for 50 samples.

1. Stable isotope-labeled internal standards (IS) (*see* **Note 4**):

   Mixture I—1×10^{-8} M (0.01 pmol/μL; 500 μL) of [$^{13}C_5$]*cis*-zeatin (*c*Z), [$^{13}C_5$]*trans*-zeatin (*t*Z), [$^{2}H_3$]dihydrozeatin (DHZ), [$^{2}H_6$]*N6*-(Δ^2-isopentenyl)adenine (iP), [$^{2}H_5$]*t*Z riboside (*t*ZR), [$^{2}H_3$]DHZ riboside (DHZR), [$^{2}H_6$]iP riboside (iPR), [$^{2}H_5$]*t*Z-7-*N*-glucoside (*t*Z7G), [$^{2}H_6$]iP-7-*N*-glucoside (iP7G):

   Mixture II—2×10^{-8} M (0.02 pmol/μL; 500 μL) of [$^{2}H_5$]*t*Z-*O*-glucoside (*t*ZOG), [$^{2}H_7$]DHZ-*O*-glucoside (DHZOG), [$^{2}H_5$]*t*Z-9-*N*-glucoside (*t*Z9G), [$^{2}H_3$]DHZ-9-*N*-glucoside (DHZ9G), [$^{2}H_6$]iP-9-*N*-glucoside (iP9G):

   Mixture III—1×10^{-7} M (0.1 pmol/μL; 500 μL) of [$^{2}H_5$]*t*ZR-*O*-glucoside (*t*ZROG), [$^{2}H_5$]*t*ZR 5′-monophosphate (*t*ZRMP), [$^{2}H_3$]DHZR 5′-monophosphate (DHZRMP), and [$^{2}H_6$]iPR 5′-monophosphate (iPRMP).

2. Multi-StageTips (Fig. 3; *see* **Note 5**)—microSPE columns packed with Empore™ High Performance Extraction Disks (3-layers of each sorbent type C_{18}, SDB-RPS and Cation-SR).
3. Microcentrifuge tube, 2 mL with lid (×2).
4. 100% acetone (2.5 mL).
5. 100% methanol (HPLC grade) (5 mL).

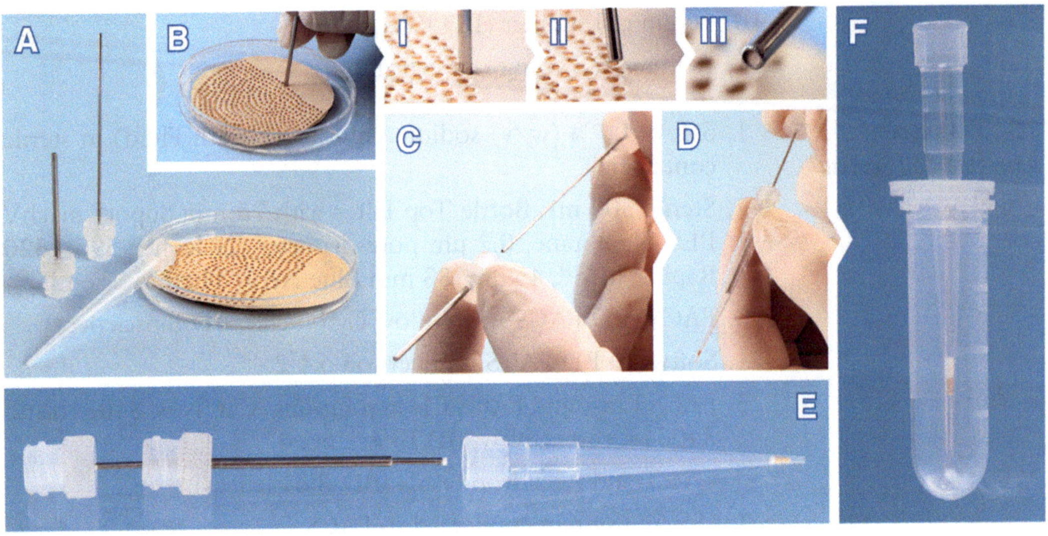

Fig. 3 Step-by-step guide for preparing a multi-StageTip. (**a**) The pipette tip, Empore™ High Performance Extraction Disk on a petri dish, cutter (blunt-ended syringe needle) and plunger (rod needle); (**b**) cutting a small disk, with approximately 1.0 mm diameter and 0.5 mm thickness (**I–III**, the cutter is gently pressed into the Empore disk, thus introducing a small disk of the sorbent material into the needle); (**c–d**) insertion of the disk into the pipette tip (200 μL) using the cutter and plunger fitted into the needle; (**e**) placement of an additional disk onto the first disk; (**f**) prepared multi-StageTips in a lid of the microcentrifuge tube (2.0 mL)

6. 50 % (v/v) nitric acid (2.5 mL): add 1.92 mL of 65 % HNO_3 to 0.58 mL water.
7. 1 M hydrochloric acid (375 μL): add 33 μL of 35 % HCl to 342 μL water.
8. 0.35 M ammonium hydroxide in 60 % methanol (2.5 mL): mix 1.5 mL methanol with 0.91 mL water, then add 0.089 mL concentrated aqueous ammonia (≥25 % in H_2O).
9. Refrigerated centrifuge.
10. Centrifuge vacuum concentrator (e.g., SpeedVac).
11. Screw top autosampler vials (volume 2 mL) and glass inserts (volume 250 μL).

2.4 UHPLC-MS/MS Method

1. Acquity UPLC® CSH C18 Column, 130 Å, 1.7 μm, 2.1 × 150 mm (Waters, Milford, MA, USA).
2. Methanol (LC-MS grade).
3. 15 mM ammonium formate (pH 3.95): add 566 μL formic acid (~98 %, LC-MS grade) to 1 L ultrapure water and adjust to pH 3.95 with ammonium hydroxide (concentration ≥25 % in H_2O).
4. UHPLC-MS/MS system and associated equipment. The LC instrument should include at least binary ultrahigh-pressure pumps, a column heater and an autosampler with temperature-

controlled system. The MS system should include a tandem quadrupole mass spectrometer.

3 Methods

The initial step is collection of samples (GFP^+ and corresponding GFP^- cell populations, *see* **Note 6**). The time required for sorting will depend mainly on the GFP expression pattern and may be substantial (e.g., 2–3 weeks for 6–10 biological replicates of a transgenic line expressing GFP in 20% of the cells, *see* **Note 7**). When the samples have been collected, CK purification using the high-throughput micropurification process presented here can be performed within six hours (for up to 48 samples). Finally with the rapid UHPLC-MS/MS method described in Subheading 3.4, less than one day is required to analyze 72 samples.

3.1 Material Preparation for One Sorting Experiment (3–4 Days)

1. Wrap mesh nylon squares in double aluminum foil and sterilize at 70 °C for 2–3 h.
2. Prepare, autoclave, and pour 3 L of MS growth medium in square petri dishes (approximately 60 ml/dish). When the medium has just solidified, transfer the mesh with forceps under sterile conditions onto its surface (*see* **Note 8**). The mesh should instantly darken slightly as it absorbs moisture from the medium (the use of mesh in the process is optional). Store the prepared dishes at least overnight at 4 °C.
3. Sow sterilized seeds of the transgenic GFP-expressing line by pipetting them in a water suspension, in three dense rows on the petri dishes containing the MS medium covered with mesh (approximately 100 seeds per row).
4. Store for 2–3 days in the dark at 4 °C to stimulate and synchronize germination.
5. Transfer the plates to the growth chamber and grow the seedlings under long day conditions for 8 days (*see* **Note 9**).

3.2 Protoplast Isolation for One Cell Sorting Experiment (3–4 h)

Protoplasts are fragile and susceptible to mechanical damage, therefore gentle handling during all steps and use of a swing-out centrifuge is highly recommended.

1. On the ninth day (*see* **Note 10**) make fresh PBI buffer and keep at room temperature in the dark until use. Harvest the roots rapidly, by opening one petri dish at a time and quickly snipping each row of seedlings' roots at the same point (*see* **Note 11**).
2. After snipping each row of roots, gently collect them with the forceps (*see* **Note 12**) and quickly place them in PB buffer, ensuring that the whole root tissue is covered by the solution.

3. Transfer the PBI buffer into a 100 mL flask and ensure that the enzymes added are fully dissolved (i.e., there is no precipitate). After rinsing the harvested roots with PB, remove the excess buffer and transfer them to the flask containing PBI (*see* **Note 13**).
4. Incubate for 2 h in the dark at room temperature, with shaking at 100 rpm.
5. Filter the protoplasts through a 40-μm cell strainer into a 50 mL Falcon tube (*see* **Note 14**). Rinse the incubation flask and all equipment involved with 25 mL PBS buffer and collect in the same Falcon tube.
6. Centrifuge (1000 ×*g*, 4 °C) the collected PBS-protoplast solution (50 mL) for 3 min and carefully discard as much supernatant as possible (*see* **Note 15**).
7. Gently redissolve protoplasts in 1 mL of 0.7% w/v sodium chloride (sheath fluid for cell sorting) and maintain at 4 °C until loading into the FACS.

3.3 Cell-Sorting (3–4 h)

There are very few generally applicable parameters in the cell-sorting process since most of them depend on the FACS equipment available, the fluorescent transgenic plant line (e.g., percentage of fluorescent cells) and the tissue being sorted (e.g., root cells have lower autofluorescence levels than shoot cells). Some adjustment of parameters may even be required when processing the same plant line and tissue (biological replicates) on different days (*see* **Note 16**). Therefore, here we describe the process according to our set-up using a BD FACS Aria I Flow Cytometer (BD Biosciences) with a 100 μm nozzle, as described in Subheading 3.2, focusing on the general parameters and tips for a successful sorting experiment. Terms in *italic* font are equipment-specific commands for the FACS Aria I instrument. However, the purpose of the commands is also explained to facilitate application of the procedures with any system.

Before starting the tissue harvesting (Subheading 3.2, **steps 1–3**).

1. Turn on the FACS instrument so that the cytometer lasers start to warm up.

   During the enzymatic digestion of cell walls in the root tissue (Subheading 3.2, **step 4**), which takes 2 h, perform the following set-up procedures:
2. *Perform Long-clean* with 70% EtOH to clean the system (*see* **Note 17**).
3. Run *fluidics start up* twice to ensure complete ethanol removal from the system.
4. Turn on the stream (sheath fluid: 0.7% sodium chloride filtered through a 0.2 μm Bottle Top Filter) and run quality control tests to check that the cytometer is performing adequately (*CS&T* application).

5. Let the stream run for 15–30 min to allow the flow to stabilize while collecting the protoplasts (Subheading 3.2, **steps 5**–7).
6. Adjust the *drop breakoff* by fine-tuning the *amplitude* and *frequency* settings, which will automatically modify the *Gap* and *Drop 1* values representing the output of *breakoff* parameters (*actual values*; *see* **Note 18**). The optimal *breakoff* parameters in our system are: *frequency* ~30, *amplitude* ~40–60, *Drop 1* ~ 100–300 and *Gap* ~10.
7. Enter the modified *Gap* and *Drop 1 actual values* as the manually imported *target values* and turn on the *Sweet Spot*. This function promotes stability of the flow and breakoff position with the given digital parameters (*Drop 1* and *Gap*), via slight automatic alterations of the *amplitude* if and when necessary.
8. When the flow has been stabilized via the *Sweet Spot* function, set the *Drop Delay*. This is the exact time required for a cell within the stream to move from the interrogation point to the droplet break-off position (*see* **Note 19**), as determined by an *Accudrop* beads solution.
9. Finally perform a *test sort* by turning on the *voltage* to charge the deflection plates (*see* **Note 20**) and the *waste drawer* so the main and side streams can be visually observed when entering the collection tubes (do not touch the deflection plates, there is high voltage!). If all streams reach the appropriate destinations (waste for the main stream and collection tubes for the side streams) sorting of the sample can start.

Cell-sorting of GFP^+ and GFP^- protoplasts:

10. After filtering the redissolved isolated protoplasts (Subheading 3.2, **step 7**) through a 0.5 μm cup filter, collect them in a round bottom tube and load them into the FACS. Make sure that the system is kept at 4 °C (from the collection chamber to the collection apparatus) and that the sample is agitated to avoid protoplasts settling in the bottom of the tube.
11. Set the criteria (*gates*) for sorting and collecting the desired cell populations. When the laser hits a cell or particle, at the so-called interrogation point, the light will be reflected and scattered at all angles (Fig. 1a). The *forward scatter* provides information about the cell's or particle's size and the *side scatter* about its granularity and complexity (*see* **Note 21**). A two-dimensional forward-side scatter plot allows identification of different cell populations and/or distinction between intact cells and smaller particles (*see* **Note 22**) in the injected sample (Fig. 1b). The protoplasts are manually selected/*gated* as the *P1* population (Fig. 1c), then further examined for the presence or absence of an internal fluorophore (autofluorescence and/or GFP). While a cell is being interrogated, the laser may also strike an internal fluorophore and then a fluorescent signal is emitted with a specific

wavelength. In such cases a two-dimensional plot showing Red and Green Fluorescence will present one cell population (non-GFP fluorescent line, Fig. 1d) or two cell populations (GFP fluorescent line, Fig. 1e). Then the GFP^- and GFP^+ cells can be selected in gates *P2* and *P3*, respectively (Fig. 1e, *see* **Note 23**).

12. After all the parameters have been set, quality control tests have been performed and the desired cell populations have been gated and assigned a collection tube position, sorting can be initiated. Cell sorting is always performed in *high purity mode* and at a *flow rate* not higher than *6*.
13. After 2 h of collection (replacing the collection tubes with new ones when 200,000 cells or some other desired number of cells has been collected, *see* **Note 6**) freeze the samples in liquid nitrogen and store them at −80 °C until further analysis.
14. Clean the equipment prior to shutting down by sequentially running the instrument with *FACS Clean*, *FACS Rinse*, and finally ultrapure water (for 5–10 min with each fluid). Turn off the *stream* then remove and clean the nozzle. Once a week also clean the flow cell and sonicate the nozzle. Finally, perform *Fluidics Shut Down* and turn off the FACS.

3.4 Cytokinin Purification from Sorted Cells (6 h)

The microSPE protocol can be applied after the cell sorting experiments have been completed and all samples containing GFP-expressing and non-GFP-expressing cell populations have been collected. The total number of samples purified by the one-step isolation protocol depends on the number of positions in the centrifuge used.

1. Prepare two microcentrifuge tubes which will be used for the loading, washing and elution steps (Subheading 3.4, **step 8**). Remove the lids from both tubes and pierce a hole in the center of one lid using forceps (the other lid can be discarded).
2. Insert the self-packed multi-StageTip containing C_{18}/SDB-RPS/Cation-SR layers (*see* **Note 24**; Fig. 3) into the hole of the lid, install it on the tube and place the tube in a centrifuge (*see* **Note 25**).
3. Let the samples thaw on ice (*see* **Note 26**) and dilute 1 mL of isolated protoplasts in 0.7% NaCl 3:1 (v/v) by adding 333 μL of H_2O. Finally, acidify the samples to pH 2.7 with 1 M HCl (7.5 μL) to enable CK binding to the Stage Tips.
4. Add 10 μL of working solutions *I–III* of isotope-labeled CK standards to each sample (*see* **Note 4**).
5. Before loading the sample, activate the multi-StageTip sorbents sequentially with 50 μL acetone, methanol, water, 50% (v/v) nitric acid, and water (step-by-step centrifugation at 434 ×*g*; 15 min for each solution; keep the centrifuge temperature at 4 °C) (*see* **Note 24**).

6. Bind CK analytes in the diluted samples by sequentially applying seven 200 μL portions to the in-tip microSPE column and centrifuging for 30 min at 678 ×*g* and 4 °C (*see* **Note 27**).
7. Rinse the tips sequentially with 50 μL of water and methanol (by loading and centrifuging for 20 min at 525 ×*g* and 4 °C).
8. Collect bound CK analytes in new clean microcentrifuge tubes, prepared in **step 1** (Subheading 3.4), by elution with 50 μL of 0.5 M NH_4OH in 60% MeOH (with centrifugation for 20 min at 525 ×*g* and 4 °C).
9. Transfer the eluates to chromatography vials with microvolume inserts (150–200 μL recommended).
10. Evaporate to dryness in a SpeedVac concentrator (at 37 °C) and store at −20 °C until UHPLC-MS/MS analyses.

3.5 Cytokinin Quantification

CK metabolites can be separated and determined using various LC-MS/MS systems. Our protocol was developed for use with a 1290 Infinity Binary LC System coupled to a 6490 Triple Quad LC/MS System with Jet Stream and Dual Ion Funnel technologies operating in positive mode (Agilent Technologies). Each technical parameter should be modified and optimized according to the apparatus used (*see* **Note 28**). However, generally:

1. Dissolve the pre-purified samples on multi-StageTips in 40 μL of 10% methanol and place them in an autosampler tray precooled to 4 °C.
2. Inject 10 μL of each sample onto a reversed-phase column (Acquity UPLC® BEH C18, 1.7 μm, 2.1 × 150 mm). The column thermostat should be set at 45 °C (*see* **Note 29**).
3. Separate the samples using the following 20 min gradient of methanol (A) and 15 mM ammonium formate (pH 3.95, B) at a flow rate of 0.35 mL /min: 0 min, 10:90 (A:B)—10.0 min, 23:77 (A:B)—15.0 min, 36:64 (A:B). At the end of the elution the column was washed with 100% A (2 min) and equilibrated to initial conditions (10:90, A:B) for 3 min. A chromatographic separation of 26 CK metabolites is shown in Fig. 4.
4. Determine the endogenous CKs in protoplasts by multiple reaction monitoring (MRM) of the protonated precursor and appropriate product ions. The MRM transitions, ionization and collision energies, and retention times, are summarized in Table 1. The optimized settings for the 6490 Triple Quad LC/MS System are also specified in **Note 29**.
5. Process data by mass spectrometry software (e.g., Agilent MassHunter Workstation Software—Quantitative Analysis, version B.05.02) and determine concentrations of the CKs using the stable isotope dilution method (*see* **Note 30**). Further data normalization is performed by calculation of the GFP^+/GFP^- ratio of the CK concentrations as described in [1, 2] (*see* **Note 6**).

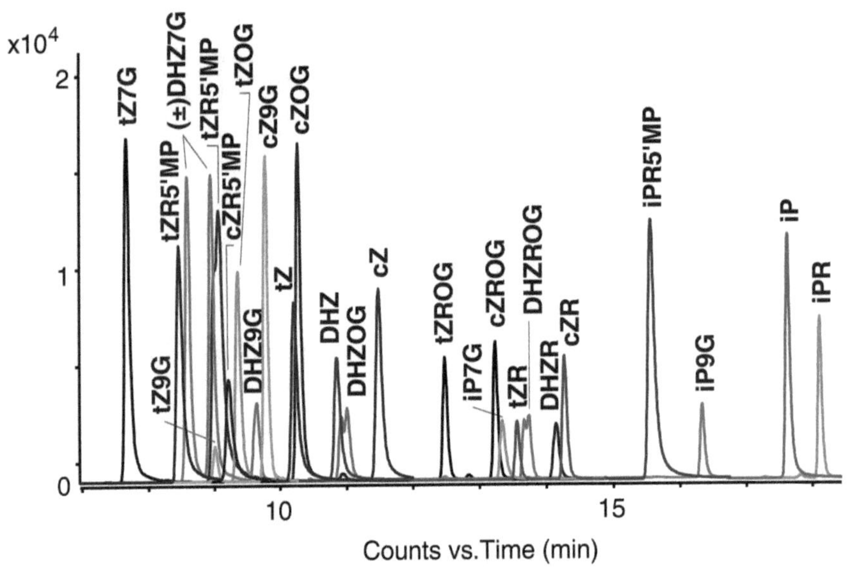

Fig. 4 Representative multi-MRM chromatogram of 26 isoprenoid cytokinins. 100 fmol of each derivative was injected and separated by UHPLC-MS/MS

4 Notes

1. The FACS samples are maintained at 4 °C, from loading in the machine to collection of the sorted cells. This will inhibit enzymatic processes and thus help ensure that levels of CKs subsequently measured correspond to the endogenous levels. However, appropriate controls should be included, which depend mainly on the experimental purposes.
2. To a 200 mL beaker add 50 mL of Milli-Q water, 10.93 g mannitol (mw = 182.17), 0.02 g $CaCl_2$ (mw = 110.99), 0.074 g KCl (mw = 74.56), 0.0386 g MES (mw = 195.2), 0.04 g $MgCl_2$ (mw = 203.3) and 0.1% w/v BSA. Stir at room temperature until all chemicals are dissolved and adjust the pH to 5.7 with 0.1 M KOH. Transfer the buffer into a 100 mL graduated cylinder and add Milli-Q water to 100 mL. The PB solution can be stored at 4 °C in darkness for up to 5 days.
3. The PB solution should be at room temperature during addition of the cell wall-degrading enzymes (cellulysin and pectolyase). The resulting PBI buffer should be kept in the dark at all times.
4. The unlabeled and labeled CK standards can be purchased from OlChemim Ltd. (Olomouc, Czech Republic, www.olchemim.cz). Dissolve the compounds in methanol to a final concentration of 1 mM (stock solutions) and store at −20 °C. Working solutions (*I–III*) containing each of the stable isotope-labeled standards can be made by diluting the stock solutions of the standards to the following final concentrations

Table 1
MRM transitions, retention times, and optimized instrument settings

Compound	RT[a] (min)	MRM transition		IS[b]	RT (min)	MRM transition		Fragmentor (V)	CE[c] (V)
		Precursor	Products			Precursor	Products		
*t*Z	10.15	220.1	136.1	$[^{13}C_5]$*t*Z	10.14	225.1	141.1	380	19
*c*Z	11.28	220.1	136.1	$[^{13}C_5]$*c*Z	11.27	225.1	141.1	380	19
DHZ	10.91	222.1	136.1	$[^2H_3]$DHZ	10.82	225.1	136.1	380	23
iP	17.88	204.1	136.1	$[^2H_6]$iP	17.79	210.1	136.1	380	16
*t*ZR	12.92	352.2	220.1	$[^2H_5]$*t*ZR	12.83	357.2	225.1	380	20
*c*ZR	13.68	352.2	220.1					380	20
DHZR	13.64	354.2	222.1	$[^2H_3]$DHZR	13.58	357.2	225.1	380	22
iPR	18.15	336.2	204.1	$[^2H_6]$iPR	18.10	342.2	210.1	380	20
*t*Z7G	7.71	382.2	220.1	$[^2H_5]$*t*Z7G	7.61	387.2	225.1	380	21
(±)DHZ7G	8.61/8.92	384.2	222.1					380	23
iP7G	13.16	366.2	204.1	$[^2H_6]$iP7G	13.05	372.2	210.1	380	21
*t*Z9G	8.93	382.2	220.1	$[^2H_5]$*t*Z9G	8.83	387.2	225.1	380	21
cZ9G	9.63	382.2	220.1					380	21
DHZ9G	9.61	384.2	222.1	$[^2H_3]$ DHZ9G	9.54	387.2	225.1	380	23
iP9G	16.44	366.2	204.1	$[^2H_6]$iP7G	16.37	372.2	210.1	380	22
*t*ZOG	9.52	382.2	220.1	$[^2H_5]$*t*ZOG	9.42	387.2	225.1	380	21
*c*ZOG	10.35	382.2	220.1					380	18
DHZOG	10.97	384.2	222.1	$[^2H_7]$ DHZOG	10.80	387.2	225.1	380	21
*t*ZROG	11.96	514.2	382.2	$[^2H_5]$*t*ZROG	11.88	519.2	387.2	380	21
*c*ZROG	12.71	514.2	382.2					380	21
DHZROG	13.29	516.2	384.2					380	22
*t*ZRMP	8.36	432.2	220.1	$[^2H_5]$*t*ZRMP	8.24	437.2	225.1	380	21
*c*ZRMP	9.01	432.2	220.1					380	21
DHZRMP	8.87	434.2	222.1	$[^2H_3]$ DHZRMP	8.79	437.2	225.1	380	23
iPRMP	15.31	416.2	204.1	$[^2H_6]$iPRMP	15.22	422.2	210.1	380	22

[a]Retention time
[b]Internal standards used for stable isotope dilution
[c]Collision energy

(of each compound): *Mixture I* (1×10^{-8} M), *Mixture II* (2×10^{-8} M) and *Mixture III* (1×10^{-7} M).

5. Self-packed multi-StageTips are prepared manually from Empore™ High Performance Disks (3 M Center, St. Paul, MN, USA) as described by Rappsilber et al. [13]. Briefly, an Empore disk is placed on a clean surface (petri dish) then very small disks (approximately 1.0 mm diameter, 0.5 mm thick) are manually cut out and introduced into a hollow blunt-ended syringe needle, by gently pressing the needle into the disk. The needle is then placed in a pipette tip (volume 200 μL) and the small disk is released and pressed gently into position using a plunger (rod) fitted into the needle. After removing the cutter and plunger, a one-layer StageTip has been created. Another two layers are added in the same manner to produce single-StageTips. To isolate the cytokinin metabolites, multi-StageTips are employed [9]. These include three layers of each of three commercially available poly-tetrafluoroethylene matrices, containing: reversed-phase octadecyl-bonded silica (C_{18}) groups, poly(styrene-divinylbenzene) (SDB) copolymer modified with sulfonic acid groups to increase hydrophilicity (SDB-RPS Disk), and ion-exchange sorbent including sulfonic acid as a cation exchanger (Cation-SR Disk) (Fig. 3).
6. While GFP^+ cells are usually of most research interest, the GFP^- cells present in each sorted sample should be collected and analyzed too, as they provide an internal reference population that is crucial for data normalization and interpretation.
7. One to two cell-sorting experiments should be sufficient for collecting 20,000–200,000 GFP+ cells (enough for cytokinin analysis) from a transgenic line in which 20% of the cells are fluorescent. To obtain a complete data set from cell sorting experiments 6–10 biological replicates are required, thus two or three weeks should be allowed for sample collection with approximately three sorting experiments (days) per week.
8. At this stage bubbles should be eliminated by being pushed out, or replacing the mesh if necessary, as plants will be unable to grow on them.
9. Healthy plants give rise to healthy root protoplasts with relatively high resistance to mechanical stress caused by the required handling for protoplast isolation. This not only increases the efficiency of the method but also reduces biological variation between samples.
10. Cells should be harvested at the same time every day to avoid possible effects of circadian regulation on the results.
11. Assist the cutting by keeping the cotyledons in place by holding them gently with the full length of a forceps. This process is also facilitated by the presence of the mesh and the high sowing density (*see* Subheading 3.1, **steps 2** and **3**, respectively).

12. Since the seedlings have been grown at high density the roots can be collected as a group rather than individually, thereby reducing the extent of tissue wounding and accelerating the process.
13. Gentle manual shaking every 30 min during incubation will improve the efficiency of the protoplast isolation procedure.
14. The undigested tissue will stay on the cell strainer and the collected PBI-protoplasts mixture should be transparent.
15. After centrifugation the protoplasts will accumulate at the bottom of the tube, but they should be very easy to resuspend (especially if they are in good condition). Therefore, the supernatant should be slowly pipetted out using a 10 mL glass pipette, and ca. 0.5 mL of supernatant should be left in the tube to avoid discarding protoplasts that have already started to resuspend. The cell wall-digestion enzymes have already been diluted (1:2) by addition of the 25 mL PB for rinsing (Subheading 3.2, **step 5**), and from that point on the samples will be kept at 4 °C, thus enzymatic activity will be reduced.
16. For example, the flow rate parameter should be adjusted according to the density of protoplasts isolated and injected in the FACS, which can vary from day to day. The *amplitude* parameter regulating the formation of single droplets during FACS may also require adjustment, which may affect other parameters, such as the *drop delay*.
17. The FACS instrument is cleaned prior to shut down. However additional rinsing with 70% ethanol upon start-up further ensures cleanliness of the machine, which is crucial for its high quality performance.
18. *Drop breakoff* is the point at which the first drop is detached from the main stream. The formation of individual drops at a stable and optimum physical space within the FACS and a stable flow is essential for high quality and purity cell-sorting, and (hence) reproducible data. *Frequency* and *amplitude* are adjusted to create the vibration wave that will provide the optimum *droplet breakoff*. *Amplitude* controls the strength of the steam vibration while the *frequency* regulates the number of droplets generated per second, measured in kHz. *Drop 1* represents a random point of the first separated droplet measured in pixels, but it is required as a known time offset from the start. *Gap* is the distance between drops (measured in pixels) and is used for stream supervision.
19. Knowledge of the stable period between the times when a cell or particle is being interrogated by the lasers and it reaches the last attached droplet in the stream (breakoff position) is essential for the sorting and thus isolation of the desired cell population. This is because when a desired interrogated cell arrives in

the last droplet connected to the stream, a charge is applied to the stream which is retained by the desired droplet/cell after leaving the stream. This charge facilitates the sorting of the drop/cell in the correct collection tube through enabling its attraction by an oppositely charged deflection plate oriented almost parallel to the main stream.

20. Sorting a desirable cell into a specific collection tube is facilitated by applying a charge in the main stream, which is retained by the detached drop containing the desirable cell for sorting. The charged drop passes through an electrostatic field created by two deflection plates that are oriented almost parallel to the main stream and is then deflected to the left or right according to the polarity of its charge. Thus, cells contained in oppositely charged drops will be shunted into side streams moving in opposite directions and sorted into respectively positioned collection tubes while uncharged drops containing undesirable cells/particles will not be affected by the electrostatic field and pass down to the waste aspirator (main stream direction).
21. The light is perceived by detectors, converted to a voltage pulse and the information can then be visualized in plots on the computer screen (Fig. 1b–e).
22. In the suspension of protoplasts injected in the FACS "debris" will also be present. This will include all material apart from the desired intact protoplasts and may include undigested cell wall parts or organelles derived from broken protoplasts (Fig. 2c).
23. There may be some overlap between the GFP^+ and GFP^- populations. Therefore, it is essential to inspect and avoid possible overlap between gates *P2* and *P3*. This can be done by visualizing histograms of the populations' GFP fluorescence as shown in Fig. 1e.
24. Loading times of individual solutions can vary depending on experimental conditions and type of centrifuge instrument, and must be optimized.
25. At the beginning of the experiment, chill the centrifuge to 4 °C (approximately 10 min).
26. Thawing the isolated GFP^+ and GFP^- protoplasts collected by FACS (samples snap-frozen as described in Subheading 3.3, **step 13**), facilitates complete disruption of the protoplasts, thus providing access to the endogenous CKs.
27. This is by far the most time-consuming step of the purification protocol (3.5 h).
28. Reproducible chromatographic separation and retention time stability is essential for separation of *cis*- and *trans*-zeatin isomers, their ribosides, nucleotides as well as *O*- and *N*-glucosides with identical precursor and basic product ions (Table 1).

29. In our system, MS conditions in positive ion mode are as follows: Drying Gas Temperature, 200 °C; Drying Gas Flow, 16 L/min; Nebulizer Pressure, 35 Psi; Sheath Gas Temperature, 375 °C, Sheath Gas Flow, 12 L/min; Capillary, 3400 V; Nozzle Voltage, 0 V; Delta iFunnel High/Low Pressure RF, 150/60 V.
30. Calibration curves ranging from 100 amol to 1 pmol were created from analyses of CKs metabolites by plotting known concentrations of each unlabeled analyte against analyte/IS ratios. To quantify levels of endogenous CK metabolites in protoplasts, their ratios to the appropriate stable isotope-labeled standards (Table 1) can be determined and further used in final calculations, based on the known quantity of IS added (*see* **Note 4**).

Acknowledgements

This work was supported by the Ministry of Education, Youth and Sports of the Czech Republic (the National Program for Sustainability I No. LO1204 and the "Návrat" program LK21306) and by the Czech Science Foundation (No. GA14-34792S), the European Molecular Biology Organization (EMBO ASTF 297-2013), the Erasmus program (European Commission: Lifelong Learning Programme), Development-The Company of Biologists (DEVTF2012), the Plant Fellows programme (http://www.plant-fellows.ch/), the Swedish Governmental Agency for Innovation Systems (VINNOVA), and the Swedish Research Council (VR).

References

1. Petersson SV, Johansson AI, Kowalczyk M et al (2009) An auxin gradient and maximum in the Arabidopsis root apex shown by high-resolution cell-specific analysis of IAA distribution and synthesis. Plant Cell 21:1659–1668
2. Antoniadi I, Plačková L, Simonovik B et al (2015) Cell-type specific cytokinin distribution within the Arabidopsis primary root apex. Plant Cell 27:1955–1967
3. Birnbaum K, Shasha DE, Wang JY et al (2003) A gene expression map of the *Arabidopsis* root. Science 302:1956–1960
4. Yadav RK, Girke T, Pasala S et al (2009) Gene expression map of the Arabidopsis shoot apical meristem stem cell niche. Proc Natl Acad Sci U S A 106:4941–4946
5. Pěnčík A, Simonovik B, Petersson SV et al (2013) Regulation of auxin homeostasis and gradients in Arabidopsis roots through the formation of the IAA catabolite oxIAA. Plant Cell 25:3858–3870
6. Moussaieff A, Rogachev I, Brodsky L et al (2013) High-resolution metabolic mapping of cell types in plant roots. Proc Natl Acad Sci U S A 110:E1232–E1241
7. Petersson SV, Lindén P, Moritz T et al (2015) Cell-type specific metabolic profiling of *Arabidopsis thaliana* protoplasts as a tool for plant systems biology. Metabolomics 11:1679–1689
8. Tarkowská D, Novák O, Floková K et al (2014) Quo vadis plant hormone analysis? Planta 240:55–76

9. Svačinová J, Novák O, Plačková L et al (2012) A new approach for cytokinin isolation from Arabidopsis tissues using miniaturized purification: pipette tip solid-phase extraction. Plant Methods 8:17
10. Novák O, Hauserová E, Amakorová P et al (2008) Cytokinin profiling in plant tissues using ultra-performance liquid chromatography–electrospray tandem mass spectrometry. Phytochemistry 69:2214–2224
11. Liu Z, Cai BD, Feng YQ (2012) Rapid determination of endogenous cytokinins in plant samples by combination of magnetic solid phase extraction with hydrophilic interaction chromatography-tandem mass spectrometry. J Chromatogr B 891–892:27–25
12. Du FY, Ruan GH, Liang SH et al (2012) Monolithic molecularly imprinted solid-phase extraction for the selective determination of trace cytokinins in plant samples with liquid chromatography-electrospray tandem mass spectrometry. Anal Bioanal Chem 404:489–501
13. Rappsilber J, Mann M, Ishihama Y (2007) Protocol for micro-purification, enrichment, pre-fractionation and storage of peptides for proteomics using StageTips. Nat Protoc 2:1896–1906

Chapter 20

Hormone Profiling in Plant Tissues

Maren Müller and Sergi Munné-Bosch

Abstract

Plant hormones are for a long time known to act as chemical messengers in the regulation of physiological processes during a plant's life cycle, from germination to senescence. Furthermore, plant hormones simultaneously coordinate physiological responses to biotic and abiotic stresses. To study the hormonal regulation of physiological processes, three main approaches have been used (1) exogenous application of hormones, (2) correlative studies through measurements of endogenous hormone levels, and (3) use of transgenic and/or mutant plants altered in hormone metabolism or signaling. A plant hormone profiling method is useful to unravel cross talk between hormones and help unravel the hormonal regulation of physiological processes in studies using any of the aforementioned approaches. However, hormone profiling is still particularly challenging due to their very low abundance in plant tissues. In this chapter, a sensitive, rapid, and accurate method to quantify all the five "classic" classes of plant hormones plus other plant growth regulators, such as jasmonates, salicylic acid, melatonin, and brassinosteroids is described. The method includes a fast and simple extraction procedure without time consuming steps as purification or derivatization, followed by optimized ultrahigh-performance liquid chromatography coupled to electrospray ionization-tandem mass spectrometry (UHPLC-MS/MS) analysis. This protocol facilitates the high-throughput analysis of hormone profiling and is applicable to different plant tissues.

Key words Plant hormones, UPLC/ESI-MS/MS, Auxin, Abscisic acid, Brassinosteroids, Cytokinins, Ethylene, Gibberellins, Jasmonates, Melatonin, Salicylic acid

1 Introduction

Plant hormones have been classified according to their chemical structures into five "classic" classes: cytokinins (CKs), auxins [in particular indole-3-acetic acid (IAA)], abscisic acid (ABA), gibberellins (GAs), ethylene, and several other plant growth regulators, such as jasmonates [jasmonic acid (JA) and its derivatives 12-*oxo*-phyodienoic acid (OPDA) and jasmonic acid-isoleucine (JA-Ile)], salicylic acid (SA), brassinosteroids (BRs), and melatonin (MEL) [1, 2]. They are crucial in regulating plant development and plant responses to various abiotic and biotic challenges [3]. Moreover, the significance of hormonal physiology in plants is increasing as we come to understand the interactive cascades of signal transduction

Jürgen Kleine-Vehn and Michael Sauer (eds.), *Plant Hormones: Methods and Protocols*, Methods in Molecular Biology, vol. 1497, DOI 10.1007/978-1-4939-6469-7_20, © Springer Science+Business Media New York 2017

network pathways, mechanisms of tissue sensitivity, genetic bases for hormonal action, and synergistic and antagonistic interaction among hormones [4]. Hence, the interest has changed from a specific plant hormone or a certain group of plant hormones to a multiple plant hormone profile analysis [5]. Frequently these molecules act at low levels, but levels can strongly differ depending on plant tissues. For instance, pollen and seeds contain BRs amounts at the nanogram/g FW level, while amounts in roots and shoots are at the picogram/g FW level [6, 7].

In the last 15–20 years many methods have been developed to quantify plant hormones using techniques such as gas chromatography (GC) and liquid chromatography (LC) coupled to mass spectrometry (MS) [8–11]. Mass spectrometers (triple quadrupole, ion trap or time-of-flight) provide a powerful analytical tool for quantification of plant hormones due to their high sensitivity, specificity, and reproducibility. Moreover, using the multiple reaction monitoring (MRM) mode increases the sensitivity of the mass spectrometer and decreases the background noise. However, depending on the plant tissues, other compounds present in the matrix can affect analyte ionization during plant hormone analyses, which must be controlled during an analytical method development for quantification of plant hormones.

In this chapter, a method for the sensitive and quantitative analysis of 20 plant hormones the cytokinins *trans*-zeatin (*t*Z), *trans*-zeatin riboside (*t*ZR), 2-isopentenyladenine (2iP), and isopentenyladenosine (IPA), the auxin IAA, the ethylene precursor, 1-amino-cyclopropane-1-carboxylic acid (ACC), ABA, SA, the jasmonates JA, OPDA, and JA-Ile, the gibberellins GA_1, GA_4, GA_9, GA_{19}, GA_{20}, GA_{24}, MEL, and the brassinosteroids, brassinolide (BL) and castasterone (CS) is described. This protocol facilitates the high-throughput analysis of hormones without time-consuming multiple steps of sample preparation followed by a rapid ultrahigh-performance liquid chromatography–electrospray ionization tandem mass spectrometry (UPLC/ESI-MS/MS), and has been successfully used for different plant tissues including leaves, roots, seeds, flowers, and pollen.

2 Materials

2.1 Chemicals

Solvents used for extraction are methanol (MeOH), isopropanol (iPrOH) and glacial acetic acid. Solvents used for UPLC-MS/MS analysis are acetonitrile, milli-Q water (18.2 MΩ), and glacial acetic acid. ABA, ACC, IAA, IPA, JA, *t*Z, *t*ZR, and SA standards can be obtained from any supplier of fine chemicals. BL, CS, GA, 2-iP, JA-Ile, OPDA, and MEL standards, together with the internal standards $[^2H_6]$ABA, $[^2H_4]$ACC, $[^2H_3]$BL, $[^2H_3]$CS, $[^2H_2]GA_1$, $[^2H_2]GA_4$, $[^2H_2]GA_9$, $[^2H_2]GA_{19}$, $[^2H_2]GA_{20}$, $[^2H_2]GA_{24}$, $[^2H_5]$IAA,

[$^2H_6$]2-iP, [$^2H_6$]IPA, [$^2H_5$]OPDA, [$^2H_5$]JA, [$^2H_2$]JA-Ile, [$^2H_4$]SA, [$^2H_5$] *t*Z, and [$^2H_5$]*t*ZR can be purchased from OlChemim Ltd. (Olomouc, Czech Republic). [$^2H_4$] MEL can be purchased from Toronto Research Chemicals (Toronto, Ontario, Canada).

2.2 Sample Extraction

1. Extraction solvents: MeOH–glacial acetic acid, 99:1 (v/v); MeOH–iPrOH–glacial acetic acid, 50:49:1 (v/v) (*see* **Note 1**).
2. *Internal standard solution*: stock solution of [$^2H_6$]ABA, [$^2H_4$] ACC, [$^2H_3$]BL, [$^2H_3$]CS, [$^2H_5$]IAA, [$^2H_6$]2-iP, [$^2H_6$]IPA, [$^2H_5$]JA, [$^2H_4$]MEL, [$^2H_4$]SA, [$^2H_5$] *t*Z, [$^2H_5$]*t*ZR, [$^2H_2$]GA$_1$, [$^2H_2$]GA$_4$, [$^2H_2$]GA$_9$, [$^2H_2$]GA$_{19}$, [$^2H_2$]GA$_{20}$, [$^2H_2$]GA$_{24}$, [$^2H_5$]OPDA, and [$^2H_2$]JA-Ile in MeOH (*see* **Note 2**).
3. Liquid nitrogen.
4. Microcentrifuge tubes (1.5 and 2 mL).
5. Mixer mill (MM400, Retsch GmbH, Haan, Germany).
6. Pipettes and pipette tips.
7. Vortex mixer.
8. Ultrasonication bath (Branson 2510, Danbury, CT USA).
9. Centrifuge type, which allows to use 1.5 to 2 mL tubes at 4 °C and 14,000 × *g*.
10. 0.22 μm PTFE syringe filters.
11. HPLC glass vials with fused-in insert (300 μL) and caps.

2.3 LC-ESI-MS/MS Analysis

1. Quaternary pump Aquity UPLC™ System (Waters, Milford, MA USA).
2. Colum C18 Kinetex 50 × 2.1 mm, 1.7 μm (Phenomenx, Macclesfield, UK).
3. API 3000 triple quadrupole mass spectrometer (PE Sciex, Concord, Ont., Canada) operating in MRM mode (*see* **Note 3**).
4. Water containing 0.05 % glacial acetic acid.
5. Acetonitrile containing 0.05 % glacial acetic acid.
6. Syringe pump (Model 11; Harvard Apparatus, Holliston, MA USA):
7. Each natural hormone (ABA, ACC, BL, CS, IAA, 2-iP, IPA, GA$_1$, GA$_4$, GA$_9$, GA$_{19}$, GA$_{20}$, GA$_{24}$, JA, JA-Ile, MEL, OPDA, SA, *t*Z, *t*ZR) and isotopically labeled hormones ([$^2H_6$]ABA, [$^2H_4$]ACC, [$^2H_3$]BL, [$^2H_3$]CS, [$^2H_5$]IAA, [$^2H_6$]2-iP, [$^2H_6$] IPA, [$^2H_2$]GA$_1$, [$^2H_2$]GA$_4$, [$^2H_2$]GA$_9$, [$^2H_2$]GA$_{19}$, [$^2H_2$]GA$_{20}$, [$^2H_2$]GA$_{24}$, [$^2H_5$]JA, [$^2H_2$]JA-Ile, [$^2H_4$]MEL, [$^2H_5$]OPDA, [$^2H_4$]SA, [$^2H_5$] *t*Z, [$^2H_5$]*t*ZR) at a concentration of 1 μg mL^{-1} in MeOH should be infused of a constant flow rate of 15 μL min^{-1}.

2.4 Quantification

1. *Calibration solutions*: Ten point standard solutions should be prepared in MeOH, containing each of the authentic standards (ABA, ACC, BL, CS, GA_1, GA_4, GA_9, GA_{19}, GA_{20}, GA_{24}, IAA, 2-iP, IPA, JA, JA-Ile, MEL, OPDA, SA, *t*Z, *t*ZR) ranging the expected physiological concentrations and within the linear response of the mass spectrometer. For example, ten point standard solutions from 0.05 to 1000 ng mL^{-1}, however, each solution should contain a constant amount of internal standards (*see* **Note 2**).

2.5 Validation

1. As the matrix components can affect the analyte stability, extraction and ionization samples should be spiked with low levels of standards.

3 Methods

The idea was to provide a protocol that facilitates the high-throughput analysis of hormone profiling which is applicable to different plant tissues. The protocol includes a fast and simple extraction procedure to quantify plant hormones and their stable isotope-labeled standards without time consuming steps as purification or derivatization followed by UHPLC/ESI-MS/MS analysis. In our research group, this method has been used to quantify hormone profiling in leaves, flowers, fruits, roots, and seeds of different species of a number of plant families including the model plant *Arabidopsis thaliana*, Brassicaceae, Liliaceae, Lamiaceae, Velloziaceae, Arecaceae, or Rosaceae. MS chromatograms for plant hormones extracted from different plant tissues are shown in Fig. 1.

3.1 Hormone Extraction

1. Freeze immediately the obtained sample material in liquid nitrogen and store it at –80 °C until analysis.
2. Transfer a pool of each sample tissue to a 2 mL microcentrifuge tube, add a magnetic bead and grind it using a mixer mill (*see* **Note 4**). Take care that the plant material does not thaw.
3. Weigh 100 mg of frozen powder and transfer it to a 1.5 mL microcentrifuge tube and take care that the sample tissue does not thaw (*see* **Note 5**).
4. Add 200 μL of extraction solvent containing *internal standards solution* to each sample. Mix each solution using vortex and extract them between 4 and 7 °C using ultrasonication for 30 min.
5. Centrifuge at 14,000 × *g* for 10 min at 4 °C. Transfer the supernatant to a 1.5 mL microcentrifuge tube and store it at 4 °C (*see* **Note 6**).

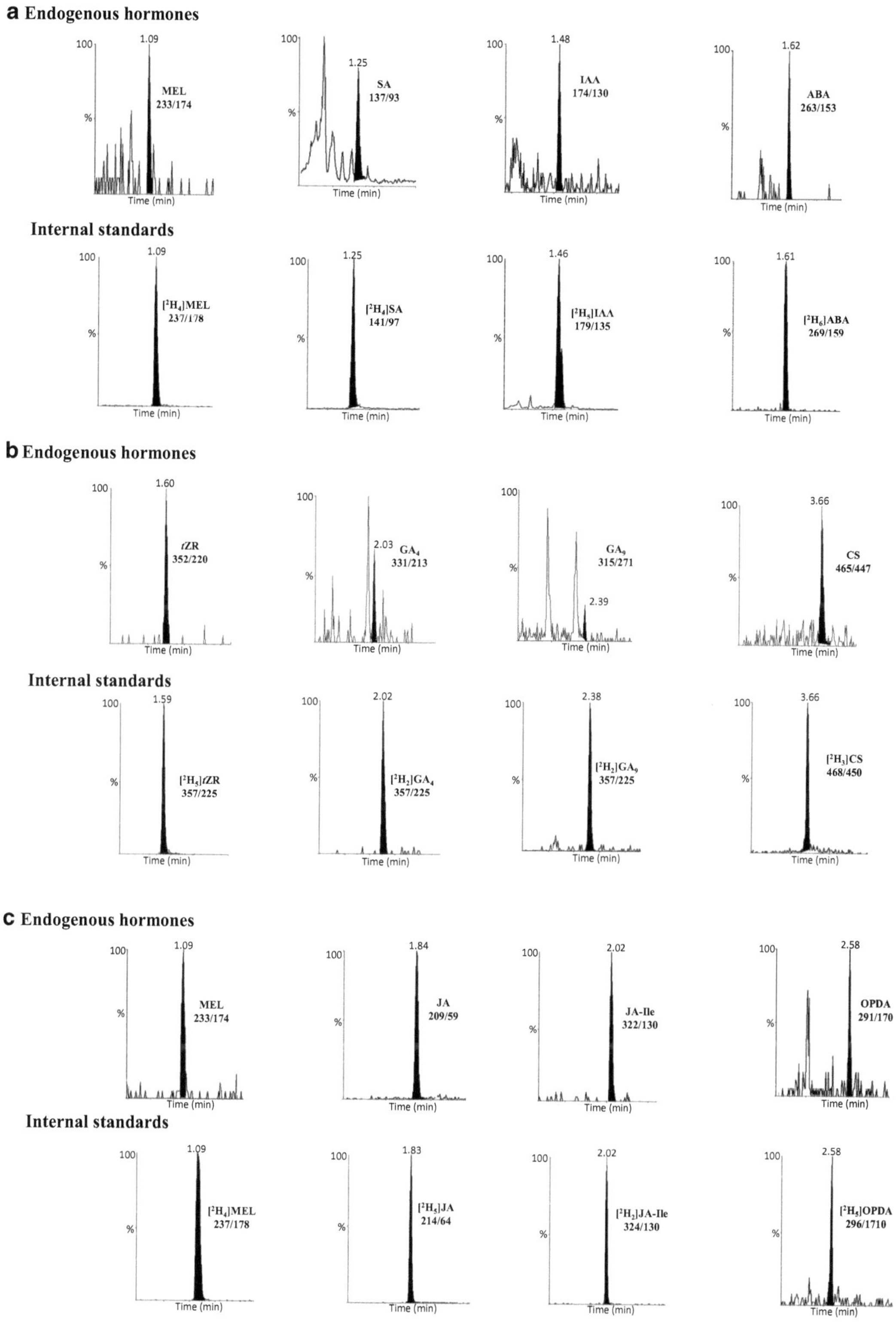

Fig. 1 Several examples of MS chromatograms for endogenous hormones and internal standards obtained by UPLC/ESI-MS/MS analysis. (**a**) MS chromatograms for MEL, SA, IAA, and ABA extracted from 100 mg *Arabidopsis thaliana* leaves. (**b**) MS chromatograms for *t*ZR, GA_4, GA_9, and CS extracted from 100 mg *Chamaerops humilis* var. *cerifera* seeds. (c) MS chromatograms for MEL, JA, JA-Ile, and OPDA extracted from 100 mg *Iris* × *hollandica* tepals

Table 1
LC conditions

Time (min)	Solvent A (%)	Solvent B (%)
0	99	1
0–4	1	99
4–4.20	99	1
4.20–5	99	1

A water containing 0.05 % glacial acetic acid
B acetonitrile containing 0.05 % glacial acetic acid

6. Re-extract the pellet three times with 100 μL of extraction solvent as described above and combine the supernatants.
7. Mix each combined supernatant using vortex and centrifuge them at 14,000 × *g* for 5 min at 4 °C.
8. Filter the supernatants through a 0.22 μm PTFE filter into a HPLC vial with a fused-in insert for analysis.

3.2 UPLC-MS/MS Analysis

1. UPLC conditions: using a C18 Kinetex column (50 × 2.1 mm, 1.7 μm) (Phenomenx, Macclesfield, UK); typical UPLC gradient conditions are listed in Table 1. The flow rate was 0.5 mL min^{-1}, the injection volume was 5 μL and the column temperature was maintained at 35 °C.
2. The mass spectrometer should operate in multiple reaction mode (MRM) due to the high selectivity using precursor-to-product ion transitions.
3. For each authentic standard and internal standard declustering potential (DP), focusing potential (FP), entrance potential (EP), collision energy (CE), collision cell exit potential (CXP) should be optimized to determine their appropriate MRM transition. Infuse each solution (1 ppm) into the tandem mass spectrometer using a syringe pump at for instance 15 μl min^{-1}. In contrast, temperature, nebulizer gas, curtain gas, collision gas, and the capillary voltage cannot be modified during analysis and average values for the different hormones should be selected (*see* **Note 7**). Optimized parameters for the API 3000 triple quadrupole mass spectrometer (PE Sciex, Concord, Ont., Canada) are given in Table 2 (*see also* **Note 8**).
4. Quantification should be performed by running all calibration solutions and construct calibration curves using spectrometer software (e.g., Analyst™ software; PE Sciex, Concord, Ont., Canada) for each authentic standard by integrating peak areas for each analyte at a given concentration and the corresponding internal standard.

Table 2
Optimized UPLC/ESI-MS/MS parameters for quantifying plant hormones

Analytes	MS method	RT[a] (min)	Precursor ion[a] (m/z)	Product ion[a] (m/z)	IS[b]	RT[b] (min)	Precursor ion[b] (m/z)	Product ion[b] (m/z)	Scan Mode	CE[a,b] (V)	CXP[a,b] (V)
ACC	1	0.26	102	56	[$^2H_4$]*t*Z	0.26	106	60	+	15	15
MEL	1	1.09	233	174	[$^2H_4$] MEL	1.09	237	178	+	20	10
*t*Z	1	1.25	220	136	[$^2H_5$]*t*Z	1.24	225	137	+	25	15
*t*ZR	1	1.60	352	220	[$^2H_5$]*t*ZR	1.59	357	225	+	25	15
2-iP	1	2.15	204	136	[$^2H_6$]2-iP	2.13	210	137	+	35	10
IPA	1	2.46	336	204	[$^2H_6$]IPA	2.45	342	210	+	25	15
BL	1	3.58	481	315	[$^2H_3$]BL	3.57	484	315	+	30	35
CS	1	3.66	465	447	[$^2H_3$]CS	3.66	468	450	+	20	15
SA	2	1.25	137	93	[$^2H_4$]SA	1.25	141	97	–	-20	-15
IAA	2	1.48	174	130	[$^2H_5$]IAA	1.46	179	135	–	-50	-10
ABA	2	1.62	263	153	[$^2H_6$] ABA	1.61	269	159	–	-30	-15
JA	2	1.84	209	59	[$^2H_5$]JA	1.83	214	64	–	-40	-15
JA-Ile	2	2.02	322	130	[$^2H_2$] JA-Ile	2.02	324	130	–	-34	-15
OPDA	2	2.58	291	165	[$^2H_5$] OPDA	2.58	296	170	–	-35	-15
GA_1	3	1.24	347	273	[$^2H_2$]GA_1	1.24	349	275	–	-40	-15
GA_{19}	3	1.57	361	273	[$^2H_2$] GA_{19}	1.57	363	275	–	-15/-25	-15
GA_{20}	3	1.63	331	287	[$^2H_2$] GA_{20}	1.62	333	289	–	-25	-15
GA_4	3	2.03	331	213	[$^2H_2$]GA_4	2.02	333	215	–	-40	-15
GA_{24}	3	2.11	345	257	[$^2H_2$] GA_{24}	2.11	347	259	–	-50	-15
GA_9	3	2.39	315	271	[$^2H_2$]GA_9	2.38	317	273	–	-50/-40	-15

MS mass spectrometry, *RT* retention time, *IS* internal standards, *CE* collision energy, *CXP* collision cell exit potential, *V* voltage

[a]Analytes

[b]Internal standards

ACC 1-amino-cyclopropane-1-carboxylic acid, *MEL* melatonin, *tZ* trans-zeatin, *tZR* trans-zeatin riboside, *2iP* isopentenyladenine, *IPA* isopentenyladenosine, *BL* brassinolide, *CS* castasterone, *SA* salicylic acid, *IAA* índole-3-acetic acid, *ABA* abscisic acid, *JA* jasmonic acid, *JA-Ile* jasmonic acid-isoleucine, *OPDA* 12-oxo-phyodienoic acid, *GA* gibberellin

3.3 Validation

The method evaluation should be conducted to demonstrate that the analytical method is applicable to obtain values, which are close to the unknown level of the analyte present in plant tissue. At least selectivity, linearity, limits of detection (LOD) and quantification (LOQ), recovery, repeatability and reproducibility should be evaluated during the validation processes.

1. Selectivity is defined as the ability to separate the analyte from the other sample components, giving pure, symmetric and resolved peaks. The method is proposed to be selective in the case that no additional peaks in the MS/MS spectrum can be detected for the band that corresponds to the analyte in the matrix compared to the original standards.
2. Linearity: calibration curves should be prepared in matrix and indicate good linearity with correlation coefficients (R) > 0.99 (*see* **Note 9**).
3. LOD and LOQ: LOD is defined as the lowest concentration of hormone, which can be distinguished from the noise in blank and is defined as a signal-to-noise ratio of 3; LOQ is defined as the lowest concentration of hormone that can be quantified precisely and accurately and is defined as a signal-to-noise ratio of 10 (*see* **Note 10**).
4. Recovery: spike six plant extracts with internal standard solution before extraction and six plant extracts with internal standard solution after extraction and calculate the recovery (%).
5. Repeatability and reproducibility: retention time and peak area reproducibility and intra- and inter-day precision in matrix should be determined and evaluated by calculating the relative standard deviation (% RSD, $n=6$). RSD values below 15 % are acceptable (*see* **Note 11**).
6. To evaluate matrix effects analyze calibration curves performed in the matrix and in solvent and calculate the ratio (*see* **Note 12**).

4 Notes

1. Due to the diverse chemical structures, physiological properties of plant hormones and matrix effects no extraction solvent has been published to isolate all plant hormones equally well from all tissue types. MeOH–glacial acetic acid, 99:1 (v/v) showed good results for plant tissues with high lipid contents such as seeds, MeOH–iPrOH–glacial acetic acid, 50:49:1 (v/v) showed good results for plant tissues such as leaves and flowers. Nevertheless, high water contents of plant tissues can influence extraction efficiency and should also be taken into account.

2. A stock solution should be prepared with a high concentration that can be stored at −20 °C and from this stock solution an *internal standard solution* can be prepared freshly before use. The optimal final concentration of the *internal standard solution* needs to be assessed empirically but 100 ng mL^{-1} [$^2H_6$]ABA, [$^2H_4$]ACC, [$^2H_3$]BL, [$^2H_3$]CS, [$^2H_5$]IAA, [$^2H_6$]2-iP, [$^2H_6$]IPA, [$^2H_5$]JA, [$^2H_4$]MEL, [$^2H_4$]SA, [$^2H_5$] *t*Z, [$^2H_5$]*t*ZR, and 20 ng/ mL [$^2H_2$]GA_1, [$^2H_2$]GA_4, [$^2H_2$]GA_9, [$^2H_2$]GA_{19}, [$^2H_2$]GA_{20}, [$^2H_2$]GA_{24}, [$^2H_5$]OPDA, [$^2H_2$]JA-Ile per sample is a good starting point.
3. The mass spectrometer should operate in MRM mode due to the high selectivity using precursor-to-product ion transitions as many compounds can present the same nominal molecular masses or peaks can overlap.
4. To avoid that plant tissue thaws it is recommended to dip microcentrifuge tubes and the tube adaptor for the mixer mill in liquid nitrogen. If the grinding process takes longer than 30 s, the samples should be dipped again in liquid nitrogen. IMPORTANT: avoid that liquid nitrogen enters in the microcentrifuge tubes, because it could destroy them.
5. To avoid the plant tissue from thawing it is recommended to dip microcentrifuge tubes and spatula in liquid nitrogen before and after weighing.
6. It is important to transfer the supernatant without any solid particles.
7. Optimized parameters obtained on the API 3000 triple quadrupole spectrometer (PE Sciex, Concord, Ont., Canada): temperature 400 °C, nebulizer gas (N_2) 10 (arbitrary units), curtain gas (N_2) 12 (arbitrary units), collision gas (N_2) 4 (arbitrary units), the capillary voltage 3.5 kV in the positive ion mode and −3.5 kV in the negative ion mode.
8. Plant hormones were analyzed using three different MS methods. CKs, ACC, and MEL were analyzed using MS method 1, ABA, SA, IAA, and jasmonates using MS method 2, and GAs using MS method 3 and given in Table 2.
9. 100 mg ground samples should be spiked with authentic standard solutions ranging for example from 0.1–500 ng mL^{-1} for CKs, IAA, ACC, GAs, MEL, and BRs; 1–500 ng mL^{-1} for ABA, SA, and jasmonates and spiked with constant levels of internal standards and extracted as described in Subheading 3.1.
10. LOD should be established using matrix samples spiked with low amounts of standards solution. In the case that none analyte-free matrix is available LODs should be determined in solvent.

11. Samples should be performed using three spiking solutions (low, medium, and high) and extracted as described in Subheading 3.1.
12. Matrix interferences can affect the efficiency of the analyte ionization resulting in suppressed or enhanced analyte ionization (decreasing or increasing the response); however, they are not well understood and seem to vary depending on analyte and matrix [12].

Acknowledgments

We thank Barbara Simancas and Javier A. Miret for help in extraction of *Arabidopsis thaliana* leaves and *Iris* × *hollandica* tepals.

References

1. Pan X, Welti R, Wang X (2008) Simultaneous quantification of major phytohormones and related compounds in crude plant extracts by liquid chromatography tandem mass spectrometry. Phytochemistry 69:1773–1781
2. Zhang N, Sun Q, Zhang H, Cao Y, Weeda S, Ren S, Guo Y-D (2015) Roles of melatonin in abiotic stress resistance in plants. J Exp Bot 66:647–656
3. Davies PJ (2010) The plant hormones: their nature, occurrence, and functions. Kluwer Academic, The Netherlands
4. Farnsworth E (2004) Hormones and shifting ecology throughout plant development. Ecology 85:5–15
5. Van Meulenbrock L, Vanden BJ, Steppe K et al (2012) Ultra-high performance liquid chromatography coupled to high resolution Orbitrab mass spectrometry for metabolomics profiling of the endogenous phytohormonal status of the tomato plant. J Chromatogr A 1260:67–80
6. Baiguz A, Tretyn A (2003) The chemical characteristic and distribution of brassinosteroids in plants. Phytochemistry 62:1027–1046
7. Takatsu S (1994) Brassinosteroids: distribution in plants, bioassays and microanalysis by gas chromatography-mass spectrometry. J Chromatogr A 658:3–15
8. Müller A, Düchting P, Weiler EW (2002) A multiplex GC-MS/MS technique for the sensitive and quantitative single-run analysis of acidic phytohormones and related compounds, and its application to *Arabidopsis thaliana*. Planta 216:44–56
9. Chiwocha SDS, Abrams SR, Ambrose SJ et al (2003) A method for profiling different classes of plant hormones and their metabolites using liquid chromatography-electrospray ionization tandem mass spectrometry: an analysis of hormone regulation of thermodormancy of lettuce (*Lactuca sativa* L.) seeds. Plant J 35:405–417
10. Pan X, Welti R, Wang X (2010) Quantitative analysis of major plant hormones in crude plant extracts by high-performance liquid chromatography mass spectrometry. Nat Protoc 5:986–992
11. Müller M, Munné-Bosch S (2011) Rapid and sensitive hormonal profiling of complex plant samples by liquid chromatography coupled to electrospray ionization tandem mass spectrometry. Plant Methods 7:37
12. Almeida Trapp M, De Souza GD, Rodrigues-Filho E et al (2014) Validated method for phytohomone quantification in plants. Front Plant Sci 5:417

Chapter 21

Use of *Xenopus laevis* Oocytes to Study Auxin Transport

Astrid Fastner, Birgit Absmanner, and Ulrich Z. Hammes

Abstract

Xenopus laevis oocytes are an expression system that is particularly well suited for the characterization of membrane transporters. Oocytes possess only very little endogenous transport systems and therefore transporters can be studied with a high signal-to-noise ratio. This book chapter provides the basic methods to use *Xenopus* oocytes for the characterization of transporters by radiotracer experiments. While the methods described here were established to study auxin transport they can easily be adapted to study other hormone transporters and their substrates.

Key words *Xenopus laevis* oocytes, Microinjection, Transport, Import export, Radiotracer, Liquid scintillation

1 Introduction

Transport proteins facilitate the uptake and also the release of substrates across membranes. The characterization of their biochemical properties is challenging because large parts of the protein are hydrophobic and reside in the two-dimensional membranes—as a consequence they cannot be expressed to high levels. Nevertheless, knowledge about the kinetics of a transport process, the interference of inhibitors, or their substrate specificity are key to the understanding of many biological processes. The vast majority of plant transporters have been studied in heterologous expression systems particularly yeast and *Xenopus laevis* oocytes.

The use of oocytes as an expression system has been pioneered more than 40 years ago [1]. It was not until the early 1990s that plant transporters were studied in *Xenopus* oocytes [2]. But since then many plant transporters were characterized in oocytes. Stage V and VI oocytes are generally considered to be particularly good expression systems for three reasons: (1) During this stage, the cells are rich in yolk and as a consequence are among the cells that have the lowest number of endogenous transport systems present in their membrane. As many as ~10^{10}–10^{11} transporters per oocyte

Jürgen Kleine-Vehn and Michael Sauer (eds.), *Plant Hormones: Methods and Protocols*, Methods in Molecular Biology, vol. 1497, DOI 10.1007/978-1-4939-6469-7_21, © Springer Science+Business Media New York 2017

are found in the oolemma (oocyte plasma membrane) upon injection of an mRNA. Therefore, transporters can be studied in many cases background free—this is especially true for charged substances [3]. (2) For obvious reasons, *Xenopus* oocytes do not naturally contain plant-specific proteins. This minimizes possible interference of transporter-modifying activities that must be postulated in plant expression systems such as plant protoplasts or BY2 cells and—to a lesser extent—also yeast and mammalian cells. (3) Oocytes are fairly big, with a diameter of about 1 mm. This allows the injection of defined amounts of substrate into the cell to produce an outward-directed electrochemical gradient in a substrate-free solution with an initial internal substrate concentration that is very similar in each cell. This is a clear advantage over the passive equilibration of substrate prior to the experiment, which inevitably leads to larger variations.

Due to these properties and advantages oocytes are mainly used to perform electrophysiological experiments using two-electrode voltage clamp (TEVC). Nevertheless, oocytes have also widely been used to study transport of radioactively labeled substrates. By this approach importers and exporters for plant hormones like gibberellic acid, jasmonate, abscisic acid, as well as several auxin transporters were characterized in oocytes [4–10]. So far hormone transporters have not been characterized electrophysiologically. This is likely due to the comparatively low transport rates for these substrates, which are in many cases below the resolution limit of TEVC.

Applying the methods described below, we were able to successfully characterize auxin import and export by expression of transporters with or without modifying kinases in oocytes. In principle, the methods can be modified and applied to other transporters and other substrates in the oocyte expression system.

2 Materials

All solutions are prepared with ultrapure nuclease-free water (Biopak®, Merck-Millipore).

2.1 In Vitro Transcription of cRNA

1. RNase Zap®: RNaseZap® RNase Decontamination Solution (Thermo Fisher Scientific, Waltham, Massachusetts, USA).
2. Phenol/chloroform/isoamyl alcohol: Roti®-phenol/chloroform/isoamyl alcohol, TE buffered, ratio 25:24:1, pH 7.5–8.0.
3. Natriumacetat: 3 M, pH 5.2.
4. Chloroform: 99.8 %, per analysis, stabilized with amylene.
5. Ethanol: ≥ 99.8 %, absolute.

6. mMESSAGE mMACHINE® SP6 Transcription Kit (Thermo Fisher Scientific, Waltham, Massachusetts, USA).
7. MEGAclear™ Transcription Clean-Up Kit (Thermo Fisher Scientific, Waltham, Massachusetts, USA).
8. NanoDrop™ 2000c Spectrophotometer or an other photometric device.

2.2 Oocyte Injection with cRNA

1. Oocytes stages V and VI are manually selected (*see* **Note 1**).
2. Barth's pH 7.4: 88 mM NaCl, 1 mM KCl, 2.4 mM $NaHCO_3$, 10 mM 2-[4-(2-hydroxyethyl)piperazin-1-yl]ethanesulfonic acid (HEPES), 0.33 mM $Ca(NO_3)_2$, 0.41 mM $CaCl_2$, 0.82 mM $MgSO_4$, pH 7.4 (with NaOH). Filter (0.45 μm) sterilize and store at 4 °C. A 5× stock solution can be prepared.
3. Gentamycin sulfate: 1000× Stock solution 50 mg/ml, store at −20 °C.
4. Nanoject II 3.5″ replacement glass capillaries (Drummond Scientific Company, Broomall, Pennsylvania, USA).
5. P-97 Flaming/Brown micropipette puller (Sutter Instrument, Novato, California, USA).
6. Stereomicroscope.
7. Plastic Pasteur pipettes with wide opening (drop volume 62 μl).
8. Disposable glass Pasteur pipettes 230 mm.
9. Pipettor pi-pump® 2500 (Glasfirn Giessen, Giessen, Germany).
10. Injection groove: Made from acrylic glass, dimensions (in cm) 5 × 5 × 0.3. Groove angle 45° (Fig. 1).
11. Nanoject II™ Auto-Nanoliter Injector (Drummond Scientific Company, Broomall, Pennsylvania, USA).
12. BSA: Bovine serum albumin Fraction V, protease free.

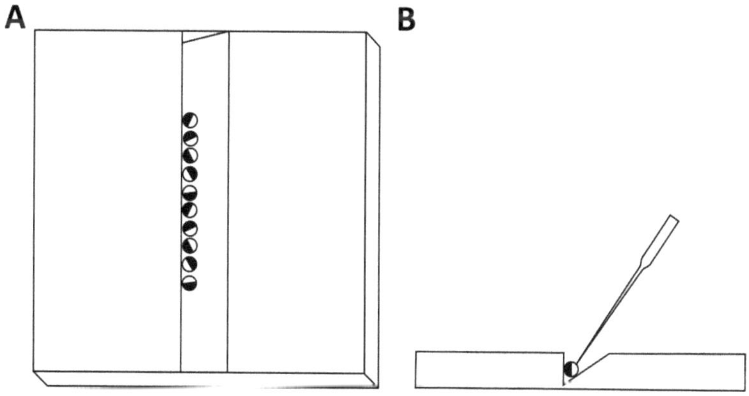

Fig. 1 Schematic representation of an injection groove. (**a**) Top view. (**b**) Lateral view with injection needle

2.3 Flux Experiments with Radioactive Labeled Substances

1. Radioactivity: Indole-3-acetic acid [5-3H], solvent ethanol, concentration 1 mCi/ml, specific activity 25 Ci/mmol, 250 μCi/vial, LOT 150522 (American Radiolabeled Chemicals, St. Louis, Missouri, USA).
2. 3-Indole acetic acid (IAA): M.W. 175.18 g/mol.
3. Barth's pH 7.4: 88 mM NaCl, 1 mM KCl, 2.4 mM $NaHCO_3$, 10 mM 2-[4-(2-hydroxyethyl)piperazin-1-yl]ethanesulfonic acid (HEPES), 0.33 mM $Ca(NO_3)_2$, 0.41 mM $CaCl_2$, 0.82 mM $MgSO_4$, pH 7.4 (with NaOH). Filter (0.45 μm) sterilize and store at 4 °C. A 5× stock solution can be prepared. Or Ringer: 115 mM NaCl, 2.5 mM KCl, 1.8 mM $CaCl_2$, 1 mM $NaHCO_3$, 10 mM 2-[4-(2-hydroxyethyl)piperazin-1-yl]ethanesulfonic acid (HEPES), 1 mM $MgCl_2$, adjust to desired pH (with HCl or NaOH). A 10× stock solution can be prepared. Filter (0.45 μm) sterilize and store at 4 °C.
4. 12-Well cell culture plate.
5. Scintillation vial 20 ml.
6. SDS: Sodium lauryl sulfate, 10% solution.
7. Scintillation cocktail.
8. Liquid scintillation counter.

2.4 Immunoblotting and Biotinylation Components

1. Homogenization buffer: 50 mM Tris(hydroxymethyl)-aminomethan (Tris) pH 7.5, 100 mM NaCl, 1 mM EDTA, 1 mM Pefabloc® SC-Protease-Inhibitor, 1 tablet/ml PhosSTOP—Phosphatase Inhibitor Cocktail Tablet (Roche, Basel, Switzerland).
2. Ultracentrifuge 100.000 g and appropriate vials.
3. PBS: 137 mM NaCl, 2.7 mM KCl, 10 mM Na_2HPO_4, 1.8 mM KH_2PO4, pH 8.0. A 10× stock solution can be prepared. Autoclave and store at room temperature.
4. Biotin: EZ-Link® Sulfo-NHS-LC-Biotin (Thermo Fisher Scientific, Waltham, Massachusetts, USA).
5. Biotin solution: 1 g EZ-Link® Sulfo-NHS-LC-Biotin/ml PBS pH 8.0 (=~2 mM biotin).
6. Quench buffer: 100 mM Glycine in PBS pH 8.0.
7. Lysis buffer: 150 mM NaCl, 20 mM Tris(hydroxymethyl)-aminomethan (Tris) pH 7.6, 1% Triton X-100, 0.5 mM phenylmethylsulfonyl fluoride (PMSF).
8. Streptavidin–agarose from *Streptomyces avidinii*, buffered aqueous solution (Sigma Aldrich, St. Louis, Missouri, USA).
9. 4× SDS-PAGE sample buffer: 250 mM Tris(hydroxymethyl)-aminomethan (Tris) pH 6.8, 8% SDS, 40% glycerol, 20% β-mercaptoethanol, 0.004% bromophenol blue.
10. Specific antibody.

3 Methods

Unless stated otherwise all steps are carried out at room temperature. For RNA work, gloves and RNase-free and freshly autoclaved tips and cups are used. Pipettes and workspace should be thoroughly cleaned with RNase ZAP®.

3.1 In Vitro Transcription of cRNA

1. Clone the CDS of the gene of interest (the transporter or the kinase) into the multiple cloning site of the vector backbone pOO2 [11] (*see* **Note 2**).
2. Linearize 20 μg of the vector containing the gene of interest (*see* **Note 3**).
3. Bring the linearization reaction to a volume of 200 μl using RNase-free water.
4. Extract with one volume phenol/chloroform/isoamyl alcohol (*see* **Note 4**).
5. Transfer the aqueous phase to a new RNase-free reaction cup.
6. Extract twice with one volume of chloroform.
7. Each time, transfer the aqueous phase to a new RNase-free reaction cup.
8. Add 1/10 volume of sodium acetate and 2.5 volumes of ethanol to the aqueous phase to precipitate the DNA on ice for 20 min.
9. Centrifuge the mixture at 4 °C for 20 min at 16,000 × *g* to recover the precipitate.
10. Remove the liquid and air-dry the pellet.
11. Resuspend the pellet in RNase-free water to a concentration of 0.5 μg/μl.
12. For in vitro transcription the mMESSAGE mMACHINE® SP6 Transcription Kit is used according to the manufacturer's instructions (*see* **Note 5**).
13. cRNA is purified using the MEGAclear™ Transcription Clean-Up Kit according to the manufacturer's instructions (*see* **Note 6**).
14. Determine the yield and purity of the cRNA synthesis using the NanoDrop.
15. Determine the quality of the cRNA obtained by an analysis on a 0.8% agarose-gel. Mix 1 μl of cRNA with RNase-free water and DNA-loading dye (*see* **Note 7**).

3.2 Oocyte Injection with cRNA

1. Use oocytes stages V and VI for cRNA injection. Inject oocytes the day after surgery.
2. Keep oocytes in sterile petri dishes in BARTH's pH 7.4 supplemented with gentamycin (50 μg/ml) at 16 °C.

3. Pull injection needles from glass capillaries. Break the needles prior to injection under a stereomicroscope to achieve an aperture of about 20 μm (*see* **Note 8**).
4. Prepare cRNA working solutions in the desired concentration (*see* **Note 9**).
5. Transfer oocytes by gravity drop into the injection groove filled with Barth's (*see* **Note 10**, Fig. 1).
6. Inject each oocyte with max. 50 nl containing the optimal amount of cRNA or water (negative control) under a stereo microscope.
7. Transfer the oocytes into a small sterile plastic petri dish containing Barth's pH 7.4 supplemented with gentamycin (50 μg/ml).
8. Incubate the oocytes until optimal expression is achieved in Barth's pH 7.4 supplemented with gentamycin (50 μg/ml) at 16 °C (*see* **Note 11**). Change the buffer every day and discard dead or damaged oocytes.

3.3 Efflux Experiments with Radioactive Labeled Substances

1. Prepare and fill one recovery plate (Fig. 2a) and one flux plate (Fig. 2b) for each construct to test, as shown in Fig. 2. Fill each well required with 2 ml Barth's (*see* **Note 12**). Place all the "recovery"-plates in the fridge (4 °C) to cool them. Store the "efflux" plates at room temperature. Before starting the injection put one "recovery" plate on ice.
2. Prepare the injection needles as described in Subheading 3.2.
3. Prepare substrate for injection (*see* **Note 13**).
4. Place the injection groove on an ice-cold metal block.
5. Place the petri dish containing the oocytes on ice.
6. Fill the injection groove with ice-cold Barth's.
7. Inject ≥10 oocytes per time point with radiolabeled substrate.

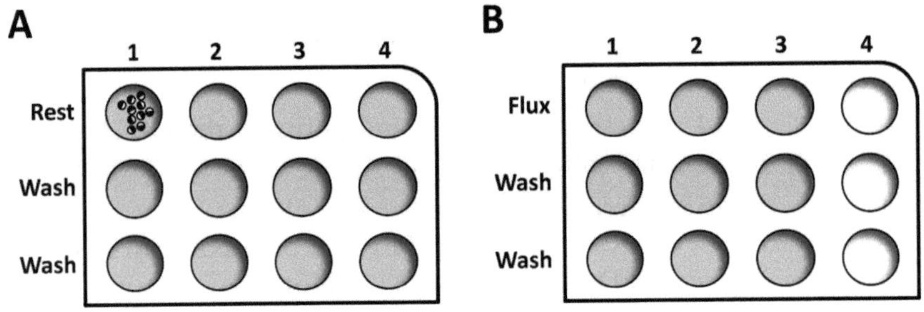

Fig. 2 Schematic representation of recovery an flux plates. (**a**) Recovery plate. This plate is placed on ice. Oocytes are placed in the top vial to allow closure of the injection spot and diffusion of substrate in the oocyte. Oocytes are transferred to the wash well by gravity drop and the to the top vial of the flux plate. (**b**) Flux plate. This plate is kept at room temperature. Oocytes are allowed to transport substrate for a defined amount of time and then transferred to the wash solution by gravity drop

8. Transfer the oocytes immediately after injection into ice-cold Barth's in the first upper well of the "recovery" plate and keep them on ice for 10 min.
9. To wash, transfer oocytes by gravity drop into the well below containing 2 ml of ice-cold Barth's.
10. For the second wash, transfer oocytes by gravity drop into the well below containing 2 ml of ice Barth's.
11. Transfer the oocytes in the first uppermost well of the flux plate and allow them to efflux for 30 min at room temperature.
12. Proceed with the other time points (15 and 7.5 min) and use for each time point a new column of the efflux plate (Fig. 2b).
13. After the efflux time is up, wash the oocytes twice by transferring them to the next well by gravity drop.
14. Transfer oocytes individually into one scintillation vial for each oocyte.
15. Oocytes representing the 0 min are only put in the recovery plate for 10 min, washed twice, and immediately separated into one scintillation vial per oocyte.
16. Lyse the oocytes in the scintillation vials with 100 μl 10% SDS and let them burst (app. 10 min).
17. Add 4 ml scintillation cocktail to the lysed oocytes and vortex vigorously.
18. Determine the amount of radioactivity in each vial by liquid scintillation counting.

3.4 Import Experiments with Radioactive Labeled Substances

1. Place 12 oocytes into the first well of a flux plate (Fig. 2b) containing 2 ml Ringer/Barth's containing the desired amount of labeled substrate (*see* **Note 12**).
2. Incubate oocytes for an appropriate time, usually between 5 and 120 min depending on the expression level of the transporter. In order to do kinetics at least four time points are required!
3. To wash oocytes, transfer them into 2 ml of ice-cold Ringer/Barth's by gravity drop.
4. Wash a second time.
5. After the second wash transfer individual oocytes to one scintillation vial per oocyte.
6. Lyse oocytes by addition of 100 μl 10% SDS.
7. Vortex vigorously to homogenize, and add 4 ml scintillation cocktail.
8. Determine the amount of accumulated radioactivity by liquid scintillation counting.

3.5 Immunoblotting of Microsomal Fractions

1. Transfer 8–25 oocytes in a reaction cup.
2. Homogenize in 40 μl homogenization buffer per oocyte with a 1 ml pipet tip by trituration, i.e., pipetting up and down vigorously (*see* **Note 14**).
3. Centrifuge at 2000 ×*g* and 4 °C for 10 min.
4. Transfer the supernatant in a reaction cup suitable for ultracentrifugation (*see* **Note 15**).
5. Fractionate the solution in an ultracentrifuge at 150,000 ×*g* and 4 °C for 30 min.
6. Transfer the supernatant, which represents the soluble protein fraction, in a new reaction cup and store it for further analysis.
7. Resuspend the pellet in 8 μl homogenization buffer supplemented with 4 % SDS.
8. Perform SDS-PAGE and Western blot according to standard protocols (*see* **Note 16**).

3.6 Biotinylation of Plasma Membrane Proteins

In order to be able to compare the amount of transporter in the membranes between experiments it is necessary to assess the amount of transporter in the oocyte plasma membrane.

1. Use 20 oocytes for each construct. As control use water-injected oocytes (*see* **Note 17**).
2. Wash oocytes twice in Barth's without antibiotics and incubate for 30 min in Barth's at 16 °C (*see* **Note 18**).
3. Transfer oocytes in a reaction cup.
4. Wash oocytes three times with 1 ml PBS at 4 °C for 1 min.
5. Incubate the oocytes for 15 min in 0.5 ml biotin solution at room temperature in the dark. Turn the reaction cup slightly from time to time to evenly biotinylate the oocytes.
6. Wash three times with 1 ml ice-cold PBS for 2 min.
7. Wash once with 1 ml quench buffer.
8. Incubate oocytes with 1 ml quench buffer on ice for 20 min.
9. Wash once with 1 ml ice-cold PBS.
10. To lyse oocytes incubate for 30 min with 100 μl lysis buffer on ice and homogenize by trituration.
11. Centrifuge at 1500 ×*g* and 4 °C for 15 min.
12. Transfer the supernatant to a new reaction cup (*see* **Note 15**).
13. Perform Bradford analysis to determine the protein content.
14. Prepare the streptavidin–agarose. Vortex vigorously before use. Pipet 50 μl with a cut pipet tip in a reaction cup. Centrifuge shortly and remove the supernatant. Wash once with 1 ml lysis buffer and remove supernatant after a centrifugation step.

15. Bind the biotinylation reaction to the streptavidin–agarose beads. Pipet a defined amount of protein to the washed streptavidin–agarose (*see* **Note 19**).
16. Incubate overnight at 4 °C while slightly shaking.
17. Centrifuge at 5000 g and 4 °C for 1 min.
18. Collect the supernatant (not biotinylated fraction).
19. Wash biotin-streptavidin-agarose pellet four times with 1 ml ice-cold PBS. Remove supernatant after a centrifugation step (5000 × *g*, 4 °C, 1 min).
20. Add 20 μl of 4× SDS-PAGE sample buffer and boil at 95 °C at 5 min (*see* **Note 20**).
21. Perform SDS-PAGE and Western blot.
22. Use a specific antibody to detect your protein of interest.

4 Notes

1. The quality of the oocytes used is absolutely crucial. The laws regulating keeping of vertebrates are becoming increasingly strict. As a consequence, keeping frogs in your own department may be complicated and starting the own facility may be impossible due to regulations. In many universities *Xenopus* facilities exist in medical faculties and/or zoology/animal physiology departments. It is certainly of great advantage to obtain the oocytes from a department in which oocytes are routinely used. As an alternative, it is possible—albeit fairly expensive—to purchase oocytes from private companies.
2. It is necessary to keep in mind that the *Nco*I site will generate an ATG start codon. pOO2 contains the 3′-UTR of the the *Xenopus laevis* beta globin gene and a poly A sequence behind the multiple cloning site to stabilize the cRNA transcript [11]. For in vitro transcription the SP6-RNA-polymerase-promotor is used.
3. Linearization of the vector forces transcript termination. This increases the yield of transcripts with an optimal size. Cut after the poly A sequence. *Mlu*I or *Pml*I can be used. If the construct contains both sites introduction of a silent mutation to eliminate the restriction site is required.
4. This step removes any residual RNase and is absolutely required because in most Midi/Maxi Kits RNase A is added during the plasmid purification process.
5. We recommend the optional DNase I treatment to remove the linearized plasmid DNA.

6. Best recovery rates are achieved when the elution is done with preheated (95 °C) elution buffer. No second elution is performed. Typically, around 25 ug cRNA is recovered. If higher cRNA concentrations are required perform a LiCl (provided with the mMESSAGE mMACHINE® SP6 Transcription Kit) precipitation according to the manufacturer's instructions.
7. To ensure that the cRNA is not degraded, standard electrophoresis on a TAE/TBE gel is generally sufficient. In rare instances more than one band can be detected. If this is the case a denaturing RNA gel electrophoresis is required. In very few cases the appearance of another band (we never observed more than one additional band) may be due to internal transcription start or premature termination. In these instances, the cRNA can be used but it must be checked subsequently by Western blotting that only the expected protein is present in oocytes.
8. The settings to pull injection needles depend on the platinum filament and vary between pullers and filaments. We recommend a rather slow pull with medium heat to yield rather long needles. Reading the "pipette cookbook" (http://www.sutter.com/PDFs/pipette_cookbook.pdf) is recommended. We break the needles open by gently pushing them against the rim of a petri dish under a stereomicroscope. Opening size is checked under a microscope. With a little bit of experience this process becomes very reproducible.
9. This point is critical but entirely empirical and will vary between transporters. When establishing this method start with 1, 0.5, and 0.1 μg/μl. If the protein is GFP labeled monitor the increase in fluorescence. Alternatively perform Western to check protein abundance. If neither of that is possible: Perform transport experiments to see if the activity is present and closely monitor changes depending in expression time. It is going to be a race between death and optimal expression. The oocyte membrane becomes instable when too much protein is inserted and the oocytes rupture. As a result some transporters must be characterized a day after RNA injection—others will take up to a week to accumulate in the membrane to an amount that allows performing the experiment. If more than one protein is expressed it is necessary to optimize the expression of all proteins. It is mandatory to establish that upon coexpression of another protein the amount of transporter does not decrease. In these cases it is necessary to adjust amount of cRNA or expression time accordingly.
10. To transfer the oocytes—particularly when handling oocytes injected with radioactivity or which are transferred out of a radioactive solution—we break glass Pasteur pipettes to yield an opening of about 3 mm and flame polish the opening to

melt sharp edges. We then insert the pipettes into the pi-pump® and suck up the oocytes. The oocytes sink to the bottom of the liquid. If the liquid level is exactly at the opening of the pipette, the oocytes will fall out when the pipette is attached to the surface of a liquid. The transfer of liquid is extremely low. We checked the carryover of radioactivity with the oocytes between wells (*see* below) and found no radioactivity in the second wash solution.

11. Lab incubators that can be used at 16 °C are fairly expensive. As a considerably cheaper alternative we use a wine cooler that can be adjusted to 16 °C (purchasing this equipment may lead to interesting discussions with the university's accounting staff).
12. Until efflux oocytes are kept cold to minimize loss of activity during handling and closure of the injection spot. Oocytes are more stable at pH 7.4. It may become necessary to vary the external pH. In this case we generally use Ringer solution. This has historical reasons. In electrophysiology Ringer is used because it has higher ionic strength. Adjusting pH in Barth's will probably work and show no difference. We never tried. Mainly because "this is how we always do it."
13. Carefully check pH! The pH inside an oocyte is 7.4 [12]. Substrates must also be adjusted to this pH! Oocytes will burst when injecting an acid! In case of auxin, start with a dilution of radioactivity that injection of 50 nl results in an end concentration of 1 μM radiolabeled auxin inside the oocyte based on an oocyte volume of 400 nl after injection [13].
14. If the phosphorylation status of the transporter upon coexpression of a kinase should be analyzed, it is necessary to use phosphatase inhibitors. For best results use fresh, not frozen oocytes. Perform the SDS-PAGE and Western blot immediately.
15. Fat from yolk will be covering the supernatant. Transfer as little fat as possible.
16. To solubilize, do not boil membrane proteins in SDS-PAGE sample buffer prior to PAGE. Incubate at 42 °C for 15 min.
17. Carefully check the pH of all buffers. During the washing steps it is crucial that the oocytes do not burst. To remove residue shake the oocytes carefully.
18. The buffers used for biotinylation must not contain primary amino groups, as biotin reacts with them.
19. Several dilutions of membrane factions must be checked to determine the binding capacity of streptavidin agarose. Do not overload and saturate the streptavidin agarose.
20. In this case it is necessary to boil to break the streptavidin from the biotinylated protein! It may run differently from the protein that is solubilized at 42 °C.

Acknowledgement

This work was supported by DFG grant HA3469/6-1.

References

1. Gurdon JB, Lane CD, Woodland HR, Marbaix G (1971) Use of frog eggs and oocytes for the study of messenger RNA and its translation in living cells. Nature 233:177–182
2. Boorer KJ, Forde BG, Leigh RA, Miller AJ (1992) Functional expression of a plant plasma membrane transporter in Xenopus oocytes. FEBS Lett 302:166–168
3. Zampighi GA, Kreman M, Boorer KJ, Loo DD, Bezanilla F, Chandy G, Hall JE, Wright EM (1995) A method for determining the unitary functional capacity of cloned channels and transporters expressed in Xenopus laevis oocytes. J Membr Biol 148:65–78
4. Krouk G et al (2010) Nitrate-regulated auxin transport by NRT1.1 defines a mechanism for nutrient sensing in plants. Dev Cell 18:927–937
5. Pellizzaro A et al (2014) The nitrate transporter MtNPF6.8 (MtNRT1.3) transports abscisic acid and mediates nitrate regulation of primary root growth in Medicago truncatula. Plant Physiol 166:2152–2165
6. Ranocha P et al (2013) *Arabidopsis* WAT1 is a vacuolar auxin transport facilitator required for auxin homoeostasis. Nat Commun 4:2625
7. Saito H et al (2015) The jasmonate-responsive GTR1 transporter is required for gibberellin-mediated stamen development in Arabidopsis. Nat Commun 6:6095
8. Swarup K et al (2008) The auxin influx carrier LAX3 promotes lateral root emergence. Nat Cell Biol 10:946–954
9. Yang Y, Hammes UZ, Taylor CG, Schachtman DP, Nielsen E (2006) High-affinity auxin transport by the AUX1 influx carrier protein. Curr Biol 16:1123–1127
10. Zourelidou M et al (2014) Auxin efflux by PIN-FORMED proteins is activated by two different protein kinases, D6 PROTEIN KINASE and PINOID. ELife 3:e02860
11. Ludewig U, von Wiren N, Frommer WB (2002) Uniport of NH_4^+ by the root hair plasma membrane ammonium transporter LeAMT1;1. J Biol Chem 277:13548–13555
12. Broer S, Schneider HP, Broer A, Rahman B, Hamprecht B, Deitmer JW (1998) Characterization of the monocarboxylate transporter 1 expressed in Xenopus laevis oocytes by changes in cytosolic pH. Biochem J 333(Pt 1):167–174
13. Stegen C, Matskevich I, Wagner CA, Paulmichl M, Lang F, Broer S (2000) Swelling-induced taurine release without chloride channel activity in *Xenopus laevis* oocytes expressing anion channels and transporters. Biochim Biophys Acta 1467:91–100

Chapter 22

Characterizing Auxin Response Circuits in *Saccharomyces cerevisiae* by Flow Cytometry

Edith Pierre-Jerome, R. Clay Wright, and Jennifer L. Nemhauser

Abstract

Recapitulation of the nuclear auxin response pathway in *Saccharomyces cerevisiae* (yeast) provides a means to functionally assay the contribution of individual signaling components to response dynamics. Here, we describe a time course assay for characterizing auxin response circuits using flow cytometry. This method allows for quantitative measurements of the dynamic response of up to 12 circuits (strains) at once. We also describe a steady-state assay and how to utilize an R package we developed to facilitate data analysis.

Key words Synthetic biology, Budding yeast, Fluorescent reporters, Signaling dynamics

1 Introduction

Auxin influences nearly every aspect of plant growth and development. The core auxin signaling pathway consists of only a handful of components, and yet is capable of eliciting a diverse array of context-specific responses to auxin. Experimental analysis of this small signaling pathway *in planta* is confounded by the ubiquity of auxin response, integration and feedback from other signaling pathways, and genetic redundancy among the core signaling components. To test the capacity of the auxin signaling pathway to generate diverse responses to auxin, we transplanted each component of the pathway from *Arabidopsis* to the budding yeast *Saccharomyces cerevisiae* [1].

A minimal yeast auxin response circuit is composed of a TIR1/AFB auxin receptor, an Aux/IAA (IAA), a TOPLESS (TPL) co-repressor, an ARF transcription factor, and an auxin-responsive promoter driving a fluorescent reporter (Fig. 1). In the absence of auxin, ARF activity is repressed through interaction with an IAA-TPL fusion. Auxin treatment promotes interaction between the IAA and the AFB receptor, an F-box protein, which catalyzes the ubiquitination and subsequent degradation of the IAA. IAA degradation relieves the repression of the ARF transcription factor

Jürgen Kleine-Vehn and Michael Sauer (eds.), *Plant Hormones: Methods and Protocols*, Methods in Molecular Biology, vol. 1497, DOI 10.1007/978-1-4939-6469-7_22, © Springer Science+Business Media New York 2017

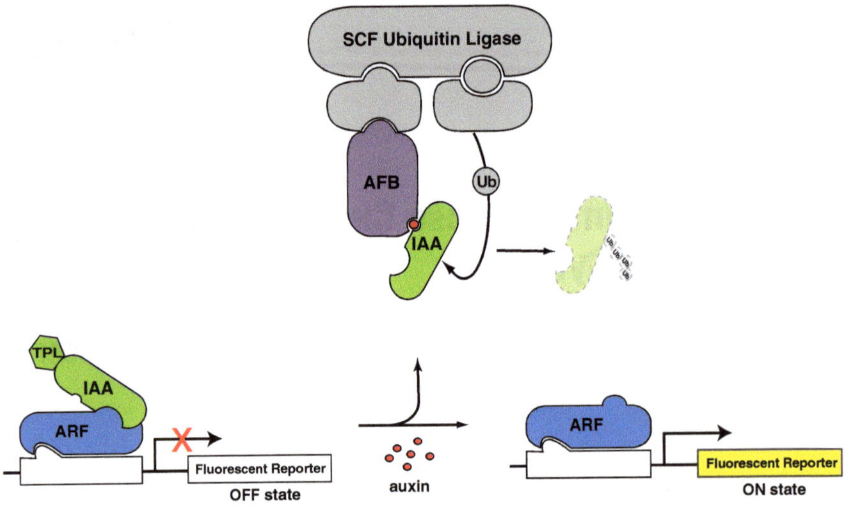

Fig. 1 A yeast auxin response circuit. An AFB auxin receptor, an IAA-TPL repressor, an ARF transcription factor, and an auxin-responsive promoter driving a fluorescent reporter are sufficient to recapitulate auxin-induced transcription in yeast

resulting in the induction of the fluorescent reporter. Flow cytometry is used to monitor levels of the fluorescent reporter over time following auxin treatment.

This user-defined auxin response circuit can be used to generate numerous circuit variants by substituting variants of any one of the core components alone or in combination. These component variants can be derived from the naturally evolved gene family members or by altering a functional domain within a component through mutation or truncation [1, 2]. The modularity of the circuit also facilitates decomposition into sub-modules to focus on specific aspects of the pathway, such as auxin-induced degradation of the IAAs. Auxin response circuit composition can therefore be tailored to suit the need of the researcher. As such, we focus here on describing a method for the quantitative characterization of auxin circuits rather than detailing their assembly.

2 Materials

2.1 Yeast Materials

1. Diploid yeast strains containing circuit variants to be assayed (*see* **Note 1**).
2. YPDA agar plates: Add about 700 mL of water to a glass beaker. Add a magnetic stir bar and set on a stir plate. Add each ingredient while stirring: 10 g yeast extract, 20 g of peptone, 80 mg of adenine. Bring volume up to 900 mL with water. Stir well, and then split into two 450 mL aliquots in 1 L bottles. Add 10 g of agar into each bottle. Autoclave for 25 min at 121 °C. After cooling, bring media up to final volume of

500 mL per bottle containing 2% glucose by supplementing with 50 mL 20% sterile filtered glucose. Pour 20 mL molten YPDA into 95 × 15 mm petri dishes as required.

3. Synthetic Complete (SC) liquid media: Add about 1 L of water to a glass beaker. Add a magnetic stir bar and set on a stir plate. Add each ingredient while stirring: 2.55 g of yeast nitrogen base without amino acids or ammonium sulfate (Sigma); 2.1 g of yeast synthetic drop-out medium without histidine, leucine, tryptophan, and uracil (Sigma); 7.5 g ammonium sulfate; 150 mg of histidine; 150 mg of leucine; 150 mg of tryptophan; 37.5 mg of uracil; and 120 mg adenine. Bring volume up to 1.35 L with water. Stir well, and then split into two 675 mL aliquots in 1 L bottles. Autoclave for 25 min at 121 °C. After cooling, bring each bottle to a final volume of 750 mL by adding 75 mL 20% sterile filtered glucose. Store at room temperature and protect liquid media from light.

2.2 Cytometry Assay Materials

1. 17 × 100 mm sterile plastic (translucent polypropylene) culture tubes with snap cap.
2. 50 mL plastic conical centrifuge tubes.
3. 25 mL serological pipettes.
4. Pipet-Lite™ XLS Adjustable-Spacer Multichannel Manual Pipette, 100–1200 μl (Rainin).
5. Multiplex culture tube lids (*see* **Note 2**).
6. 96-well PCR plates.
7. Indole-3-acetic acid (auxin).
8. 95% EtOH.
9. 2000 μl Eppendorf™ Deepwell™ Plates 96.
10. 25 mL reagent troughs.
11. 12-channel pipette with 100 μl capacity.
12. AeraSeal™ breathable film seals (Excel Scientific).

2.3 Equipment

1. Flow cytometer with plate capacity configured to detect GFP or YFP. We currently take fluorescence measurements with a BD Accuri™ SORP flow cytometer with a CSampler 96-well plate adapter using an excitation wavelength of 514 nm and an emission detection filter at 545/35 nm (*see* **Note 3**).
2. Benchtop shaker incubator amenable to rapid and repeat sampling of growing yeast cultures such as a MaxQ™ 4000 Benchtop Oribital Shaker (Thermo Scientific).

2.4 Data Analysis Software

We have developed a package called `flowTime` for the R programming language to facilitate data analysis for these assays. This package will soon be available via http://bioconductor.org (current and continuing development version available at http://github.com/wrightrc/flowTime). We suggest using RStudio (http://www.rstudio.com) to edit and run the R-code for your data analysis. Numerous resources are available on the internet to gain an introductory knowledge of the R language [3].

3 Methods

3.1 Auxin Time Course Assay

This protocol can be used to assay auxin-induced transcription of up to 12 strains in parallel. This constraint is due to the time required for auto-sampling of multiple wells by the cytometer as well as to minimize the time a yeast culture sits in the sampling plate. We therefore generally limit each read to 12 strains. This assay can also be used for monitoring auxin-induced degradation with only minor adjustments (*see* **Note 4**).

Day One

1. For each strain to be assayed, inoculate ¼ to ½ of a fresh yeast colony from a YPDA plate into 3 mL of synthetic complete (SC) media. Vortex well to mix. Aliquot 100 μl into a 96-well plate and read on cytometer to estimate cell density (in events per microliter).
2. Export data as `.FCS` files and import into R as a `flowSet` (*see* Subheadings 3.3 and 3.4, for example R-code). Create summary statistics for this `flowSet` using the **summary.cyt** function with the "only" parameter set to "`yeast`" to subset the data for all yeast. The value in the "`conc`" column of the resulting data frame is the concentration in yeast events per μL for each sample or `flowFrame`. Use this value to calculate the volume of the initial 3 mL inoculant required to prepare the requisite dilution.
3. Dilute each strain to 0.25 events per μL in 12 mL of SC in a 50 mL conical tube. Mix well by vortexing.
4. Split dilutions into duplicate 5 mL aliquots in sterile plastic culture tubes and incubate for 16 h at 30 °C with shaking at 220 rpm.

Day Two

5. After 16 h of growth, combine duplicate aliquots to homogenize cultures. Mix gently by pipetting and split back into two 5 mL aliquots.
6. Separate duplicate cultures into two groups of tubes while maintaining the strain order between groups. This will be the order samples are read on the cytometer. Return tubes to shaker and replace individual plastic caps with 3D printed

multiplex tube lids. Use of these lids facilities rapid sampling of the cultures using the Rainin Adjustable-Spacer Multichannel Pipette.

7. Set cytometer settings to measure 10,000 events for each well at a flow rate of 66 μl/min and core size of 22 μm with a time limit of 1 min. Use these settings for the entire experiment.
8. Transfer 100 μl of each culture from one replicate group to a 96-well PCR plate by expanding the multichannel pipette to sample from the culture tubes and collapsing to dispense into the PCR plate. Fill an additional well with 200 μl of sheath fluid to serve as a wash between readings (*see* **Note 5**).
9. Load PCR plate containing samples onto cytometer sampling arm and run the auto-sampling program to flow samples. The density of each culture during this initial read should be around 200 events/μl (*see* **Note 6**). Repeat **steps 8** and **9** for the second replicate of cultures.
10. For each strain, mock treat the first replicate measured with 95 % ethanol. Immediately transfer 100 μl of each mock-treated culture to the next row of the 96-well PCR plate and read samples on the cytometer to record fluorescence. These measurements serve as the 0-min time point.
11. Treat each strain of the second replicate of cultures with indole-3-acetic acid (IAA) dissolved in 95 % ethanol to a final concentration of 10 μM. Immediately after auxin addition, transfer 100 μl of each culture to a new 96-well PCR plate. Open a new file with the settings described in **step** 7 and measure fluorescence for the 0-min time point.
12. Use the same plate and file to acquire measurements for auxin-treated samples at 10-min intervals. Prepare new plates and a corresponding cytometer file as needed to acquire 5 h of data.
13. Take measurements for mock-treated samples every hour using the same plate and file used in **step 9**.
14. For each plate file, export all wells as a folder of `.FCS` files. There should be a separate plate file containing the initial and mock-treated reads and several files for the auxin-treated measurements.

3.2 Auxin Steady-State Assay

Day One

1. For each strain to be assayed, inoculate ¼ to ½ of a fresh yeast colony from a YPDA plate into a well of a 2000 μl Eppendorf™ Deepwell™ Plate 96 containing 500 μl of SC media.
2. Seal plate with a breathable film seal and incubate at 30 °C with shaking at 375 rpm for 16 h.

Day Two

3. After 16 h of growth, dilute strains 1:200 into a new well containing 500 μl of fresh SC media in a deep 96-well plate (*see* **Note 7**). Prepare two replicate wells for each strain.
4. Return plate containing diluted cultures to the shaker for 2 h of additional growth at 30 °C with shaking at 375 rpm (*see* **Note 8**).
5. After 2 h, mock (95 % EtOH) treat one replicate well and auxin (indole-3-acetic acid) treat the other (*see* **Note 8**). To assay more than 12 strains or multiple doses of auxin, treatment additions must be staggered for every 12 wells. This is important to allow time for data acquisition and maintain a consistent treatment time between reads.
6. Return plate to shaker for an additional 4 h of growth (*see* **Note 8**).
7. Set cytometer settings to measure 20,000 events for each well at a flow rate of 66 μl/min and core size of 22 μm with a time limit of 1 min.
8. Sample up to 12 strains at once using a multi-channel pipette to transfer 100 μL of culture from the deep-well plate in the shaker to a PCR plate. Fill an additional well in the PCR plate with 200 μl of sheath fluid to serve as a wash between readings.
9. Load the PCR plate onto cytometer sampling arm and run the auto-sampling program to measure fluorescence.
10. Repeat **steps 8** and **9** until all strain/treatment combinations have been measured.
11. Typically, data for at least three independent replicates for each strain/treatment is collected.
12. Export data as a folder of .FCS files.

3.3 Time Course Data Analysis in R

Here, we demonstrate how to import the .FCS files into R, gate and annotate this data with experimental metadata (e.g., the strain and treatment for each sample), generate summary statistics for each sample and time point, and finally plot the time course data (*see* **Note 9**).

3.3.1 Importing and Annotating Data

1. Import your flow cytometry data using **read.flowSet**. Here, we will import an example `flowSet`:

```
plate1 <- read.flowSet(path = system.file("extdata",
"tc_example/", package = "flowTime"), alter.names = T)
# Next, add plate numbers to the sampleNames
sampleNames(plate1) <- paste('1_',
sampleNames(plate1),
sep = "")
```

2. If you have several plates, this code can be repeated and each plate can be combined (using **rbind2**) to assemble the full dataset:

```
plate2 <- read.flowSet(path = paste(experiment, "_2/", sep = ""),
alter.names = T);
sampleNames(plate2) <- paste("2_", sampleNames(plate2), sep = "");
dat <- rbind2(plate1, plate2)
```

3. Import the table of metadata for each well. The "`sample-Names`" field of the assembled *flowSet* ("`dat`" in this example) must match that of a unique identifier column in the table of metadata. In this example and as a default name is the unique identifier column. One can also create this column from a flowSet and attach the annotation columns, as demonstrated in the code below. The order of the unique identifier column does not matter, as **annotateFlowSet** will join annotation to "dat" by matching identifiers, as specified by the "mergeBy" parameter:

```
annotation <- read.csv(system.file("extdata", "tc_example.csv",
package = "flowTime"));
sampleNames(dat) #view the sample names
sampleNames(dat) == annotation$id #replace 'id' with the unique
identifier column to test if this column is identical to the
sample names of your flowset
annotation <- cbind(annotation, names = sampleNames(dat)) #If
the sampleNames and unique identifiers are in the correct order
this command will add the sampleNames as the identifier
```

4. Attach this metadata to the `flowSet` using the **annotateFlow-Set** function:

```
adat <- annotateFlowSet(dat, annotation);
head(rownames(pData(adat)))
#> [1] "0_A08.FCS" "0_B08.FCS" "0_C08.FCS" "0_D08.FCS"
"0_E08.FCS" "0_F08.FCS"
head(pData(adat))
#>name X strain RD ARF AFB treatment
#> 0_A08.FCS 0_A08.FCS 0_A08  3 TPLRD1 19 AFB2   0
#> 0_B08.FCS 0_B08.FCS 0_B08  3 TPLRD1 19 AFB2   0
#> 0_C08.FCS 0_C08.FCS 0_C08  3 TPLRD1 19 AFB2   0
#> 0_D08.FCS 0_D08.FCS 0_D08  3 TPLRD1 19 AFB2   0
#> 0_E08.FCS 0_E08.FCS 0_E08  3 TPLRD1 19 AFB2   0
#> 0_F08.FCS 0_F08.FCS 0_F08  3 TPLRD1 19 AFB2   0
```

3.3.2 Compiling and Plotting Time Course Data

1. Use the **summary.cyt** function to compile the summary statistics from the raw data in this `flowSet`. This function will gate each `flowFrame` in the `flowSet` and compile and return a

dataframe of summary statistics in the specified channel for each flowFrame:

```
# load the gate set for BD Accuri C6 cytometer
loadGates(gatesFile = "C6Gates.RData");
dat_sum <- summary.cyt(adat, ploidy = "diploid", only =
"singlets", channel = "FL1.A")
#> [1] "Gating with diploid gates…"
#> [1] "Summarizing singlets events…"
```

2. Use the resulting dataframe to plot mean fluorescence for each mock- and auxin-treated sample over time to generate induction curves (Fig. 2):

```
qplot(x = time, y = FL1.Amean, data = dat_sum, linetype =
factor(treatment)) +
geom_line() + xlab("Time post Auxin addition (min)") +
ylab("Reporter Fluorescence (AU)") +
scale_color_discrete(name = expression(paste("Auxin (", mu,
"M)", sep = ""))) +
theme_classic(base_size = 14, base_family = "Arial")
```

3.4 Steady-State Assay Data Analysis

For steady-state assays, the entire dataset (all measured events/sample) can be readily plotted. Data import and annotation remains the same.

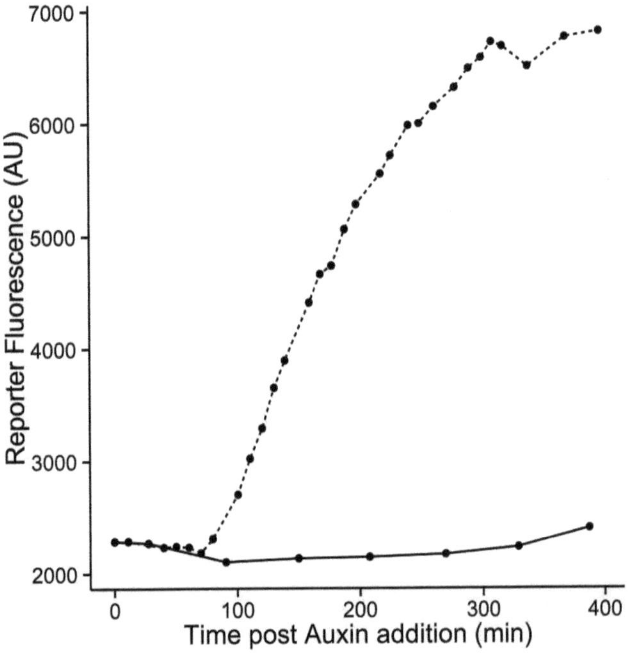

Fig. 2 Example data from a time course assay. Induction of transcription in response to auxin is shown for a mock- (*solid line*) and auxin-treated (*dashed line*) circuit

3.4.1 Compiling and Plotting Steady-State Data

1. Use the steadyState function to gate each flowFrame in the flowSet and compile and return a dataframe of the relevant data and metadata for each event:

```
loadGates(gatesFile = "SORPGates.RData")
dat.SS <- steadyState(flowset = adat, ploidy = "diploid", only =
"singlets")
#> [1] "Gating with diploid gates..."
#> [1] "Converting singlets events..."
```

2. Use the dataframe to plot the fluorescence for all events captured for each sample (Fig. 3):

```
p <- ggplot(dat.SS, aes(as.factor(treatment), FL2.A, fill = AFB))
+ geom_boxplot(outlier.size = 0) +
facet_grid(IAA ~ AFB) + theme_classic(base_family = "Arial",
base_size = 16) +
ylim(c(-1000, 10000)) + xlab(expression(paste("Auxin (", mu,
"M)", sep = ""))) +
ylab("Fluorescence (AU)") + theme(legend.position = "none");

p
```

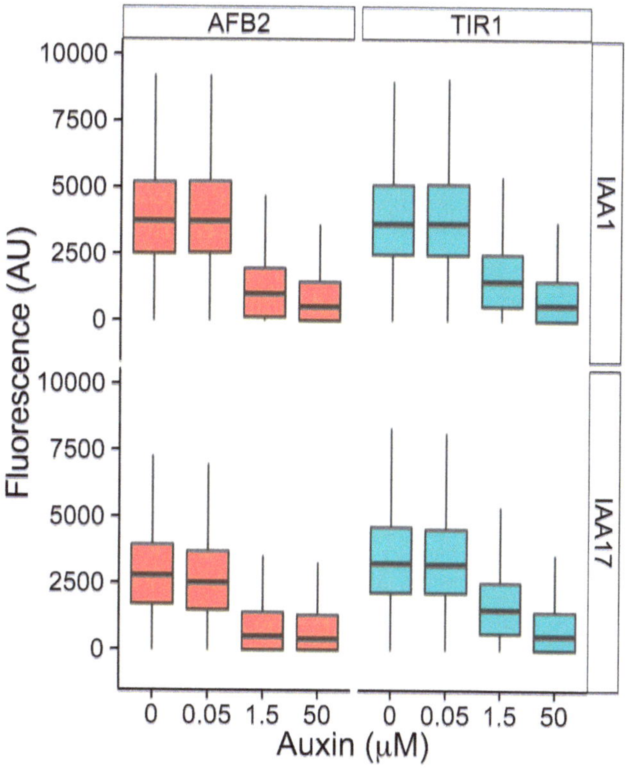

Fig. 3 Example data from a steady-state assay. Each sub-panel represents a different circuit built to assay auxin-induced degradation. Box plots display the fluorescence distribution of IAA protein levels following a dose response assay

4 Notes

1. Strain construction for auxin response circuits has been previously described [1]. Briefly, we use laboratory strains W814-29B MATα and W303-1A MATa which are both auxotrophic for uracil, histidine, tryptophan, and leucine. Each component of the pathway is singly integrated into the yeast genome at a different auxotrophic locus. Typically, a TIR1/AFB receptor is integrated at *LEU2* and an IAA-TPL fusion is integrated at *TRP1* in W814-29B MATα. An ARF is integrated at *HIS3* and the auxin-responsive fluorescent reporter (i.e., pIAA19::GFP) is integrated at *URA3* in W303-1A MATa. The two strains can then be mated to generate a diploid strain containing the entire auxin circuit. For auxin-induced degradation assays, only introduction of a TIR1/AFB receptor and a fluorescently tagged IAA is required [2].
2. The .stl design file for the 3D printed multiplex lid is publically available on Thingiverse (http://www.thingiverse.com/thing:893490). Each lid can accommodate six culture tubes in a 6 × 6 or 6 × 12 tube rack.
3. Our published work utilized a standard BD Accuri™ C6 flow cytometer with a CSampler 96-well plate adapter configured with a 488 nm laser and an emission detection filter at 533 nm to measure GFP and YFP fluorescence levels. Our current 514 nm laser configuration optimizes YFP excitation but no longer excites GFP. We tested several bench-top cytometers for YFP and GFP detection in our strains and found the Accuri to be the most sensitive.
4. For auxin-induced degradation assays, follow time course protocol with the following modifications. The first day, dilute strains to 0.5 events/μl in 10 mL of SC and split into two 4 mL aliquots. Culture density during the initial reads the next day should be ~500 events/μl. After treatments, samples are measured at 10-min intervals for 3 h.
5. A wash well should be prepared for every read containing up to 12 strains. We typically prepare the full PCR plates containing a wash well for each row so that only cultures need to be transferred to the PCR plate at 10-min intervals.
6. A starting density of ~200 events/μl after 16 h is based on the growth rate of our particular strains. We found this starting density to be optimal to ensure that yeast are in log phase for the duration of the experiment. If starting density is too low, it is best to allow the yeast to grow for an additional 30 min to an hour as fluorescence readings can be more variable.
7. For haploid strains, dilute cultures 1:100 in order to maintain the time frame outlined.

8. We typically treat with 10 μM IAA but auxin treatment can range from 0.05 to 50 μM. Dimethyl sulfoxide can be used as a solvent in place of 95 % ethanol. To conduct a dose response assay, replicate wells for each dose must be prepared in **step 3**. For auxin-induced degradation assays, treatment should be added 5 h post-dilution and fluorescence measured after 1 h.
9. To minimize noise in yeast cytometry data we typically present fluorescence measurements from only singlet yeast cells (i.e., cells without a significantly sized bud), as budding increases cell size and decreases the effective concentration of the circuit components. Unfortunately, gates isolating populations of singlet-yeast cells for both haploid and diploid strains are unique for each yeast strain, shaker setup, and flow cytometer. Determining the parameters of these gates is currently done by trial and error in R. More visual programs such as flowJo may be helpful. See the `flowCore` documentation [4] for more information about using R to create gates, and refer to [5] for an example of how to identify live and singlet cell populations.

References

1. Pierre-Jerome E, Jang SS, Havens KA, Nemhauser JL, Klavins E (2014) *Recapitulation of the forward nuclear auxin response pathway in yeast.* Proc Natl Acad Sci U S A 111(26):9407–9412
2. Havens KA, Guseman JM, Jang SS et al. (2012) *A synthetic approach reveals extensive tunability of auxin signaling.* Plant Physiol 160:135–142
3. R Core Team (2015) *R: a language and environment for statistical computing.* R Foundation for Statistical Computing, Vienna, Austria, http://www.R-project.org/
4. Ellis B, Haaland P, Hahne F, Le Meur N, Gopalakrishnan N, Spidlen J, Jiang M *flowCore: basic structures for flow cytometry data.* R package version 1.36.9. https://www.bioconductor.org/packages/release/bioc/html/flowCore.html
5. Baxter SK, Lambert AR, Scharenberg AM, Jarjour J (2013) *Flow cytometric assays for interrogating LAGLIDADG homing endonuclease DNAbinding and cleavage properties.* Methods Mol Biol 978:45–61

INDEX

Jürgen Kleine-Vehn and Michael Sauer (eds.), *Plant Hormones: Methods and Protocols*, Methods in Molecular Biology, vol. 1497, DOI 10.1007/978-1-4939-6469-7, © Springer Science+Business Media New York 2017

D

E

F

L

M

N

O

P

U

V

W

X

Y

MIX
Papier aus verantwortungsvollen Quellen
Paper from responsible sources
FSC® C105338

If you have any concerns about our products,
you can contact us on
ProductSafety@springernature.com

In case Publisher is established outside the EU,
the EU authorized representative is:
Springer Nature Customer Service Center GmbH
Europaplatz 3, 69115 Heidelberg, Germany

Printed by Libri Plureos GmbH
in Hamburg, Germany